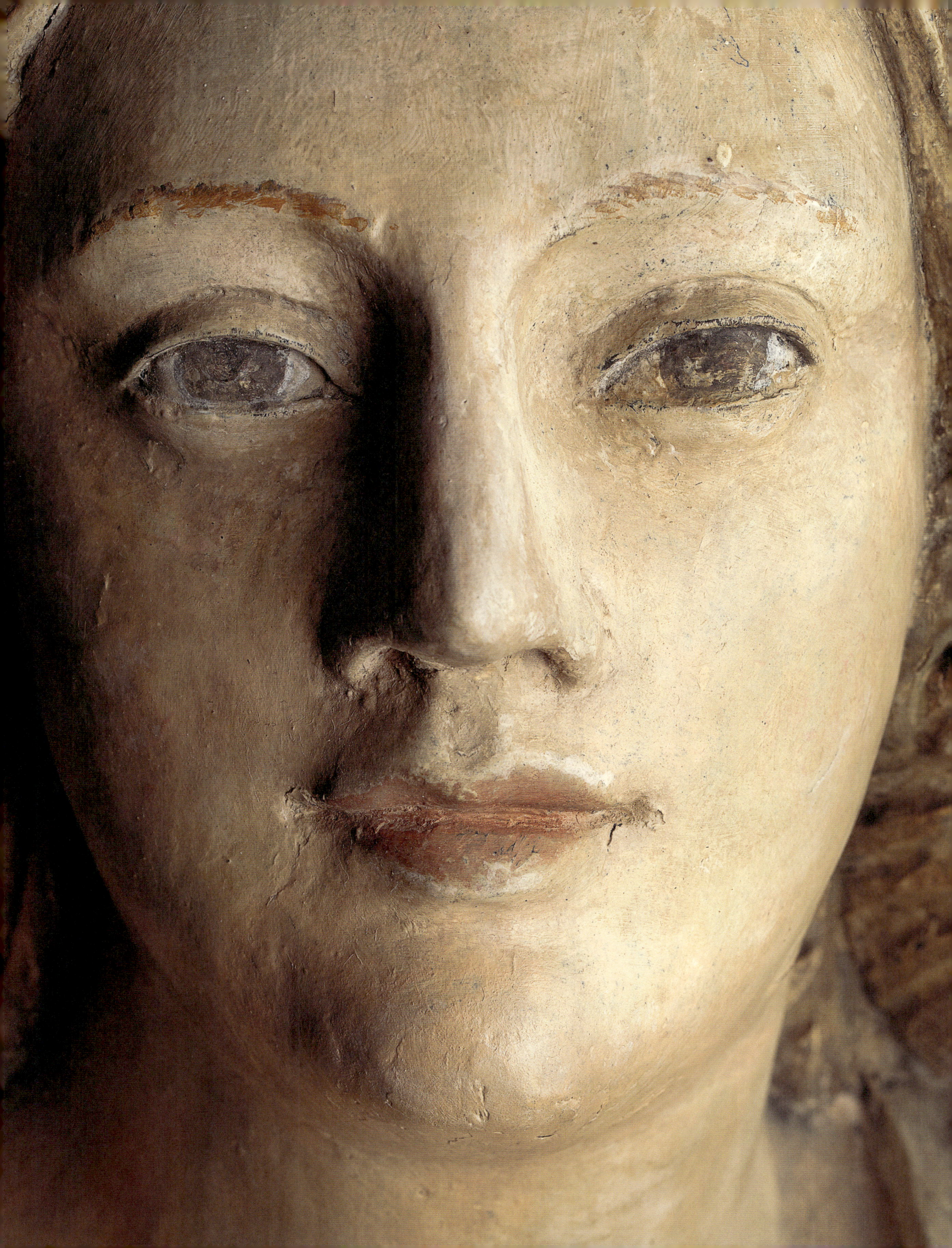

IMAGERY OF THE SEA

Figureheads and Symbols

ANNIKA BÜNZ LARS EINARSSON NIKLAS ERIKSSON FRED HOCKER
LEOS MÜLLER JOHAN RÖNNBY INGRID ULFSTEDT

TRANSLATION

WILLIAM SLEATH & SUSAN LONG

EDITED BY

KURT ALMQVIST & SVANTE HELMBAEK TIRÉN

PHOTOGRAPHER

OLA TORKELSSON

BOKFÖRLAGET STOLPE   AXEL AND MARGARET AX:SON JOHNSON FOUNDATION FOR PUBLIC BENEFIT

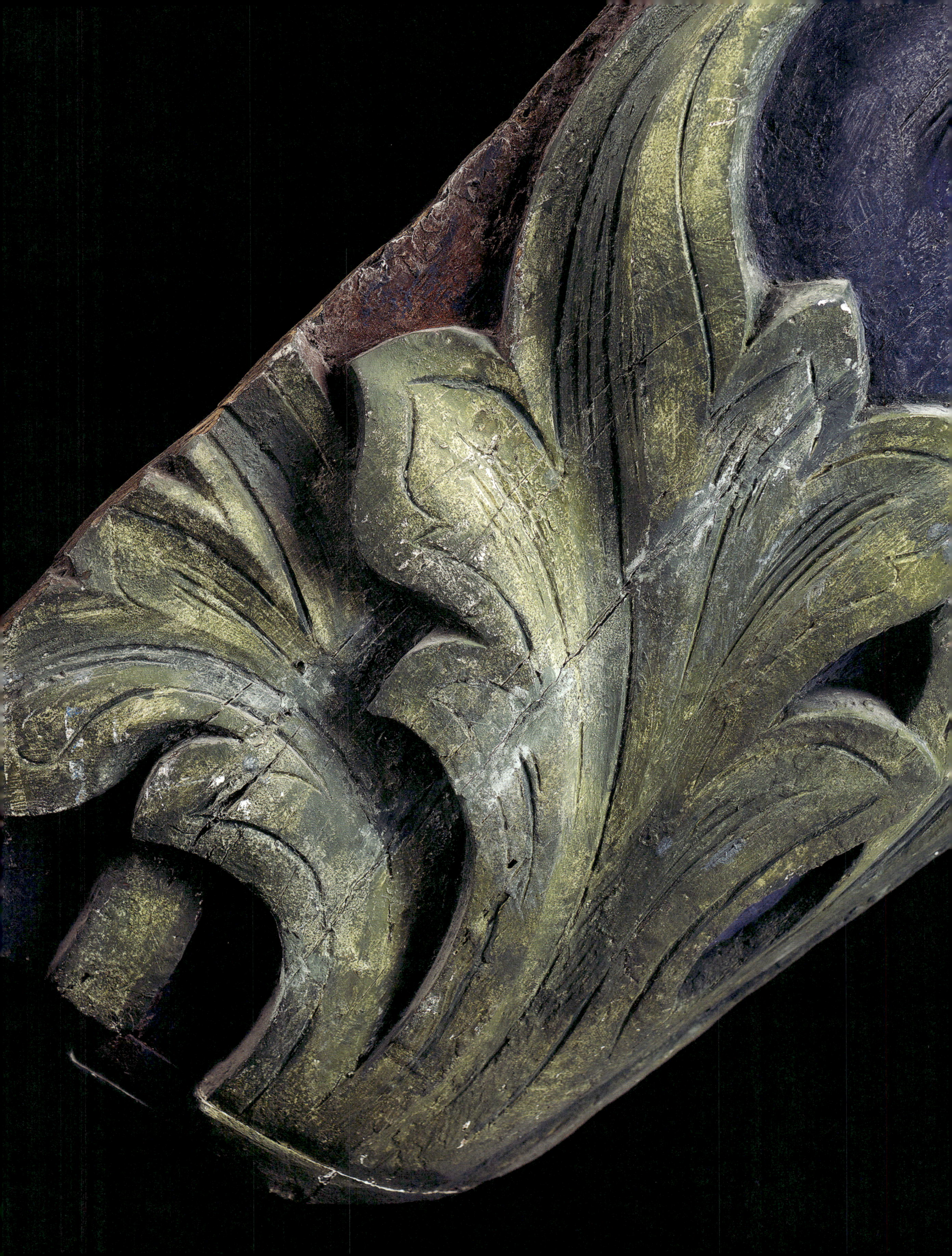

CONTENTS

PREFACE 7
Kurt Almqvist

INTRODUCTION 15
Svante Helmbaek Tirén

THE FIGUREHEAD'S ADVENTURES THROUGH TIME AND SPACE 173
Annika Bünz

THE FIGUREHEADS AT ENGELSBERG 187
Svante Helmbaek Tirén

SWEDEN AND THE SEA – FROM THE 16TH CENTURY TO THE PRESENT DAY 191
Leos Müller

THE DANISH MONSTER 199
Johan Rönnby

THE ROYAL SHIP KRONAN – PALACE AND WAR MACHINE 211
Lars Einarsson

BETWEEN LION FIGUREHEADS AND LIONS OF FASHION 223
Niklas Eriksson

VASA – THE METAPHYSICAL WARSHIP 233
Fred Hocker

VESSEL ORNAMENTS IN STOCKHOLM'S MARITIME MUSEUM COLLECTIONS 243
Ingrid Ulfstedt

CONTRIBUTORS 255
NOTES AND REFERENCES 256
IMAGE RIGHTS 261
INDEX 262

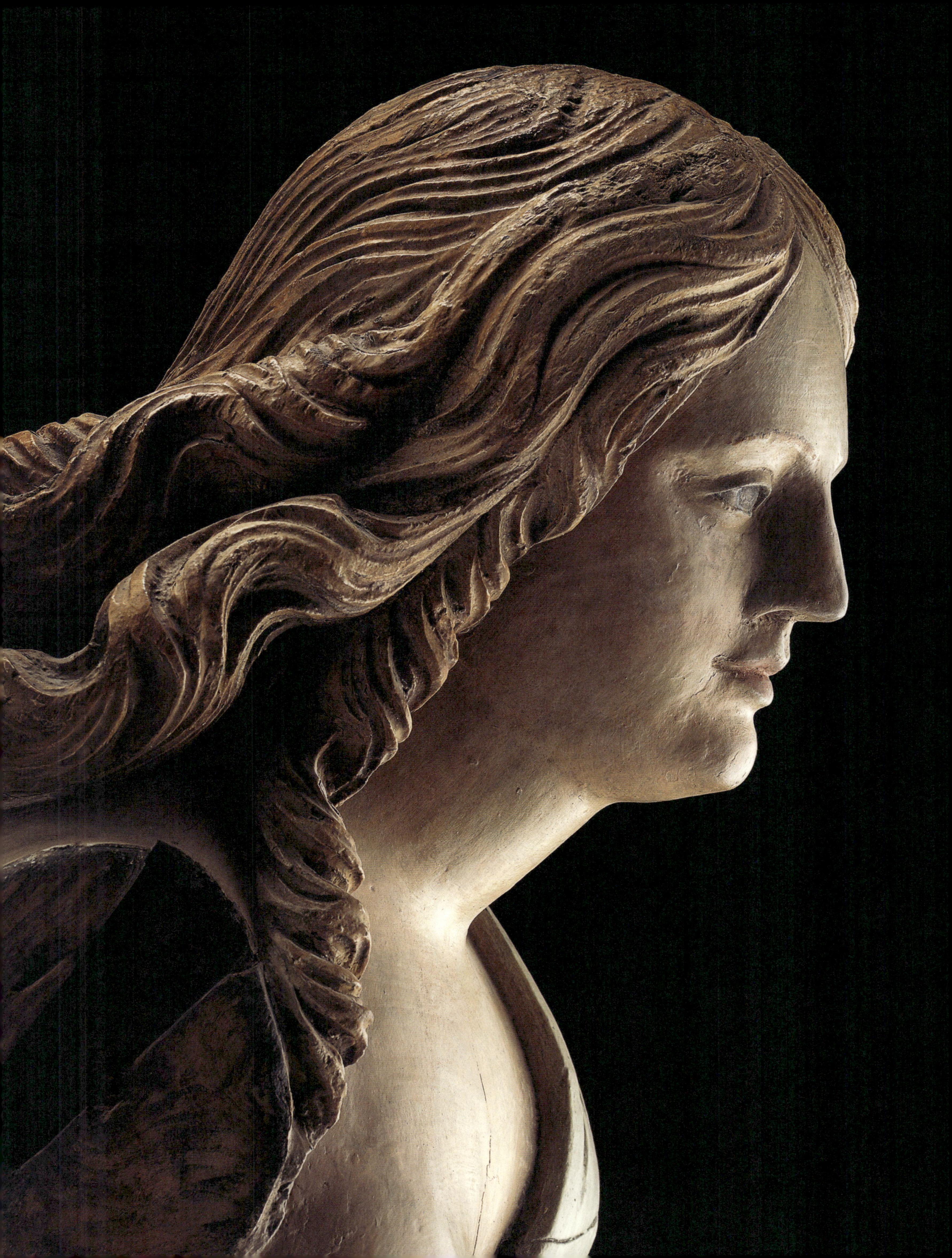

PREFACE

This book originates from the project *Imagery of the Sea* and forms part of Forum Axess. Conferences, seminars, Axess Magasin, Axess TV and book publishing all come under the umbrella of Forum Axess. The aim is to disseminate scholarly results of the so-called *tredje uppgiften*, the third assignment, that Swedish academics are obliged to do, i.e. reach out to the general public.

The Axel and Margaret Ax:son Johnson Foundation for Public Benefit is a private foundation whose main purpose is general promotion of academic research. It in particular supports the humanities and social sciences. The foundation was started up in 1947 by the Consul General Axel Ax:son Johnson (1876–1958), owner of the Nordstjernan group, together with his wife Margaret Ax:son Johnson (1887–1966).

Through collaboration with the publishing house Bokförlaget Stolpe, whose emphasis is on arts & crafts, education and beauty, the foundation's books are given a congenial form for their content.

Kurt Almqvist
President
Axel and Margaret Ax:son Johnson Foundation for Public Benefit

Full-length woman dressed in white with fair, flowing hair. Origin unknown.

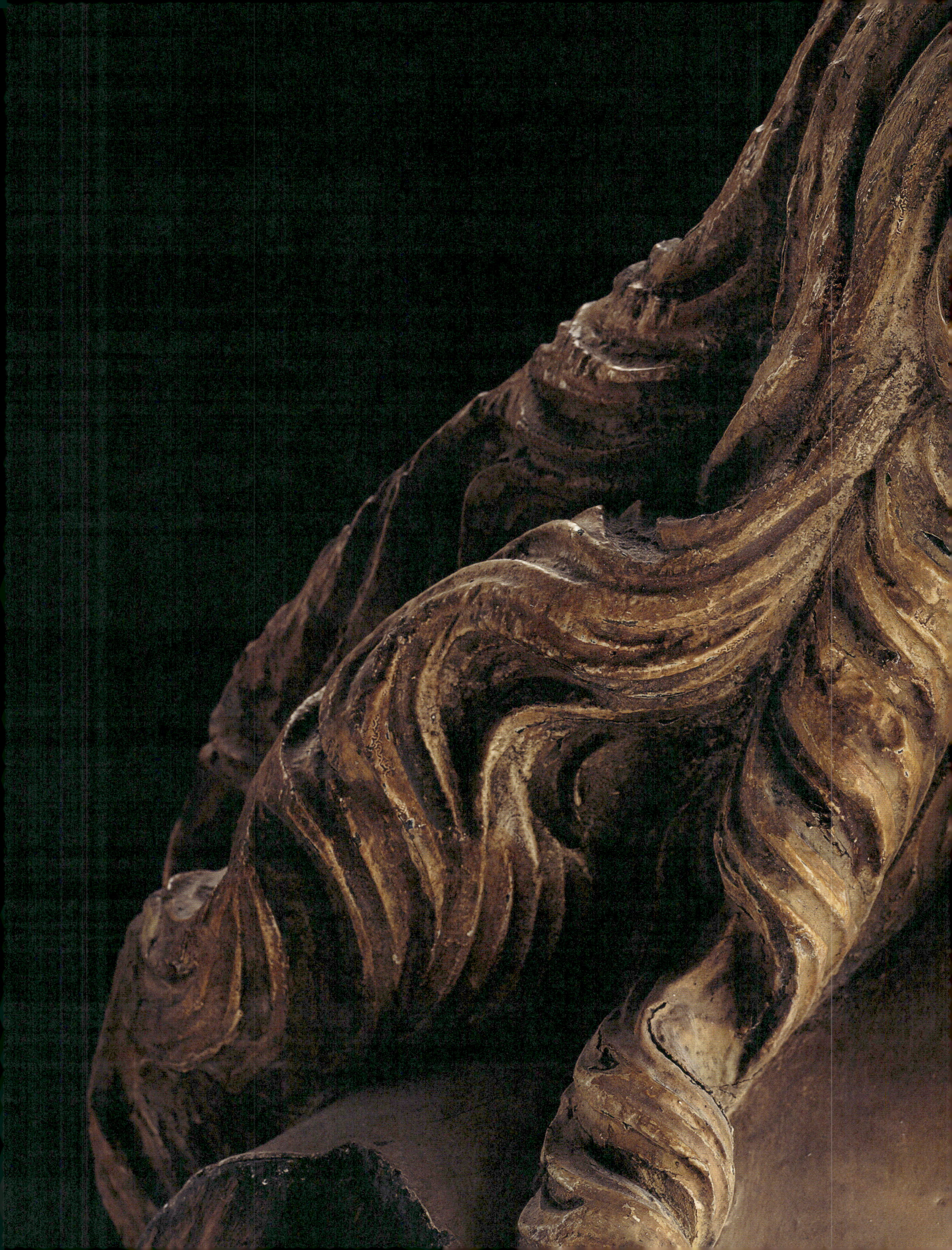

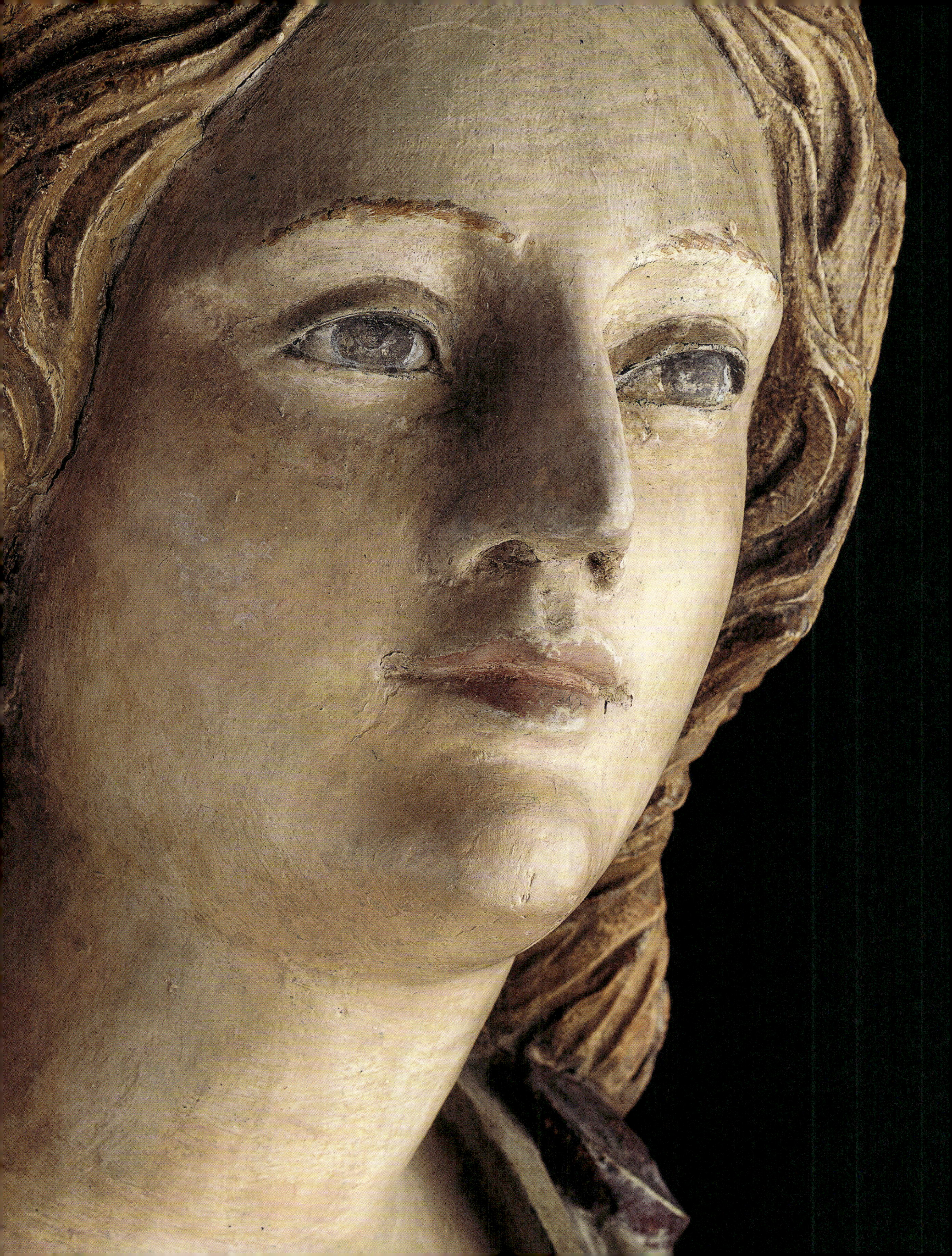

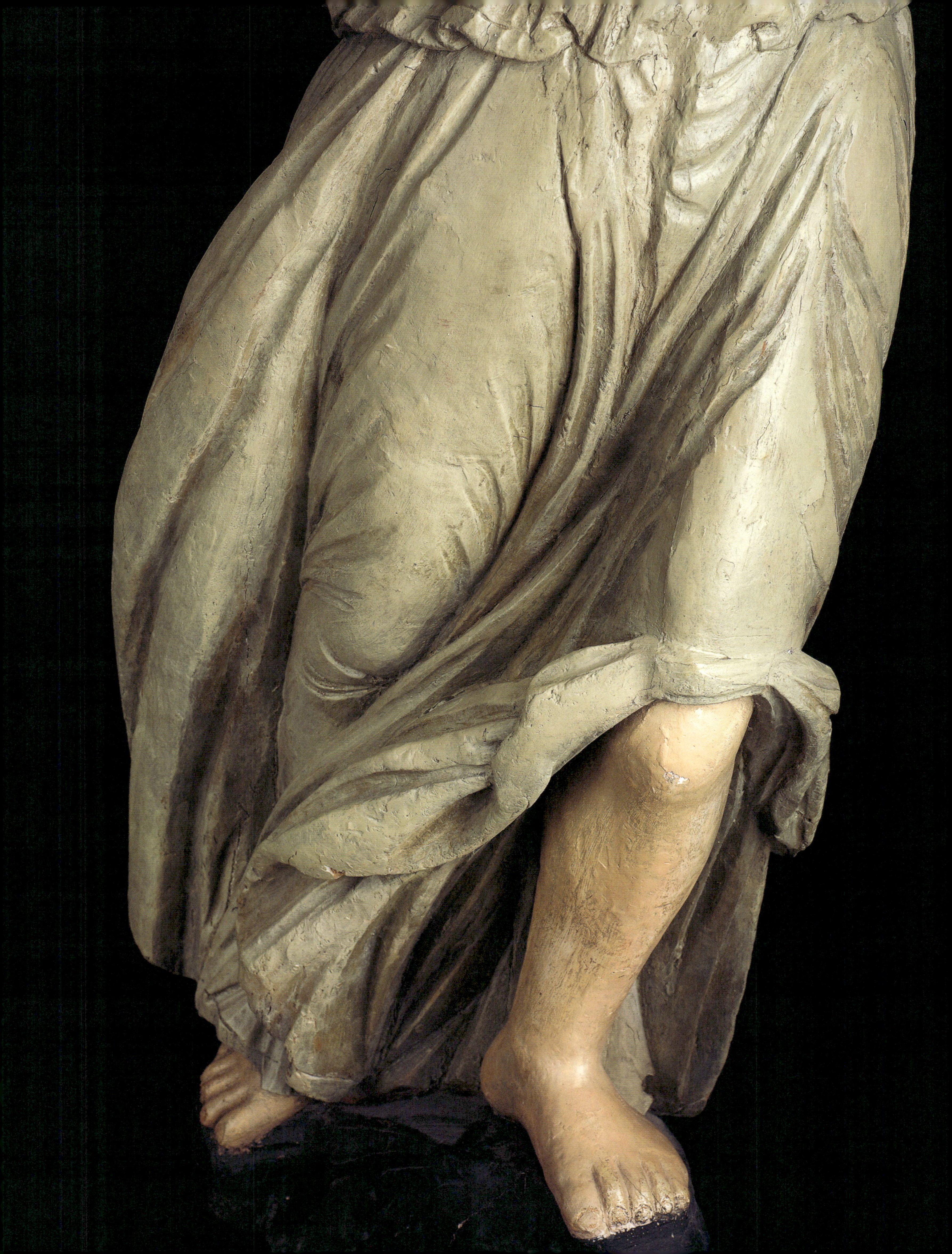

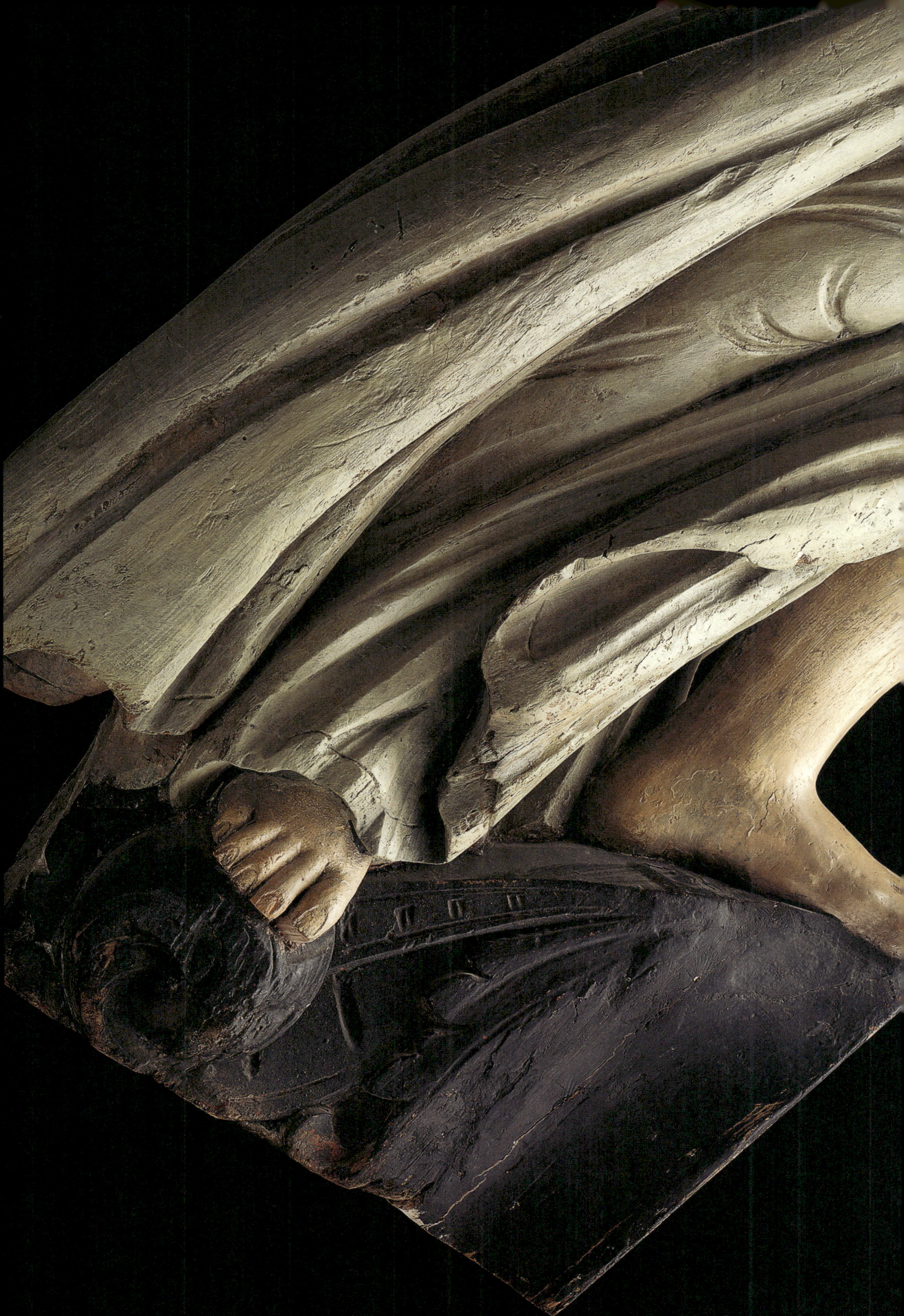

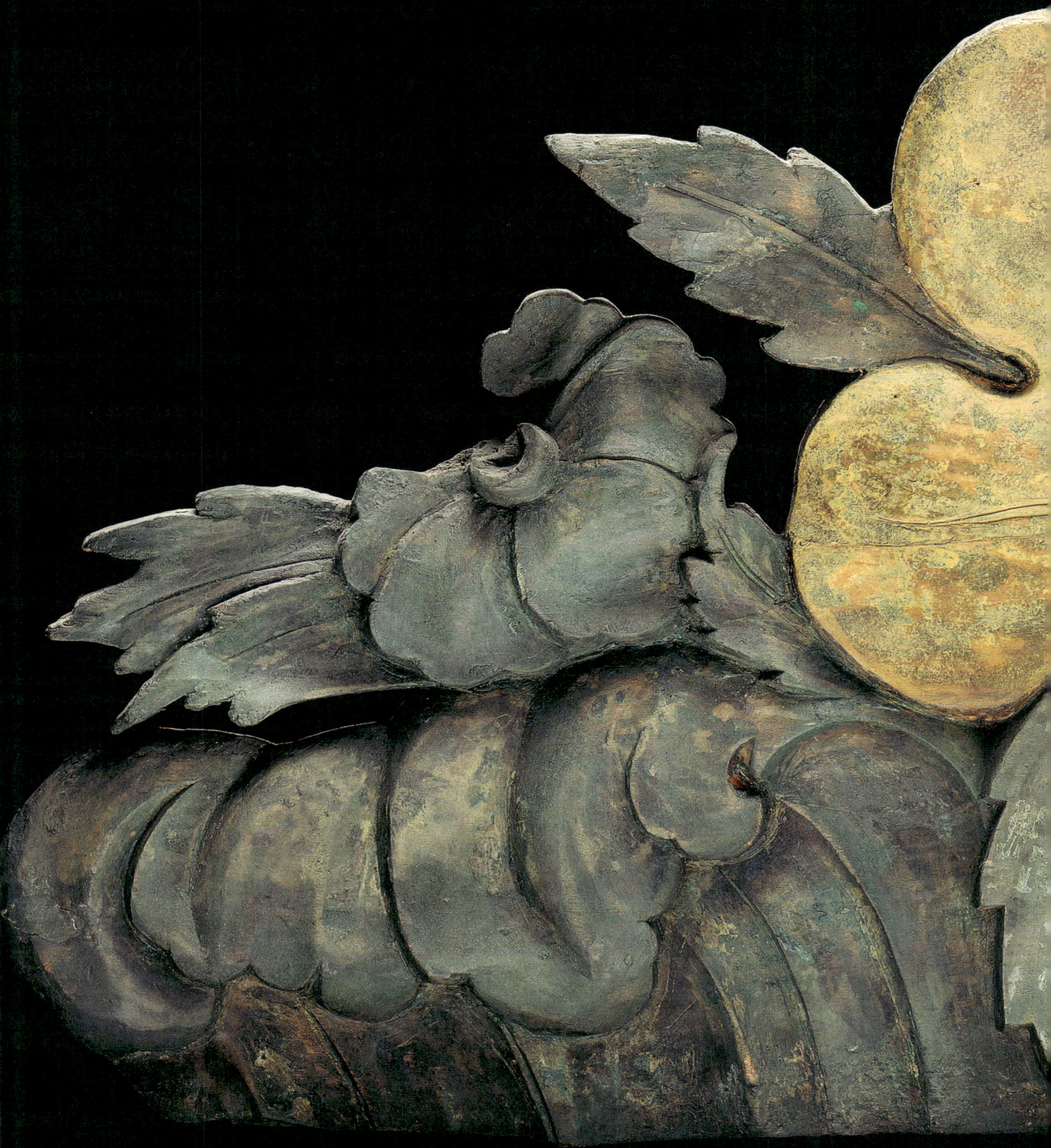

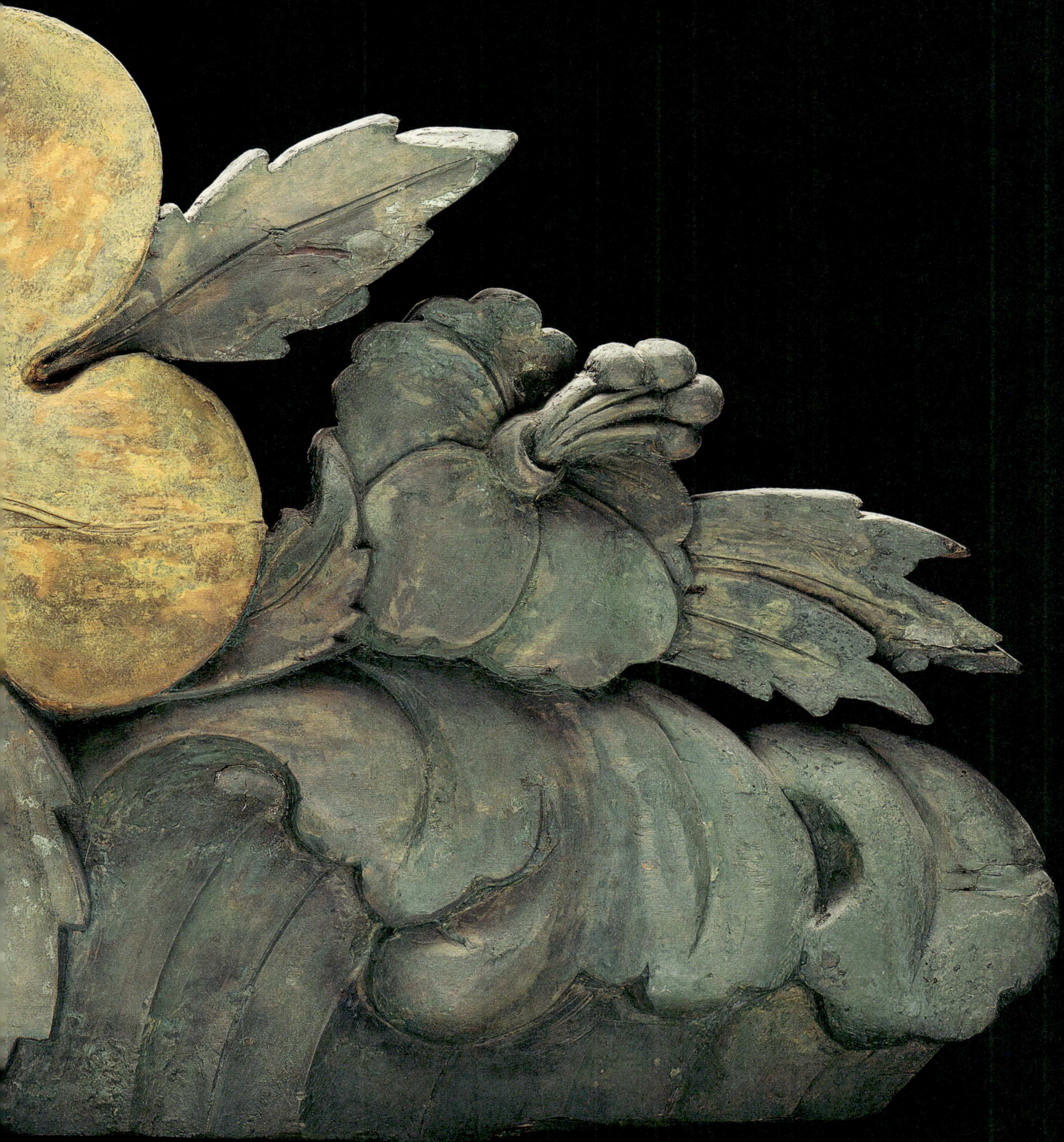

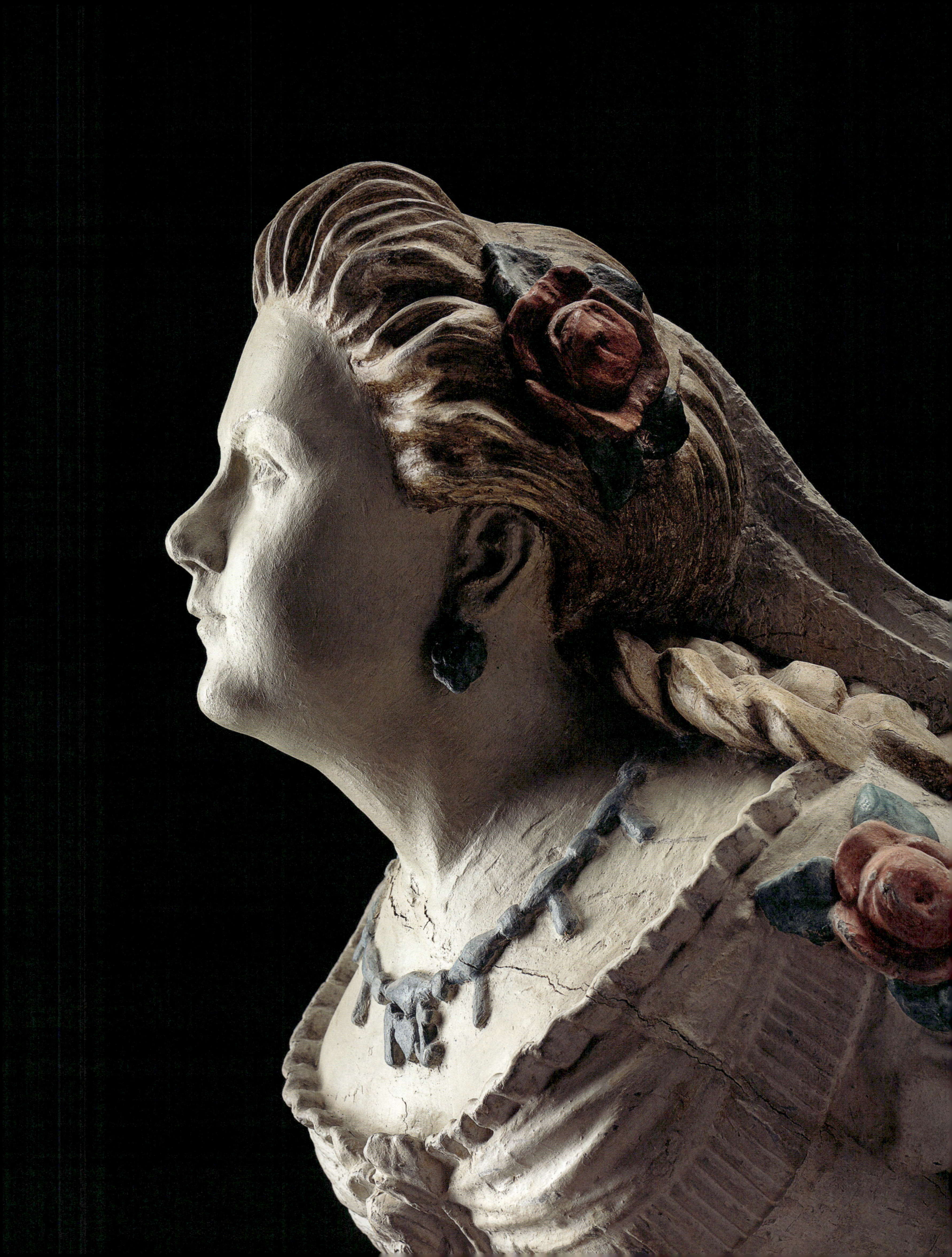

INTRODUCTION

Svante Helmbaek Tirén

THE TRADITION OF DECORATING ships and other watercraft can be traced far back in history. We see this in the carved dragon heads of Viking ships, in ornately decorated naval vessels from the period when Sweden was a great power, and in 19th-century figureheads depicting everything from portraits of shipowners and imaginary animals to gods of antiquity. While construction methods and ideals have changed, there has been a more or less constant need to convey the identity of these vessels. Watercraft from vastly different eras and contexts can ostensibly be linked together, at least superficially, through their visual images and symbols. But what did people actually want to convey with this ornamentation? Why did people devote time and energy to creating this imagery, and why does it still fascinate us today?

The basis for this book is the figureheads conserved at the UNESCO World Heritage Site Engelsberg Ironworks in Västmanland, Sweden. These figureheads, also known as the Skagen Collection, were all taken from vessels sunk off the coast of North Jutland in the late 19th century. With their great variation in motifs and artistic visual expression, they form a small cross-section of what the tradition of figureheads looked like in their heyday. It has been possible to identify some of these and connect them to specific vessels, while others have remained a mystery. Yet no matter what knowledge we have of their origins, they are all clear examples of the vital function figureheads fulfilled over many centuries. On the surface, they are aesthetically pleasing objects – but also echoes of tragedies, triumphs, hopes and human beings.

Nonetheless, the figureheads at Engelsberg are far from the only objects that tell the history of nautical art. In Sweden, there is a rich cultural heritage with objects preserved from our maritime past. Museums and private collections both hold remains of vessels and of the wide network of contacts they were once part of. Even more await discovery in the shipwrecks hidden in the depths of the sea, particularly the Baltic, with its unique conditions for preserving them.

In this book, the focus of research is on the ornately decorated naval vessels of the 16th and 17th century. The reason for this is that new technology has enabled the discovery of a large number of sensational shipwrecks, which has dramatically increased knowledge about and research material on this era. This period also coincides with one of Sweden's most ambitious artistic programmes for its vessels; almost no other age has devoted such extensive resources to nautical decoration. At the same time, it is important not to let vessels from the period when Sweden was a great power completely overshadow the imagery that both preceded and followed that era. While they have a visual advantage – with their striking design – to our present-day eyes, they are not unique in their ambition to convey the identity of their vessels. The countless methods, materials and forms of imagery used through history to embed vessels in the ideas and needs of their time all testify to the fundamental importance of images to life at sea. They in turn reflect larger conceptual worlds and questions about the importance of water and the sea in our history.

Preceding pages: Ornamental relief painted green and yellow, probably part of the sternpost decoration. Origin unknown.

Three-quarter-length figure of woman painted white wearing a veil. Origin unknown. According to tradition, this figurehead was the inspiration for the writer Karen Blixen's 1942 short story "Peter and Rosa".

So for posterity, the study of vessel ornamentation is as fascinating as it is complicated. Our relation to the sea as an interface and a landscape has changed dramatically with the great technical and social transformations of the 20th century. Naval battles, trade routes and arduous circumnavigations of the globe are no longer part of public awareness in the same way as before, although they still exist as phenomena. These visual cultures and the narrative traditions attached to them have been relegated to the past, and this historical distance often creates problems of knowledge. Awareness about the history of Sweden's importance as a seafaring nation is, to say the least, fragmentary, despite its fundamental importance to the society we live in. Seeing and understanding artefacts in their broader context requires time and attention.

This book brings together samples of research currently being conducted on the history of vessels and their visual conception. Annika Bünz introduces with a comprehensive reflection on the figureheads' travel through time and space. Leos Müller provides an overview and food for thought about the broader writing of history concerning the importance of the maritime world. Johan Rönnby describes the *Gribshunden* (The Griffin) wreck and uses its dragon-shaped figurehead as a window into the politically tinged visual world of the late medieval period. Niklas Eriksson considers the momentous development of imagery on the many merchant and naval vessels of the 17th century, thus providing a broader basis for Fred Hocker's text on the warship *Vasa* and Lars Einarsson's text on the warship *Kronan* (The Crown). Ingrid Ulfstedt takes a step forward in time and describes developments in the 18th and 19th century. The book begins with recent photos of figureheads in the Skagen Collection.

It is difficult to delimit the questions raised in interpreting the many nautical images. There is no easy way to separate maritime history from society in general; on the contrary, this history instead illustrates the inextricable links and relations between land and water. The stories about these figureheads that emerge can provide a point of departure for new insights while bringing to life the many layers of the history of the sea.

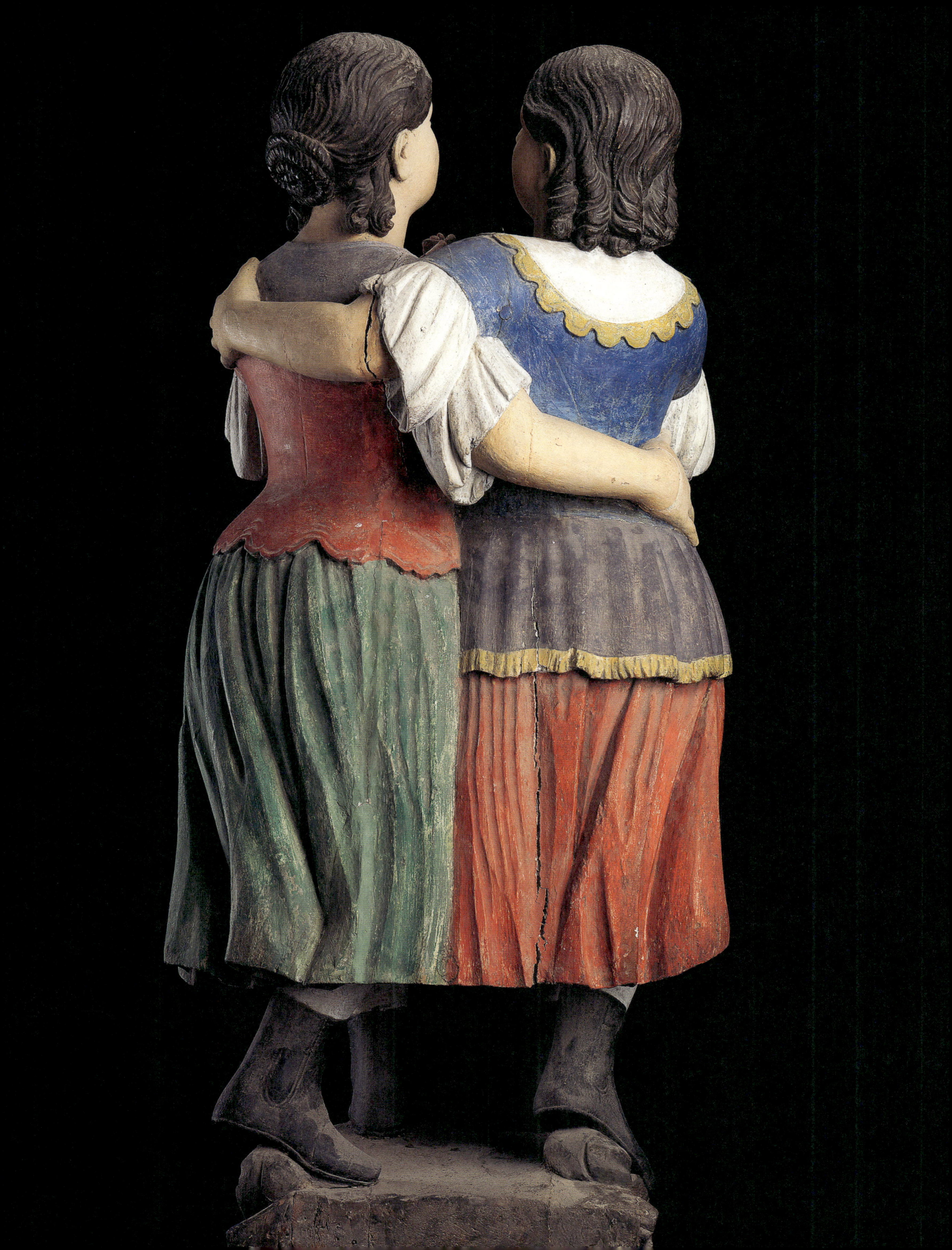

Full-length figure of two girls holding one another and a small bouquet. Believed to have belonged to the French schooner *Pauline et Marie* from Nantes, which sank near Tyborøn, Jutland in 1866.

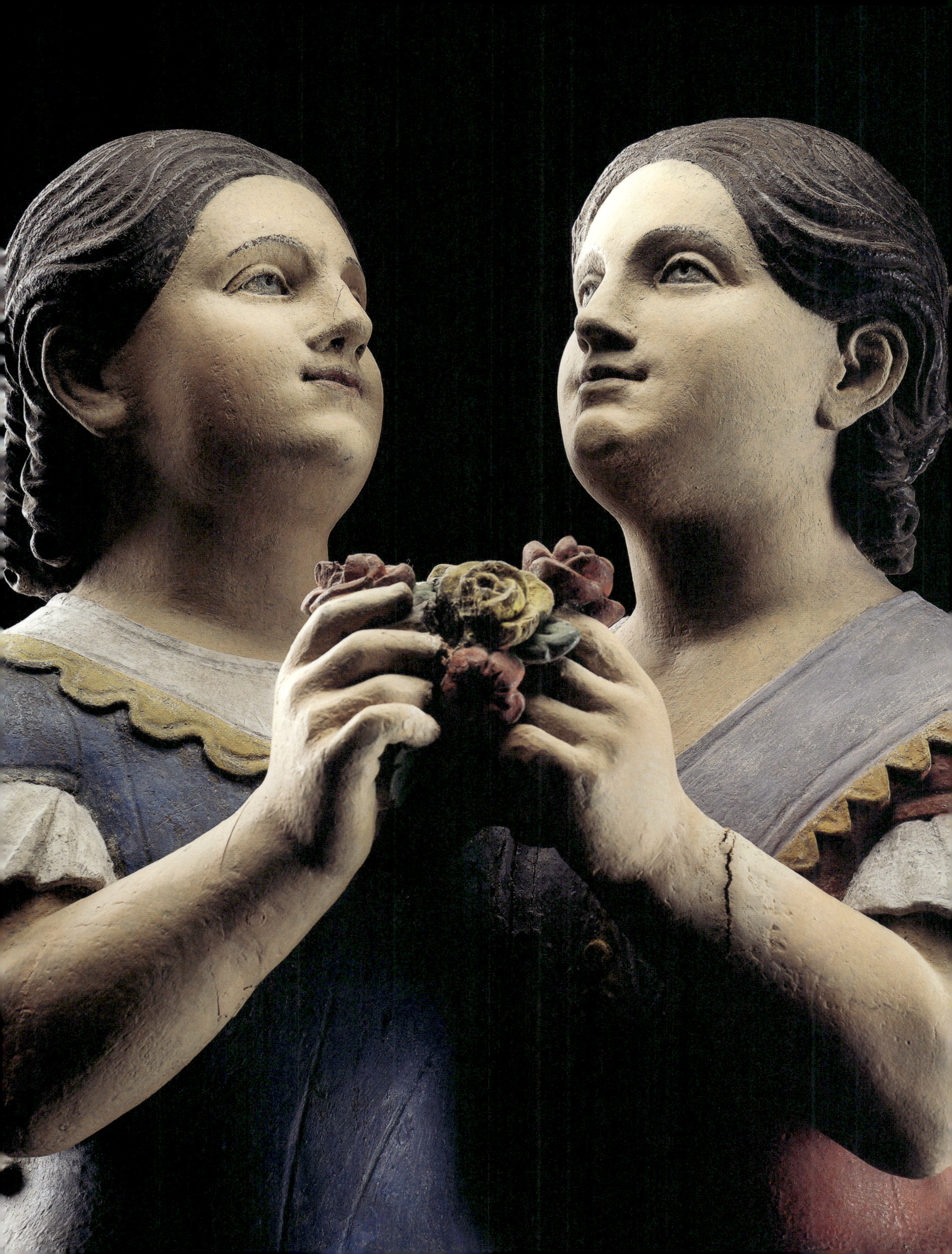

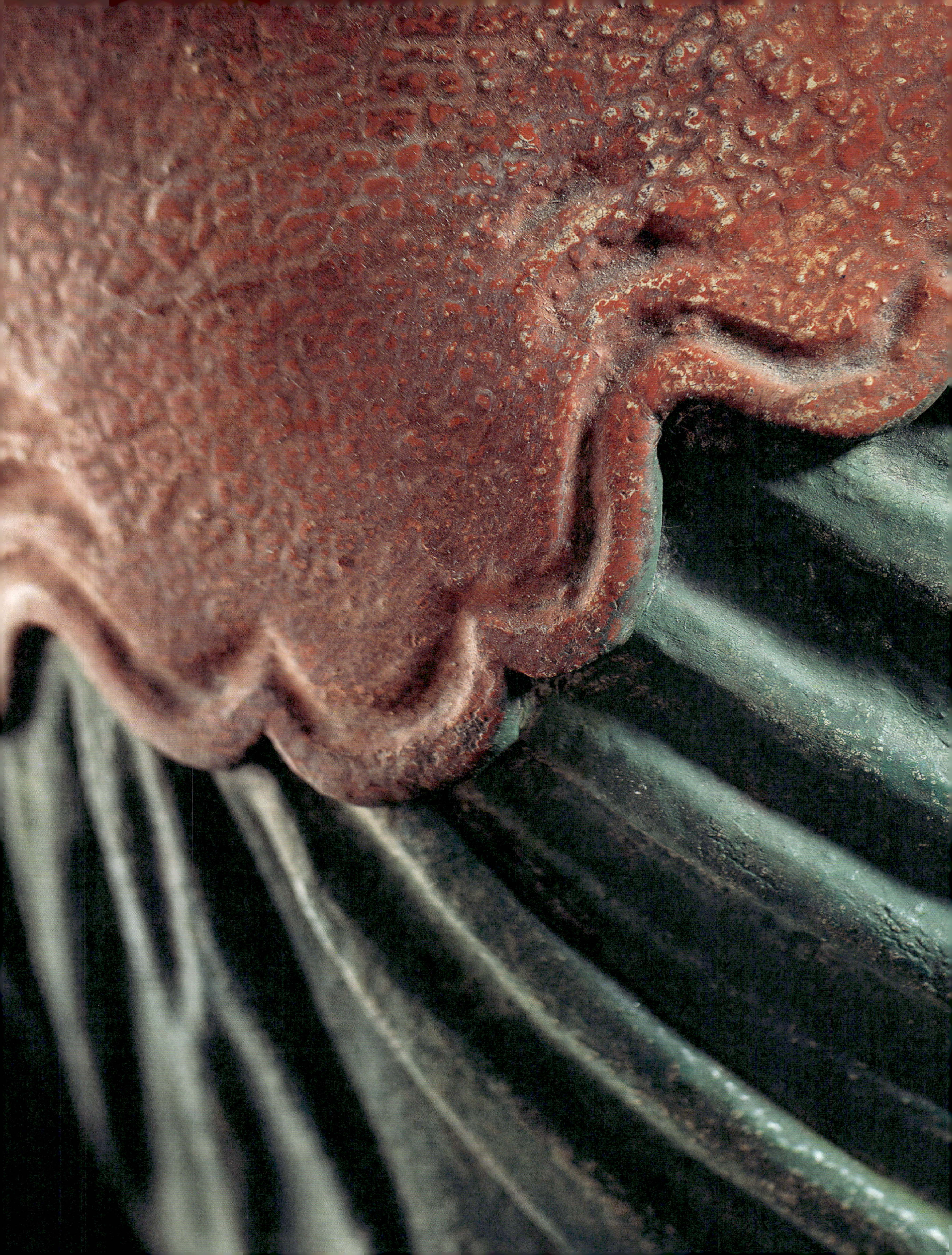

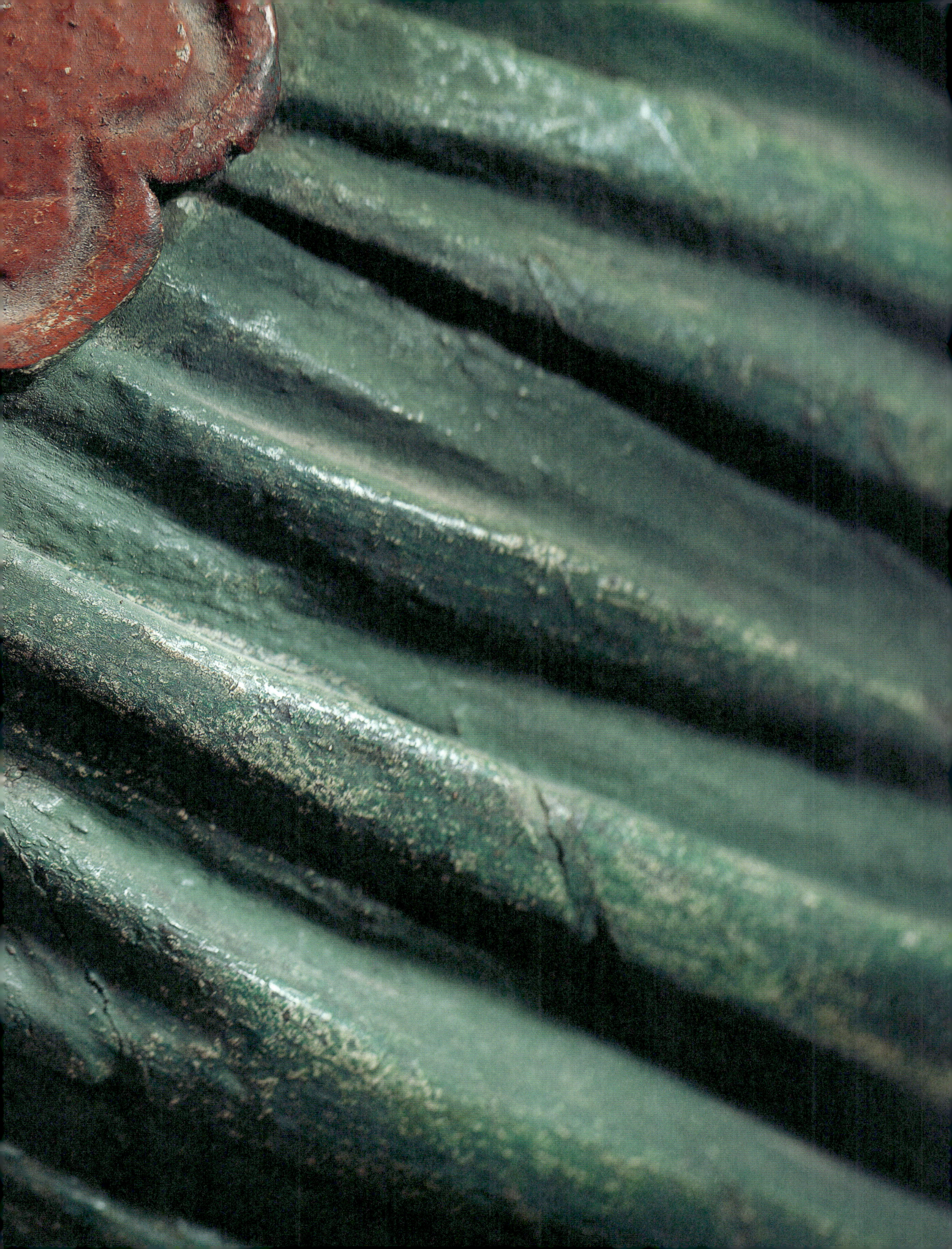

Demi-figure of the sea god Neptune with white hair and beard and gilded crown. Origin unknown.

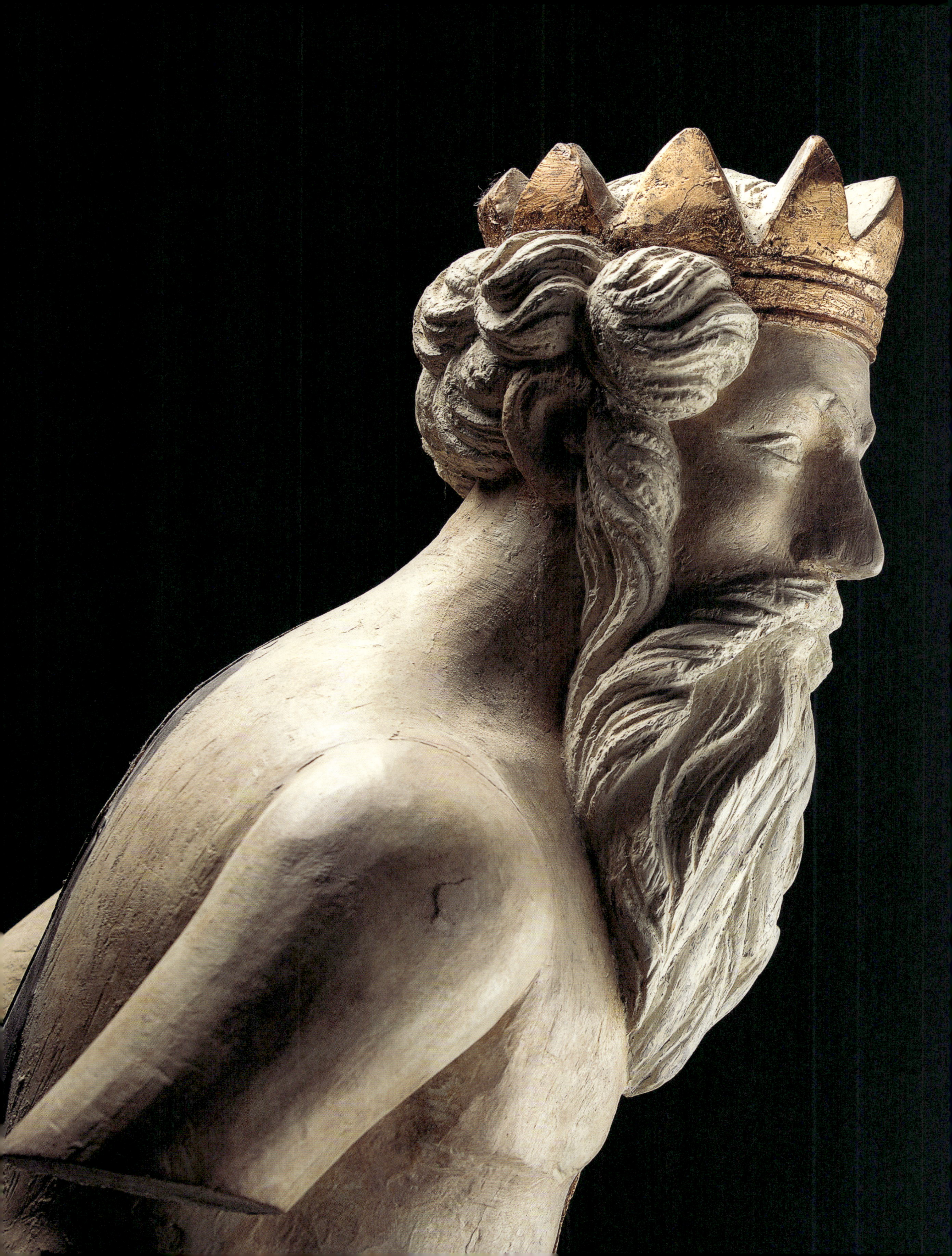

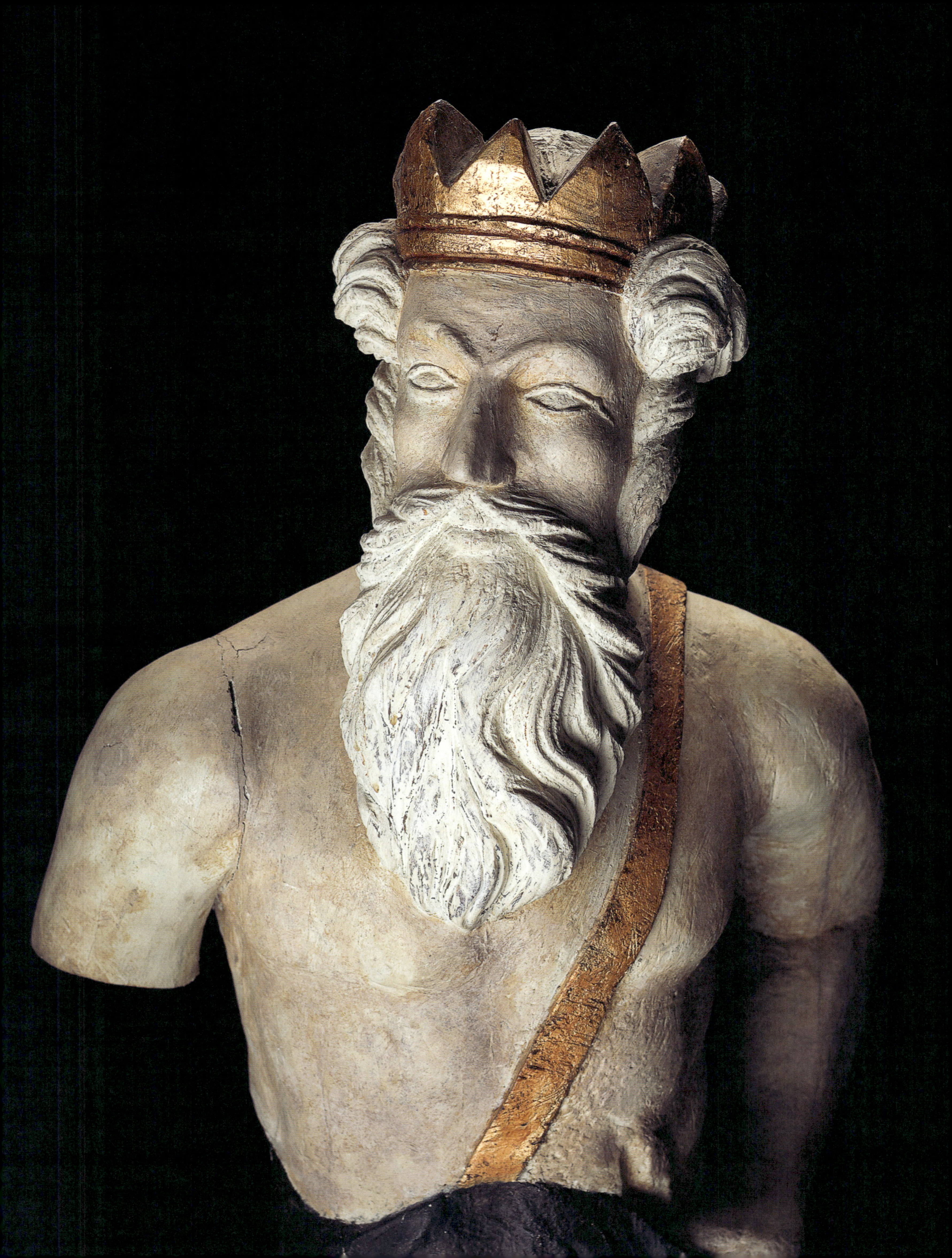

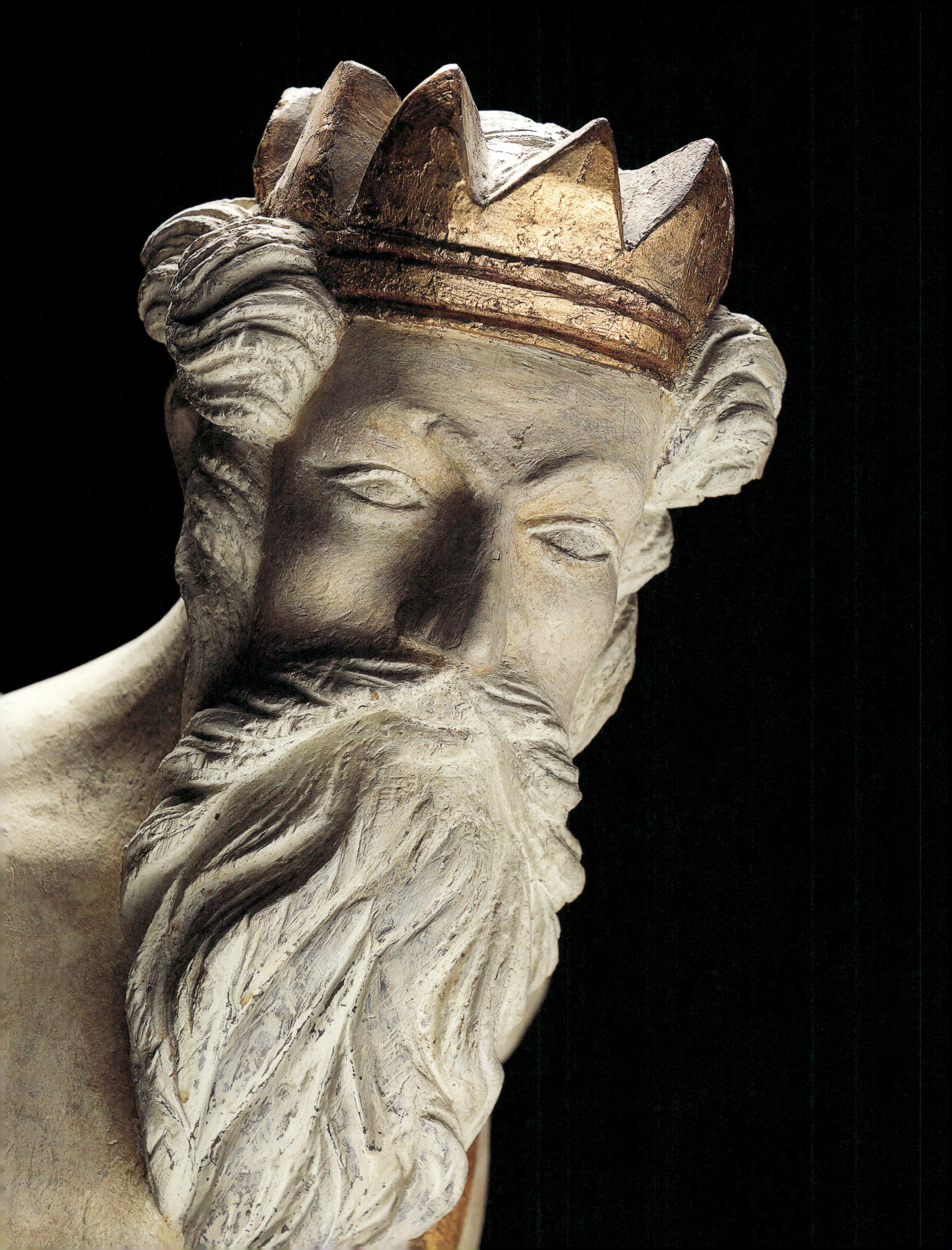

Female figure draped in reddish-brown gown with dark wavy hair, fastened at the back with a flower. Origin unknown.

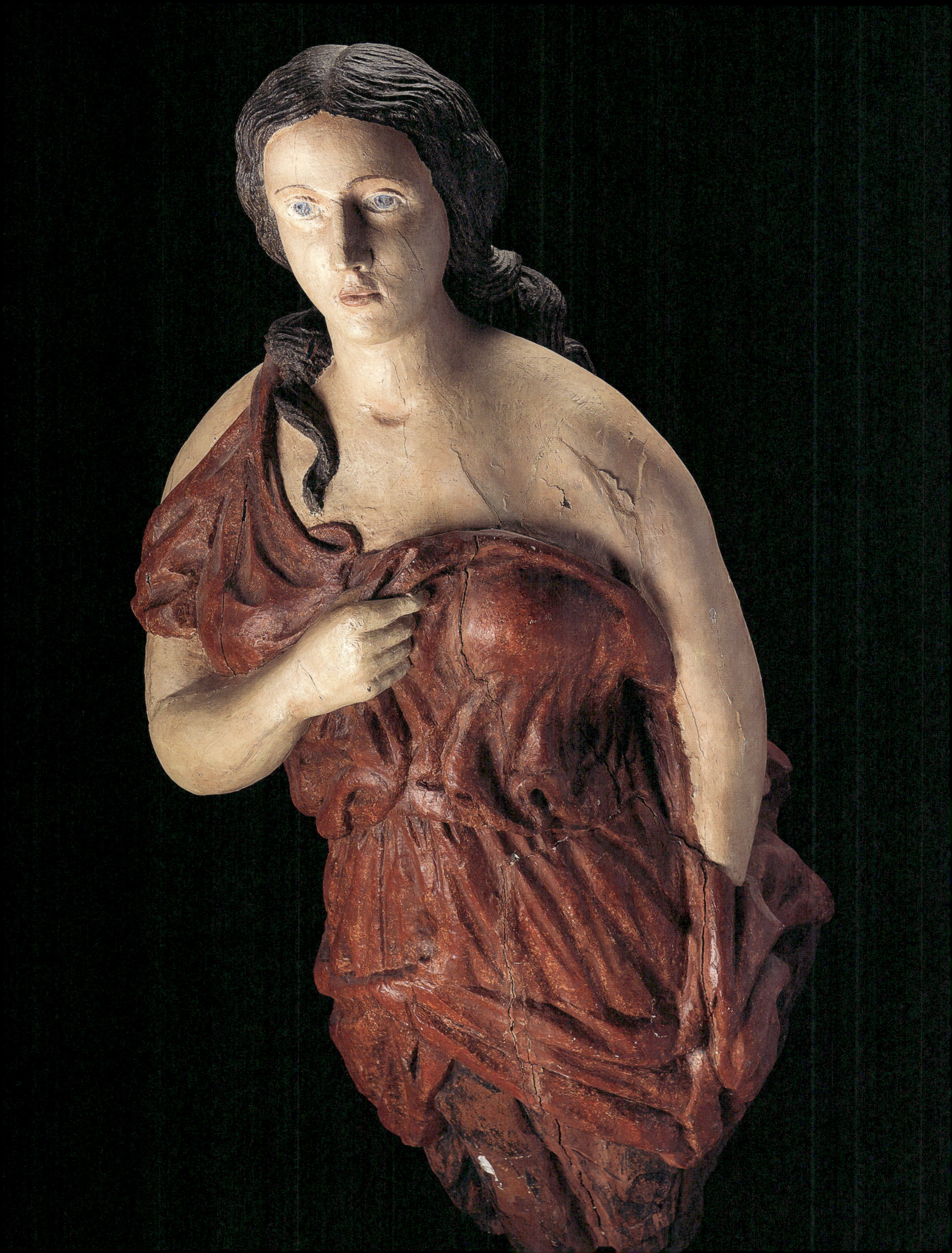

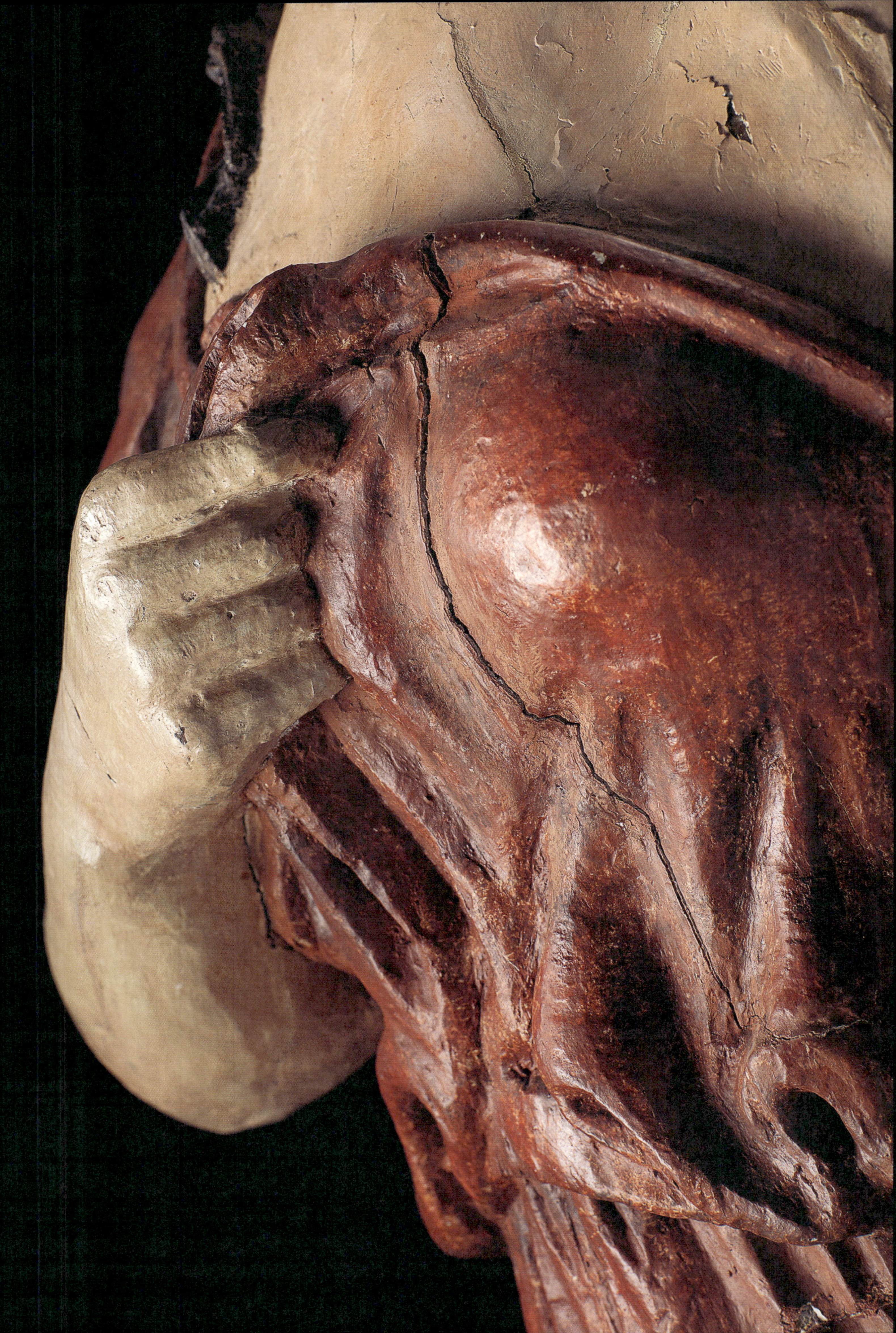

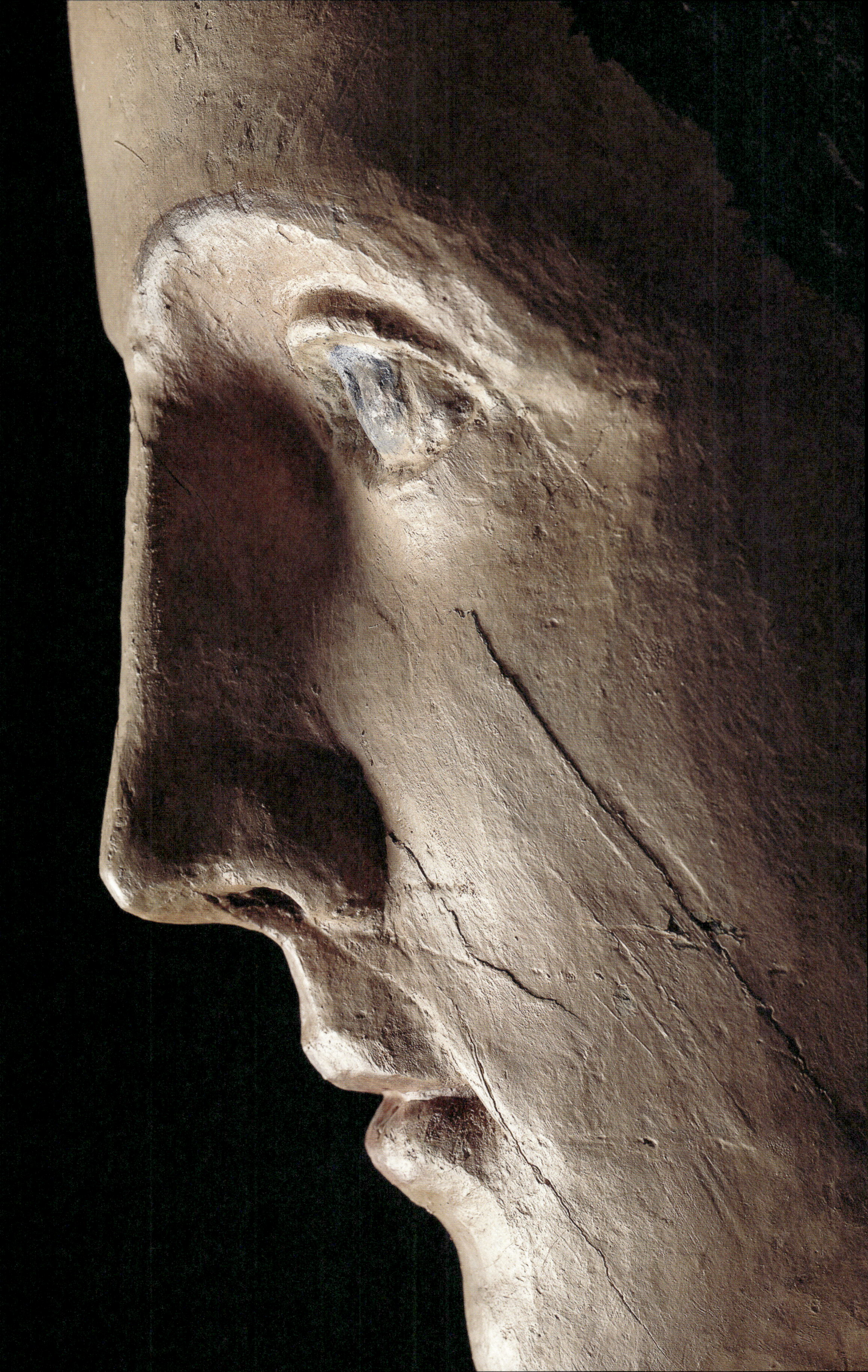

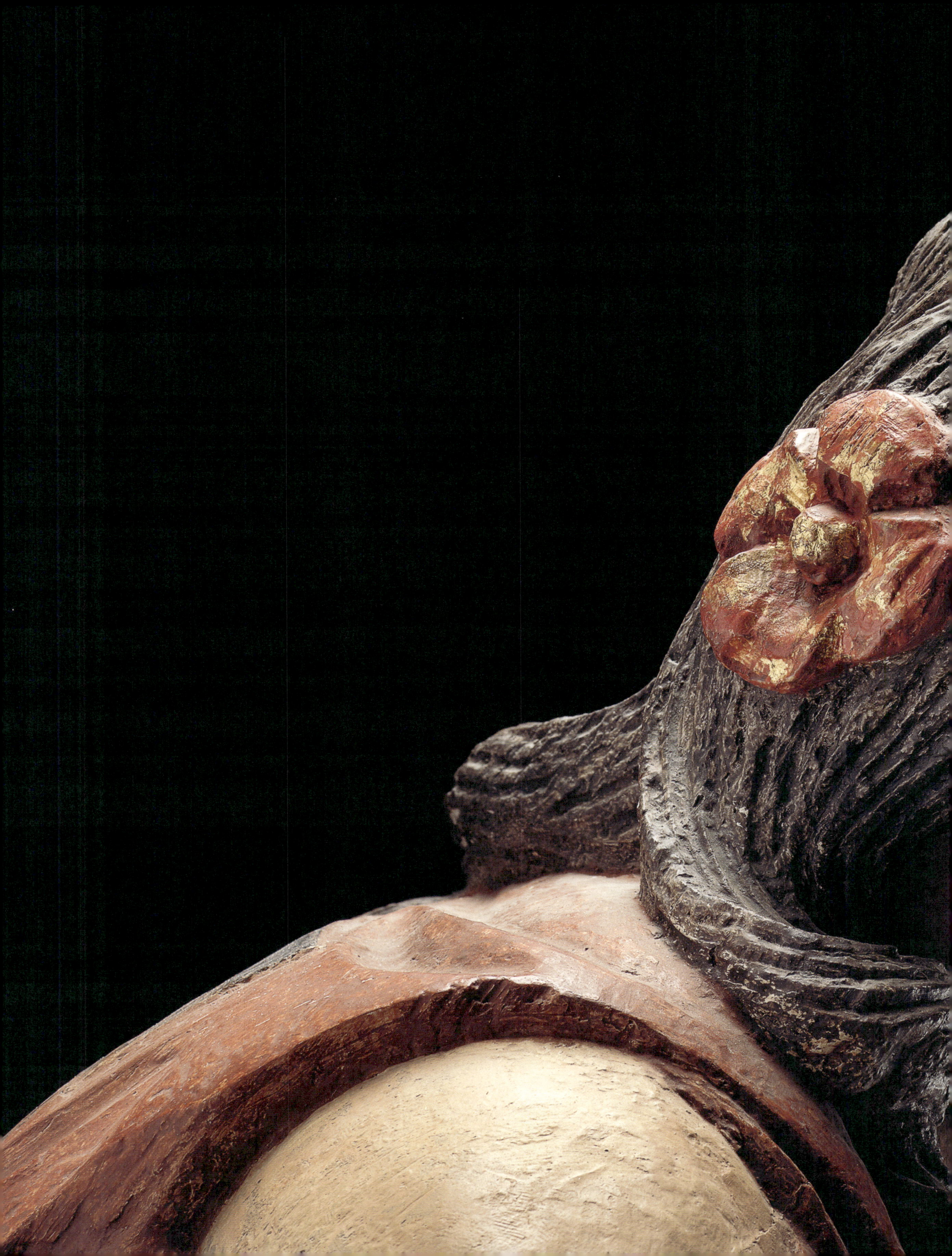

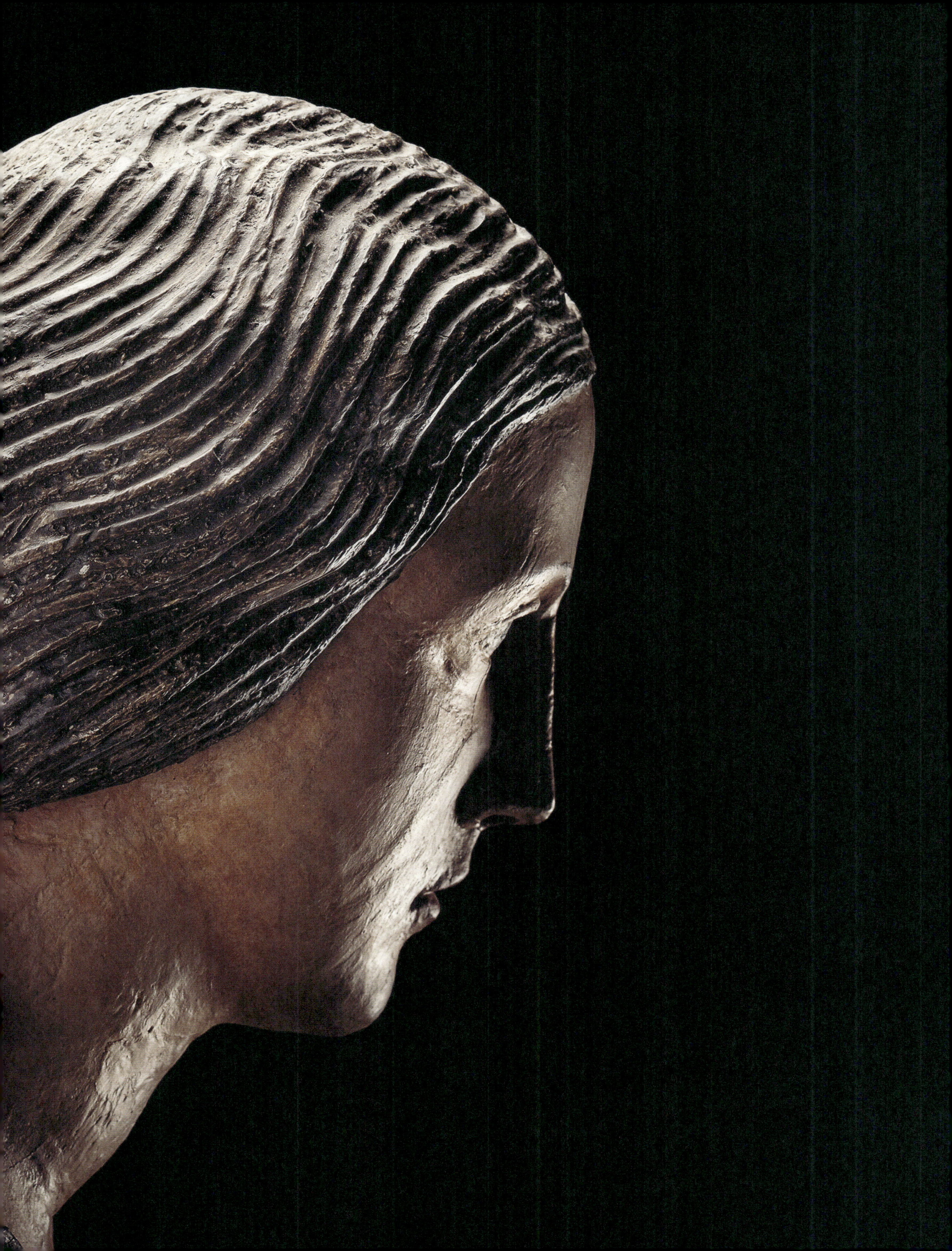

Figure of medieval king from the Norwegian barque *Haakon Haakonson* of Tvedestrand, which ran ashore in 1903.

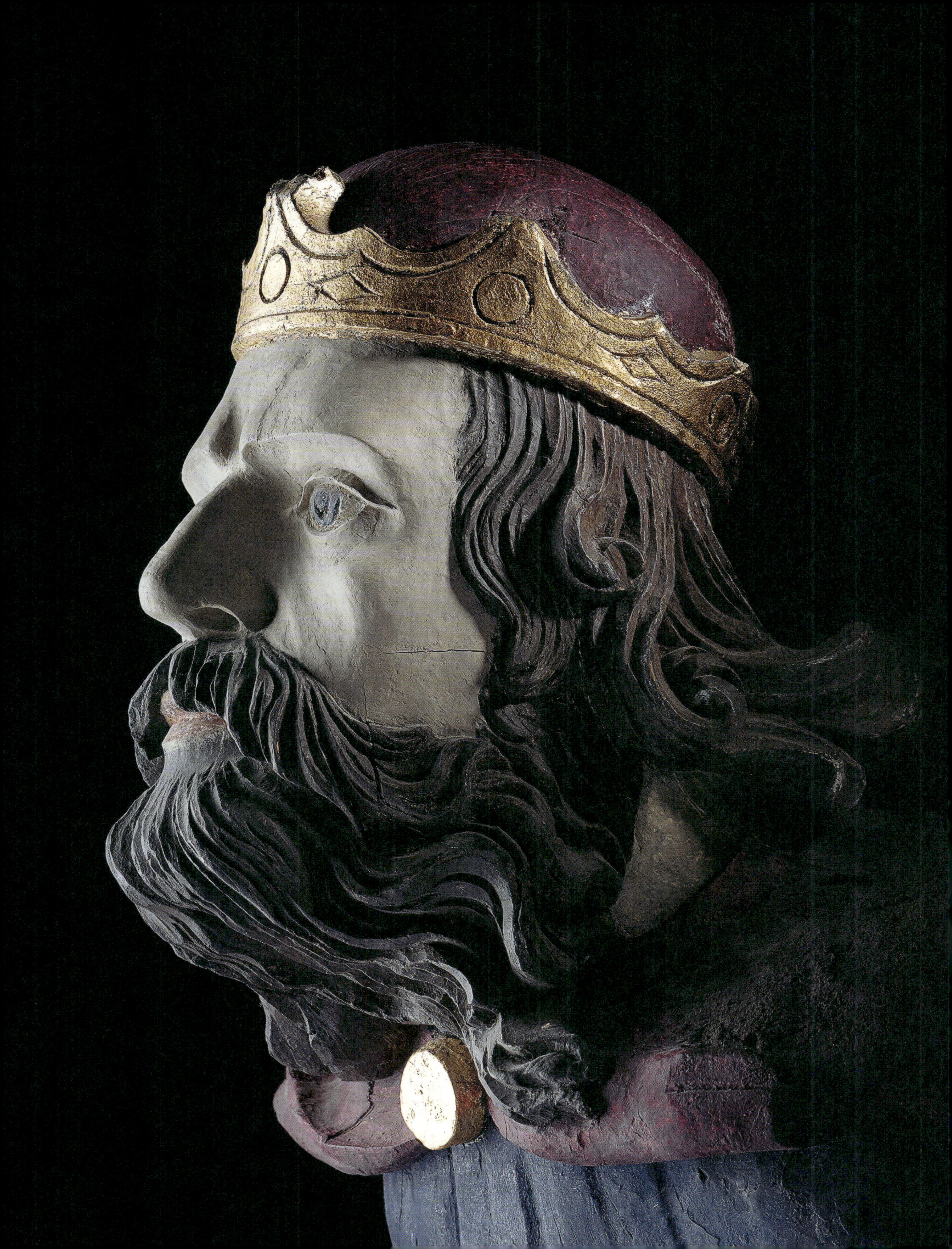

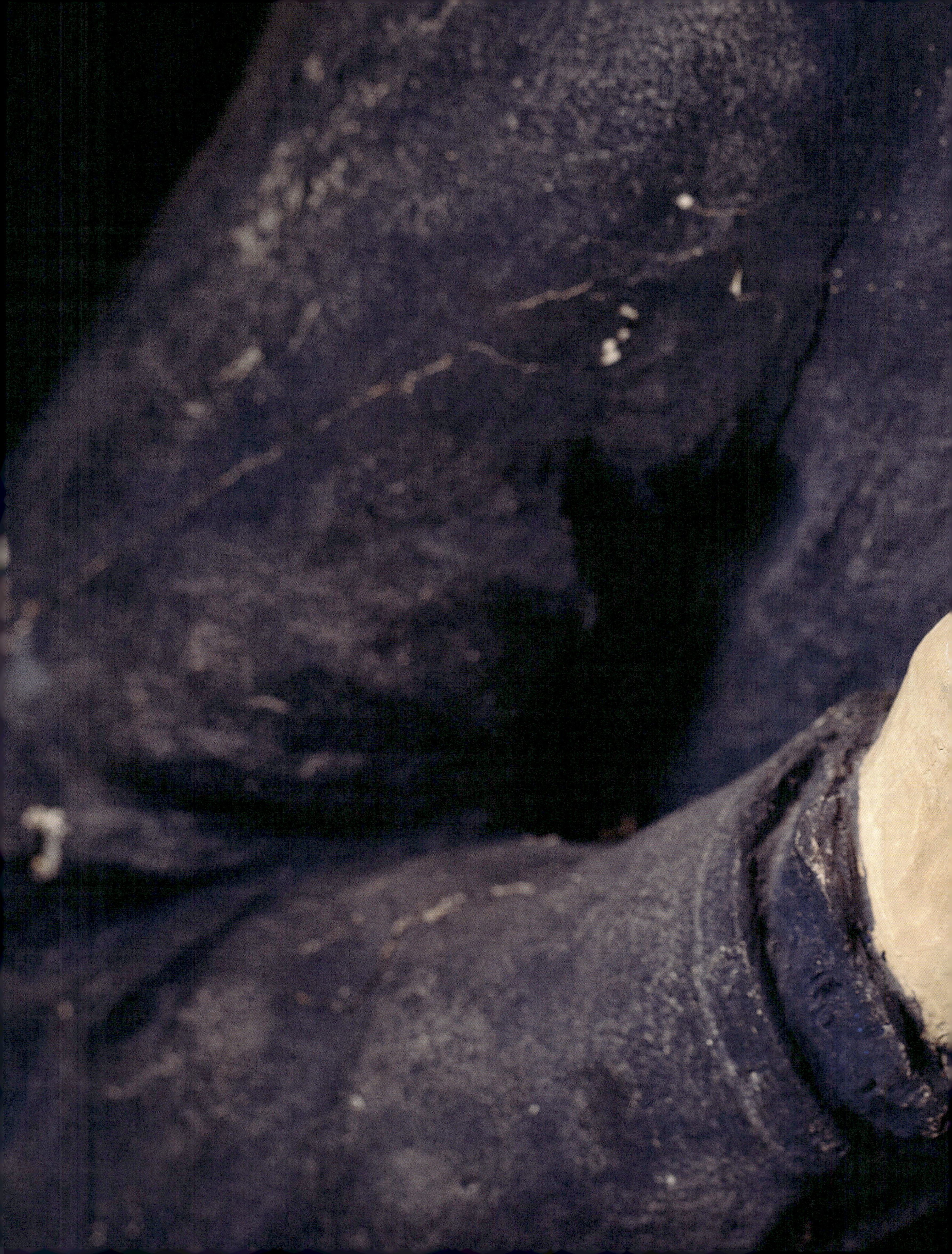

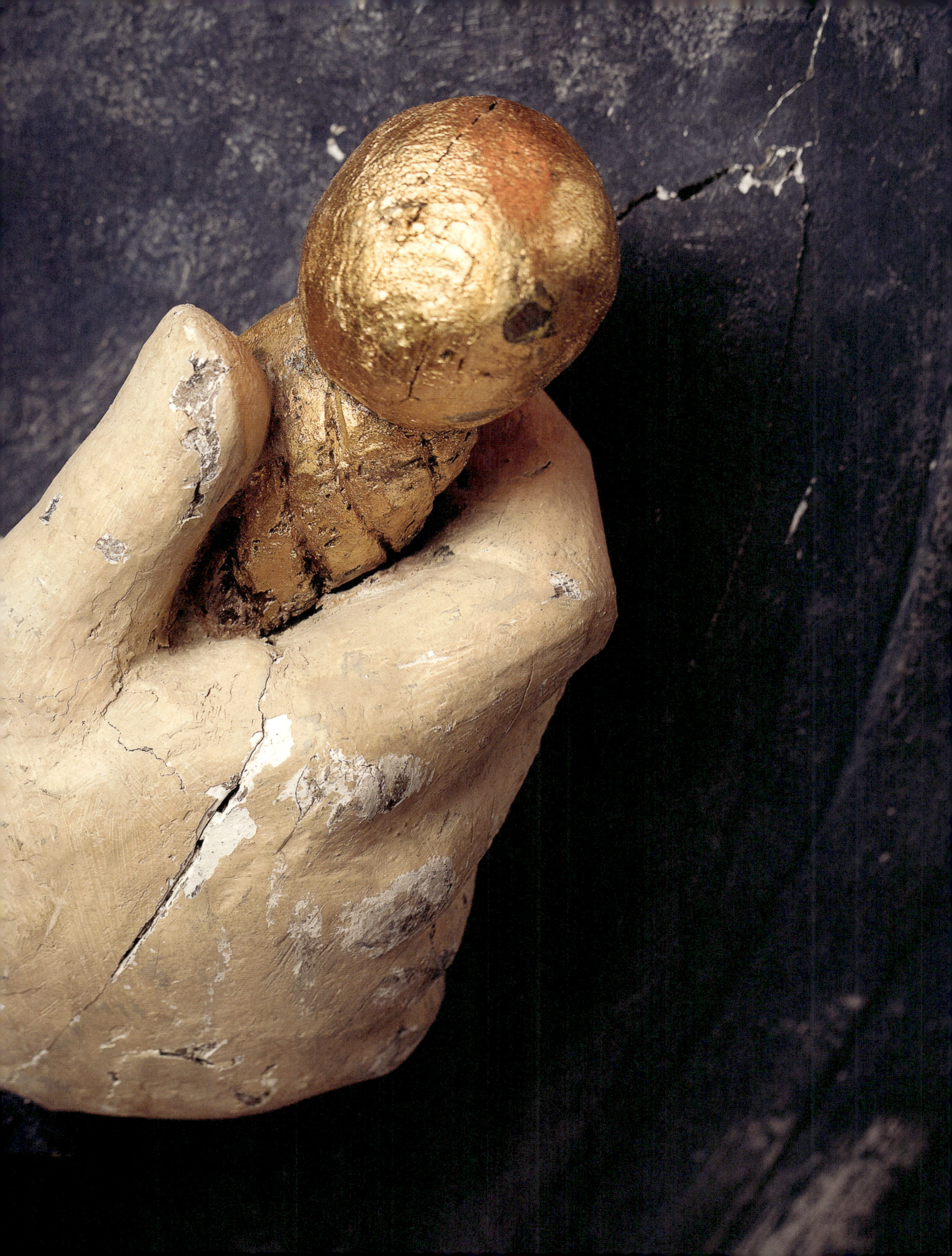

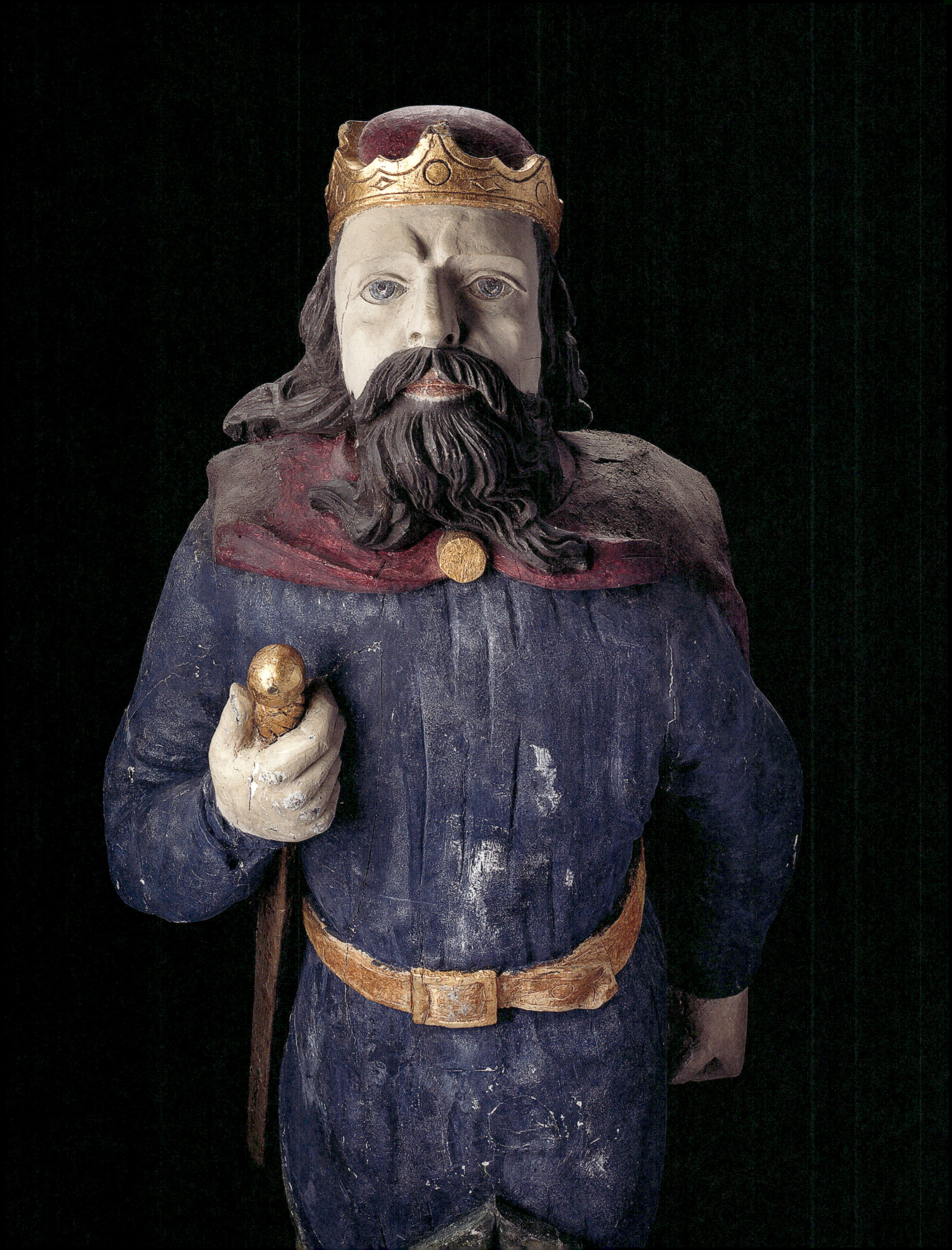

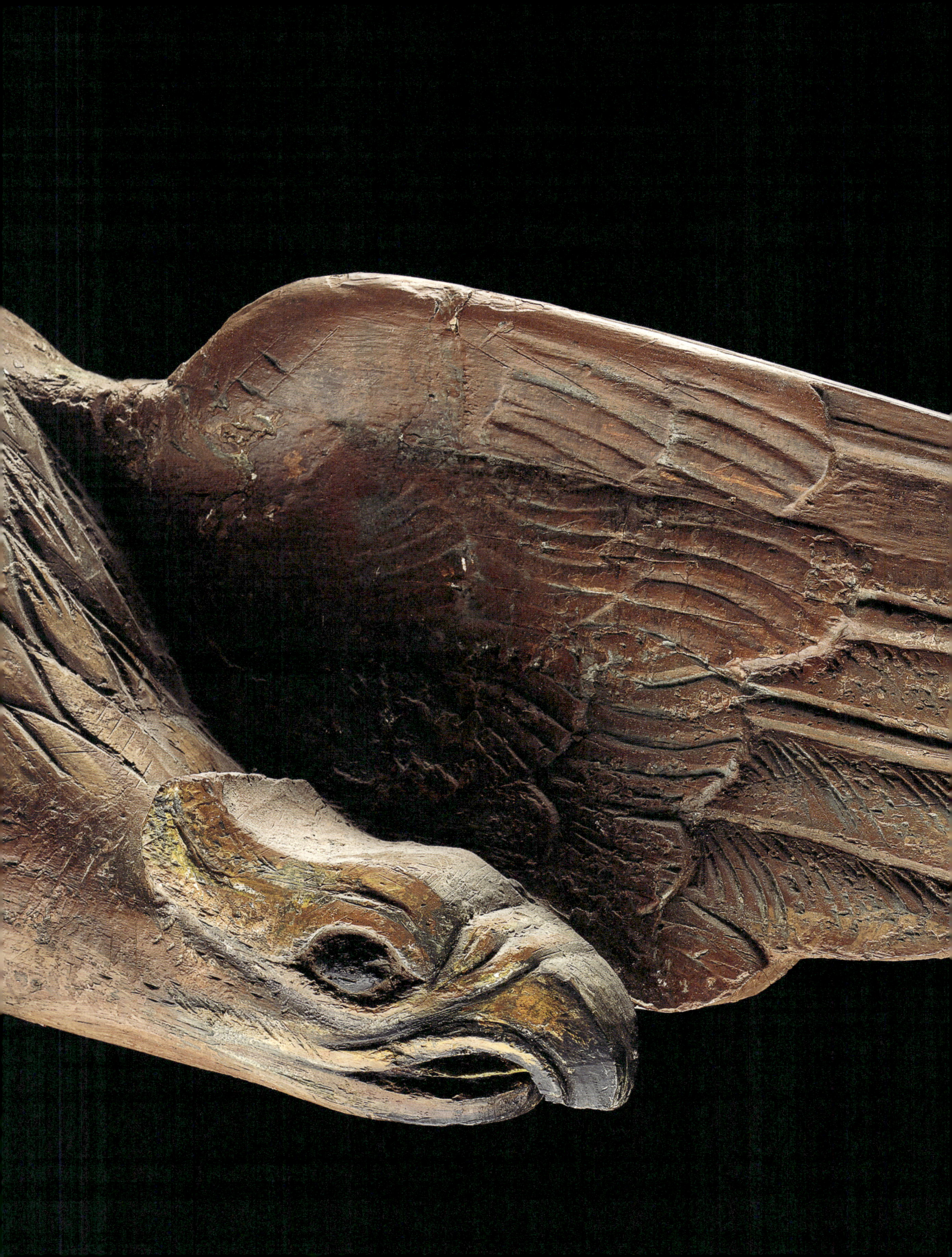

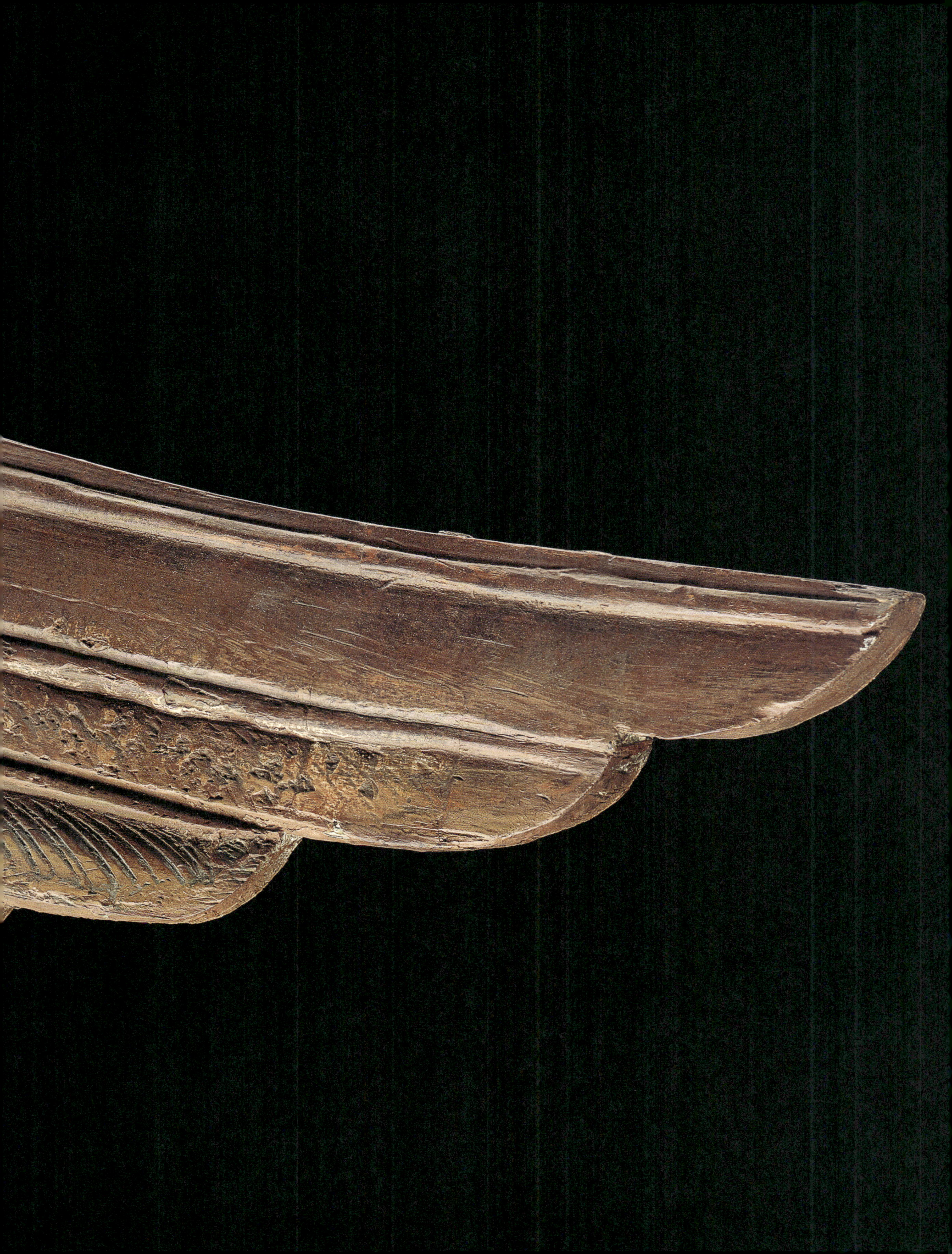

Preceding pages: Relief of eagle with spread wings, painted brown. Probably belonged to the barque *Örnskjöld* from Härnösand, which ran ashore off Skagen in 1873.

Quarter-length figure of man in Chinese attire from the English ship *China* of Bath. Ran ashore off Lyngby near Løkken in North Jutland in 1879.

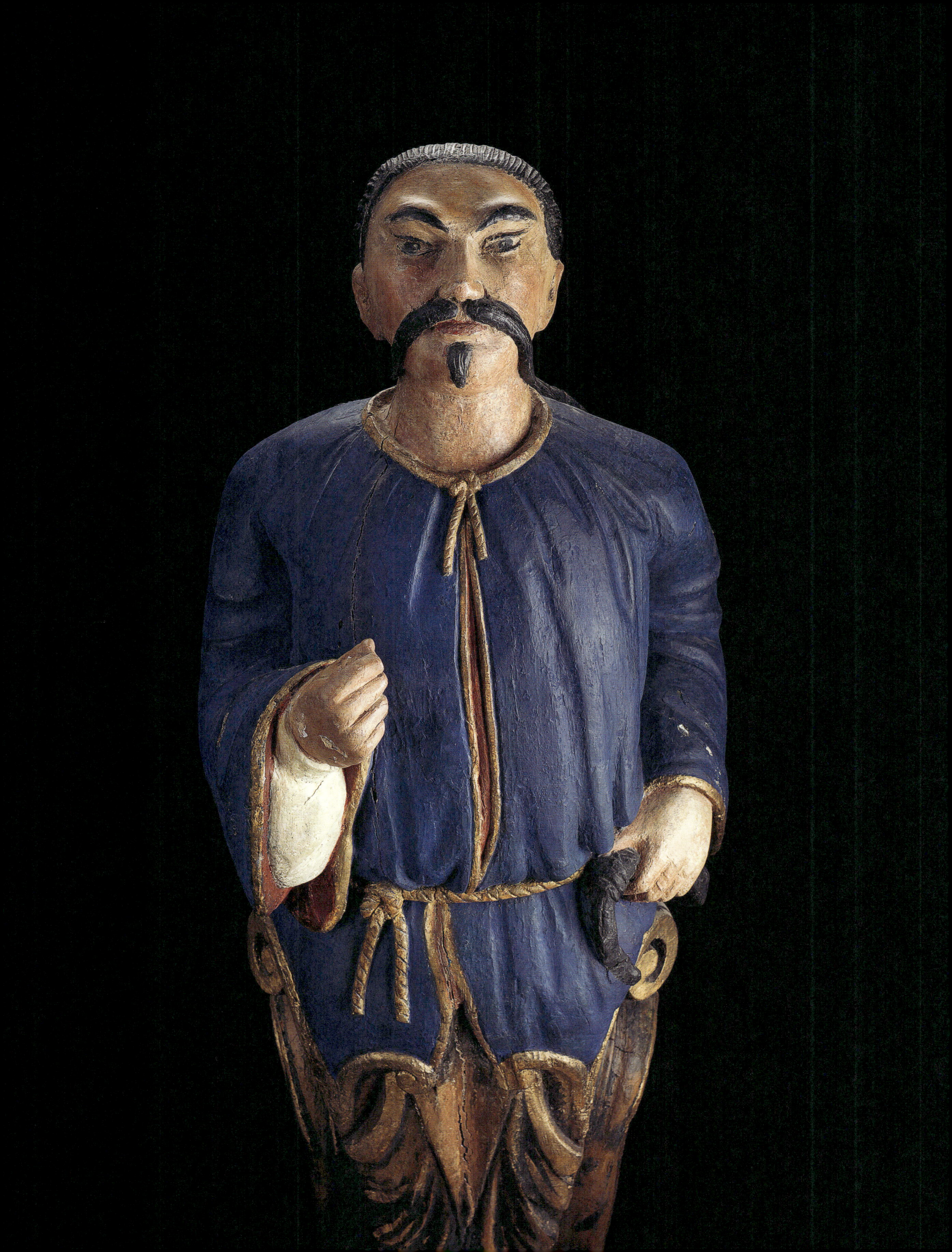

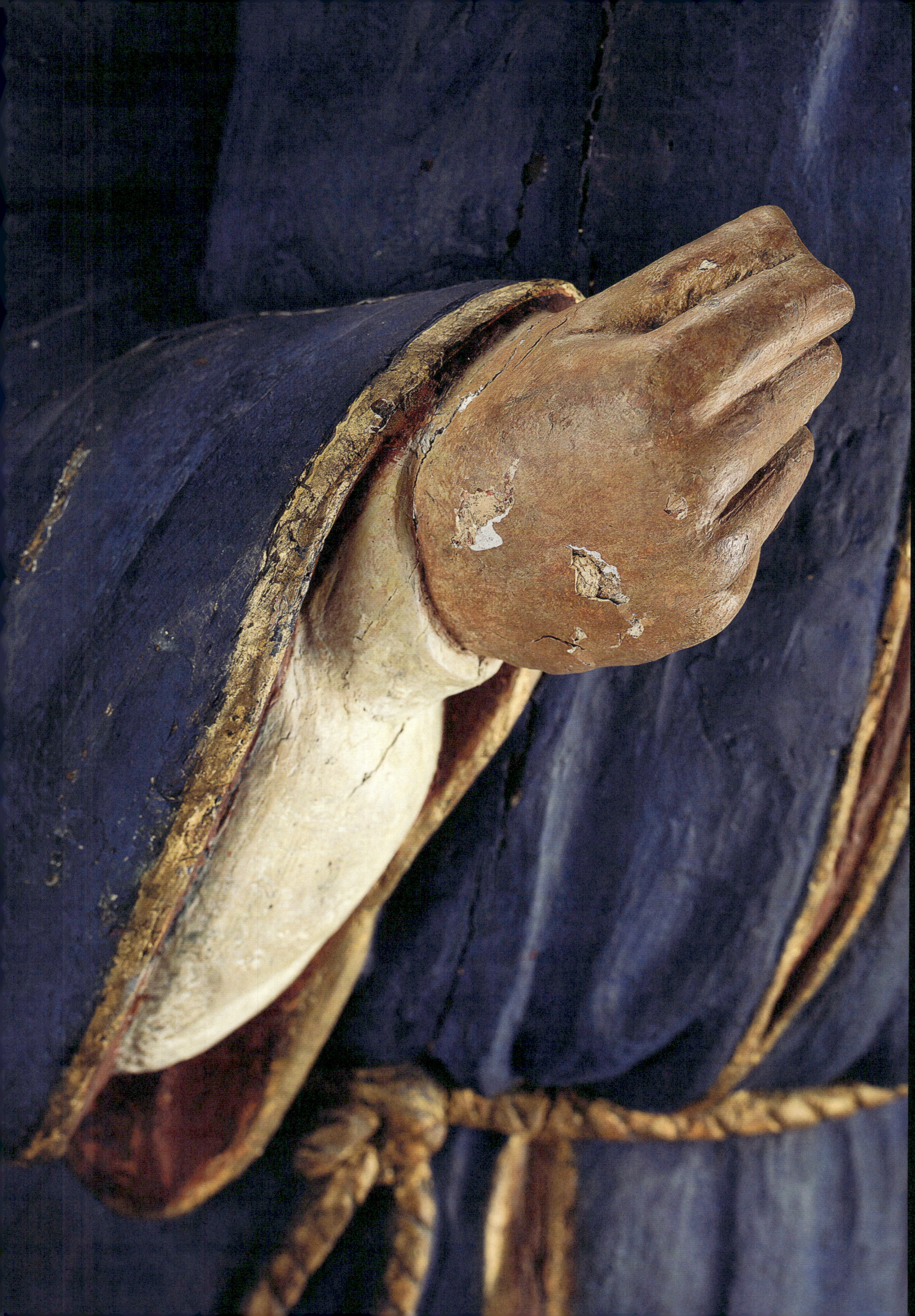

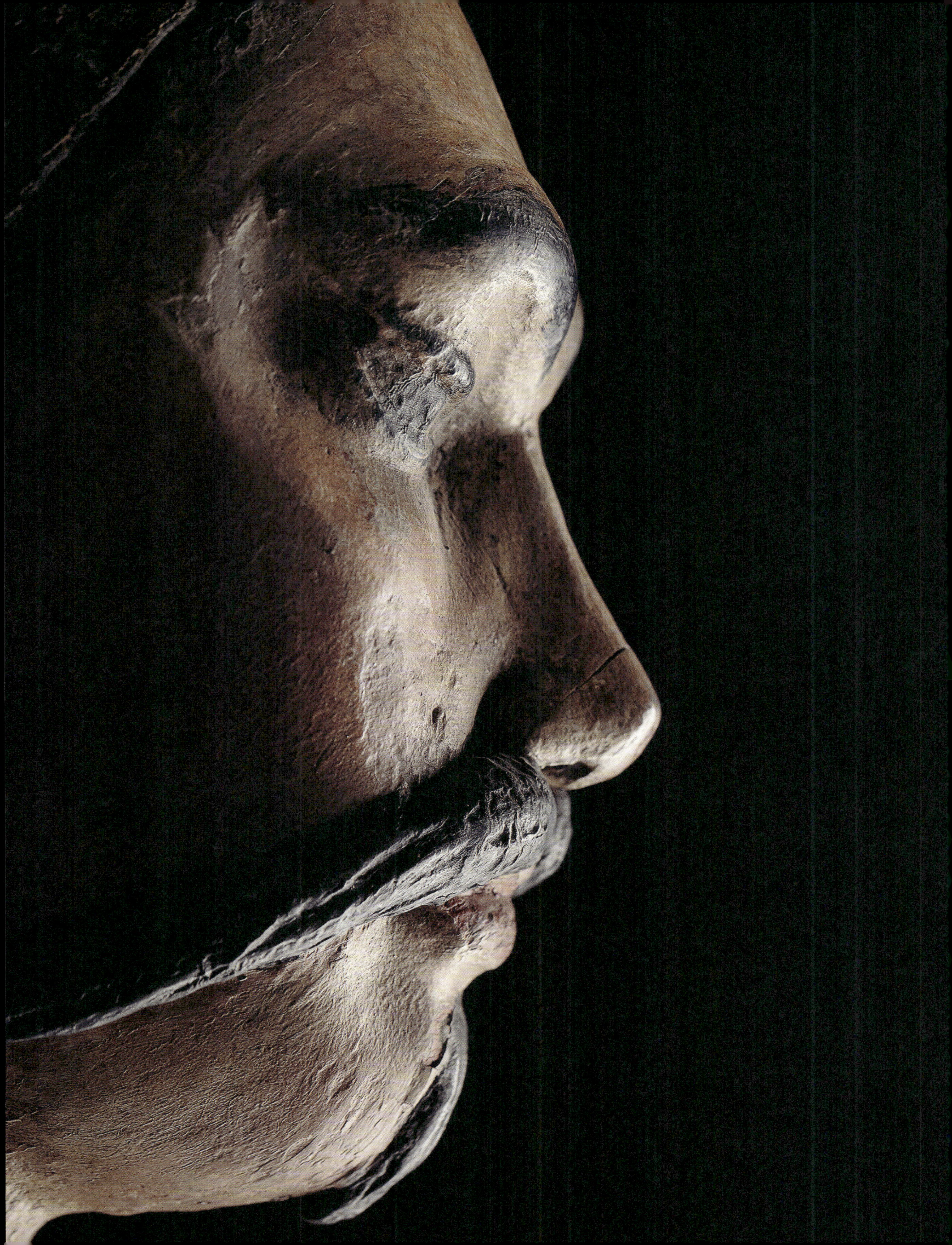

Roman youth in demi figure with dark hair, laurel wreath, blue ribbon and light-coloured drapery placed on remnants of a volute-shaped construction painted green. Origin unknown.

Following pages: Sternpost ornament with framed relief portrait of dark-haired woman with red hat and gown. Probably from the brig *Lady Stanley* of West Hartlepool, which ran ashore near Tyborøn on the west coast of Jutland in 1888.

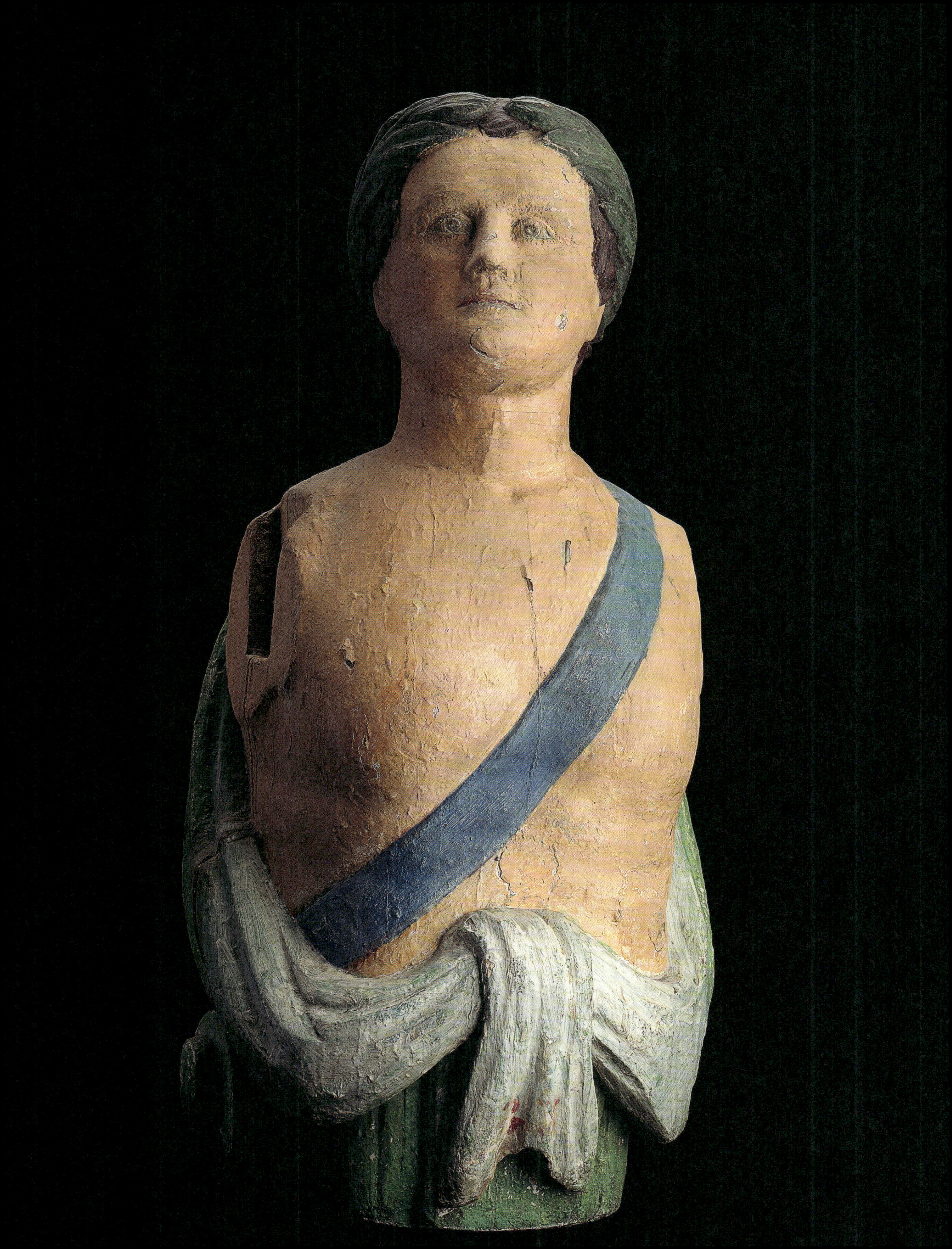

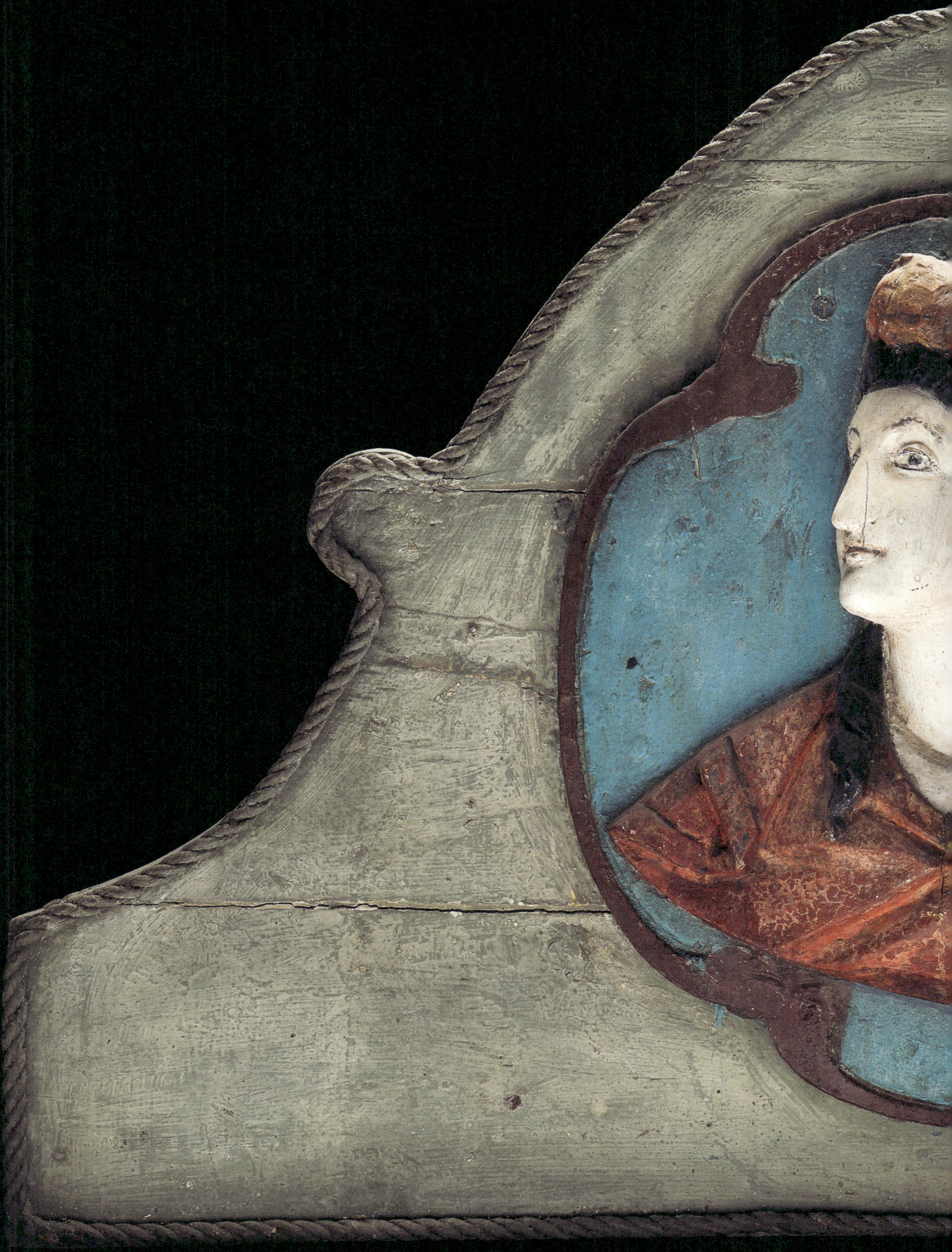

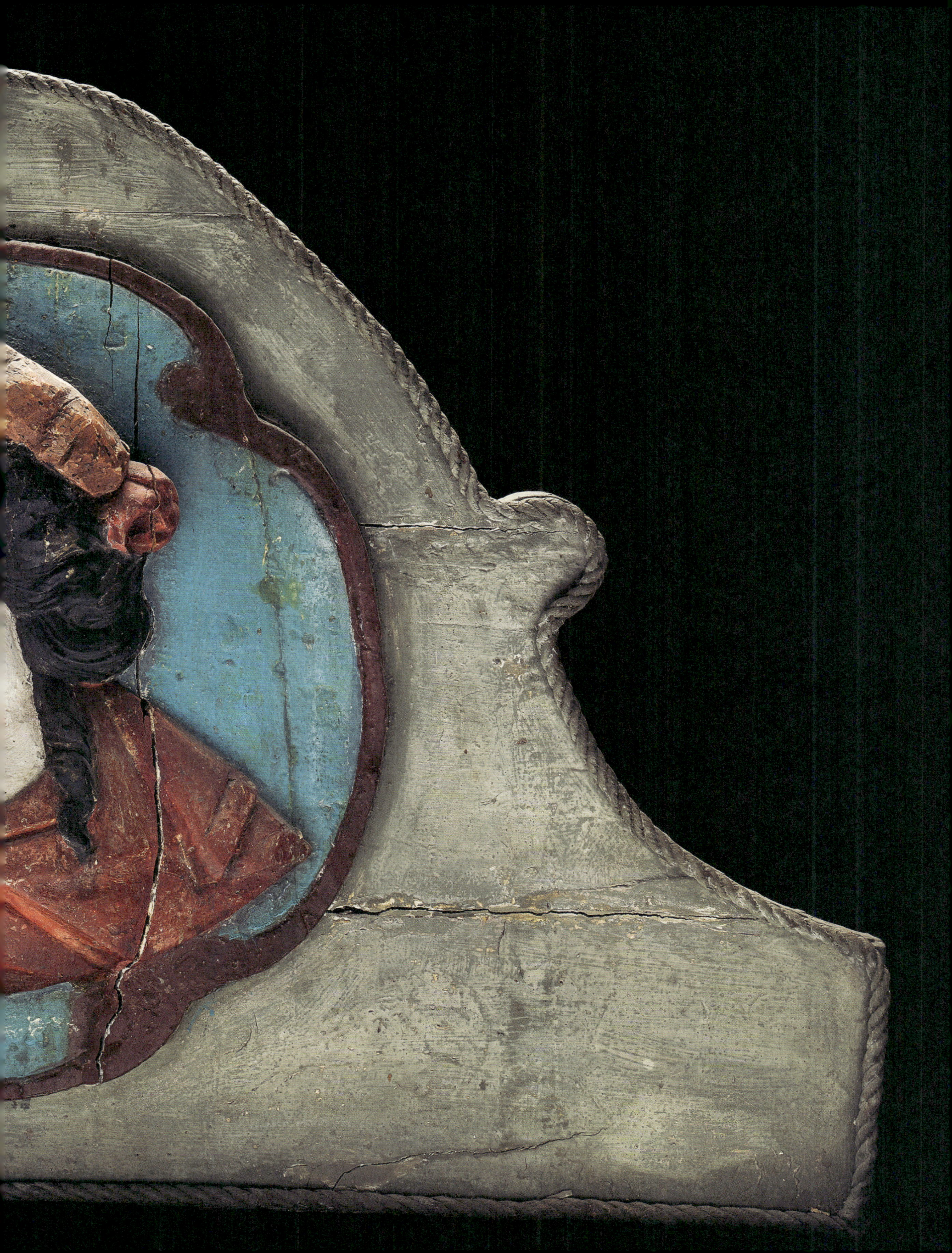

Relief of the shipowner type from the German ship *Onkel* of Geestemünde, which sank off Skagen in 1881.

Following pages: Relief of dark-haired woman in full-length figure dressed in Classical style. Origin unknown.

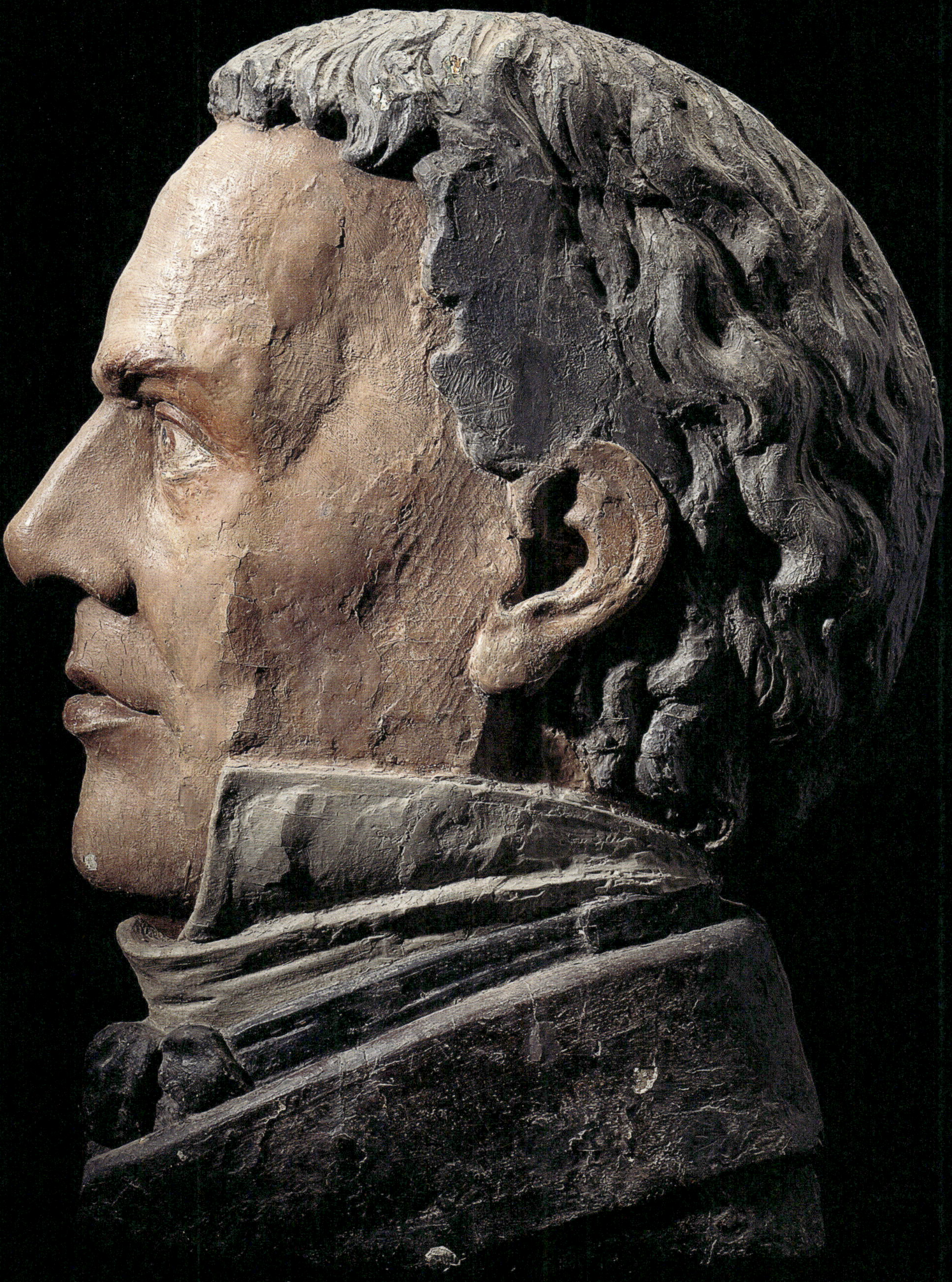

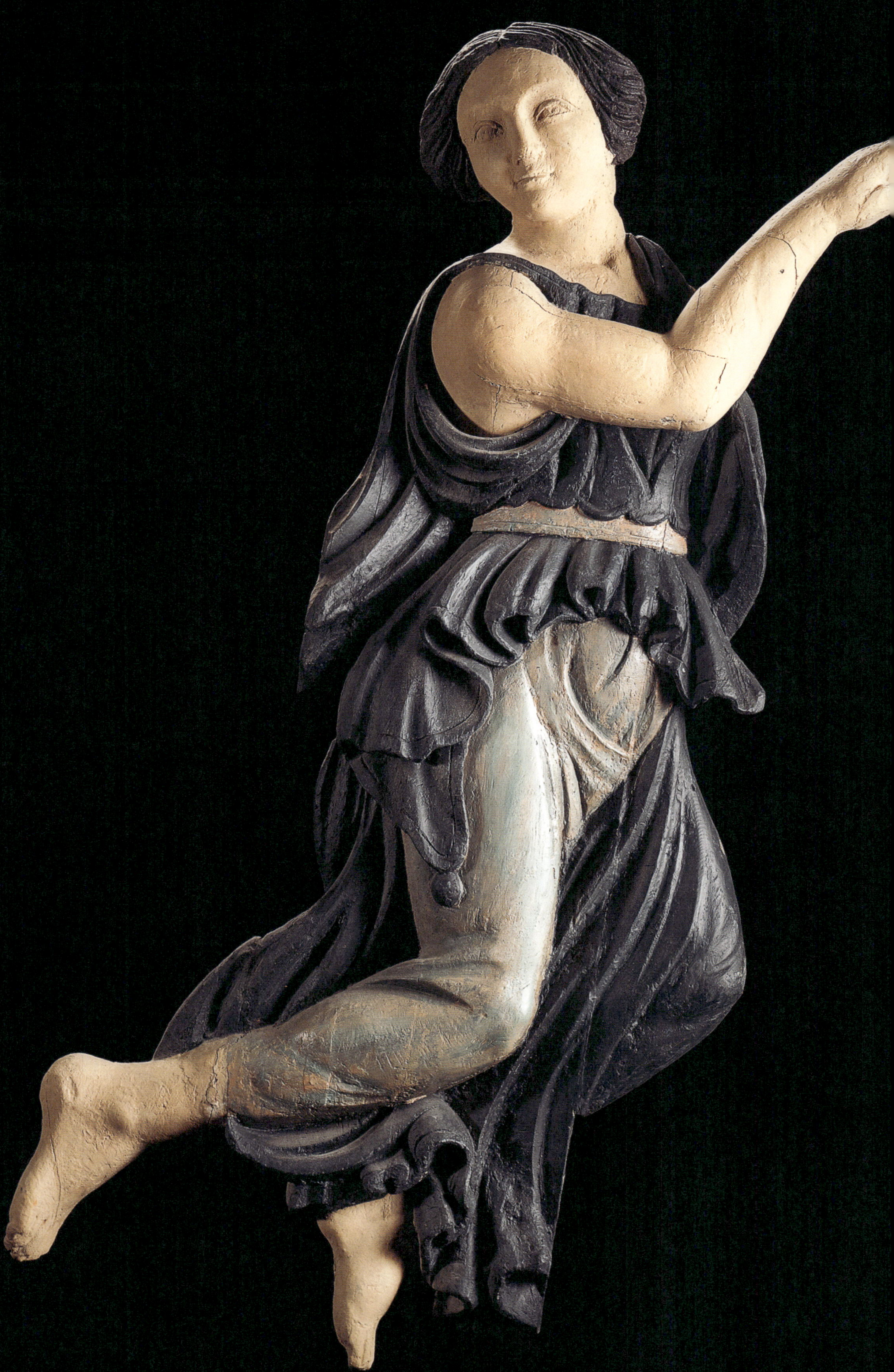

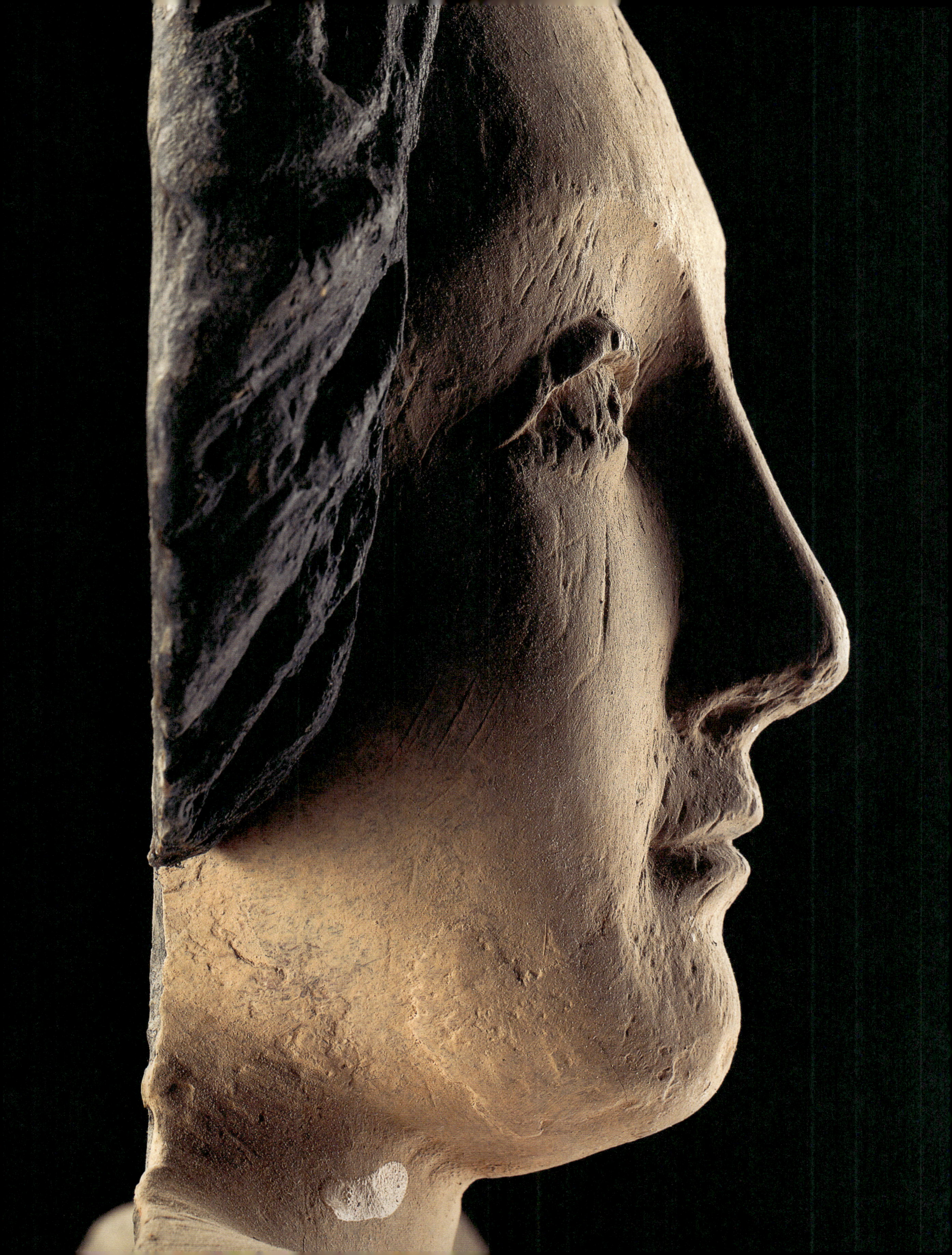

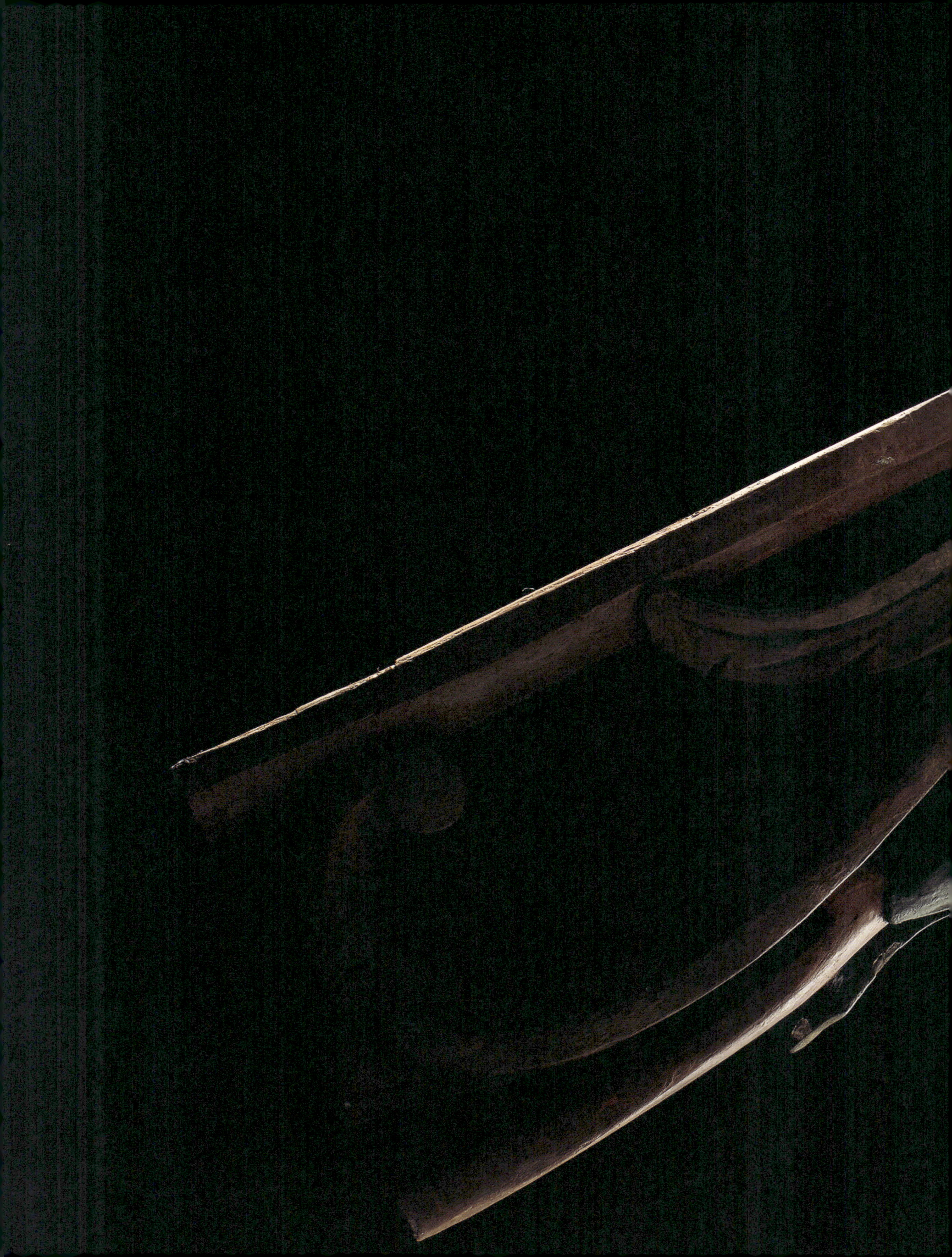

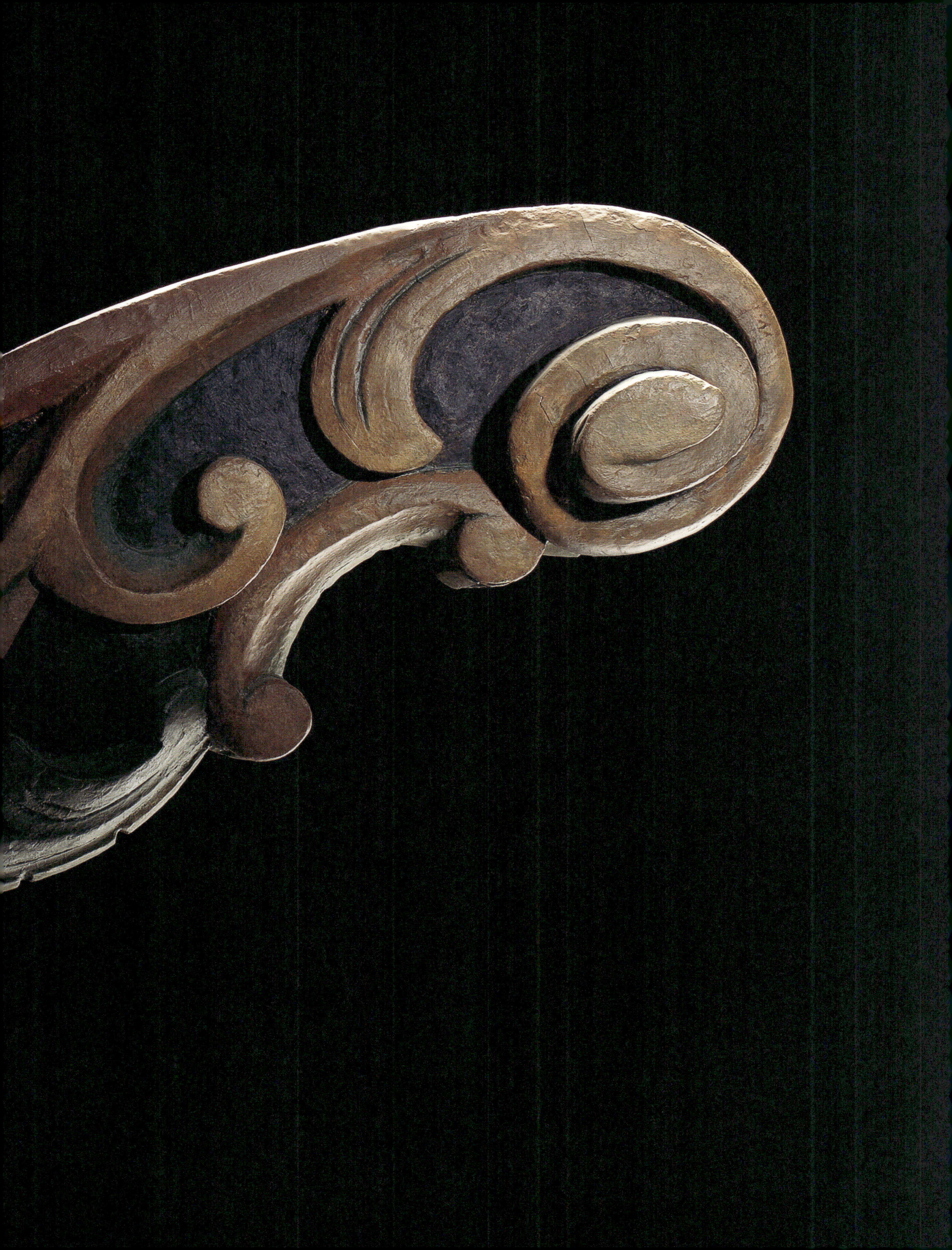

Preceding pages: Figurehead curl in the form of a volute curve, painted green and gold on black background. Origin unknown.

Three-quarter-length figure of woman in Classical gown, tiara and hair partly flowing. Origin unknown.

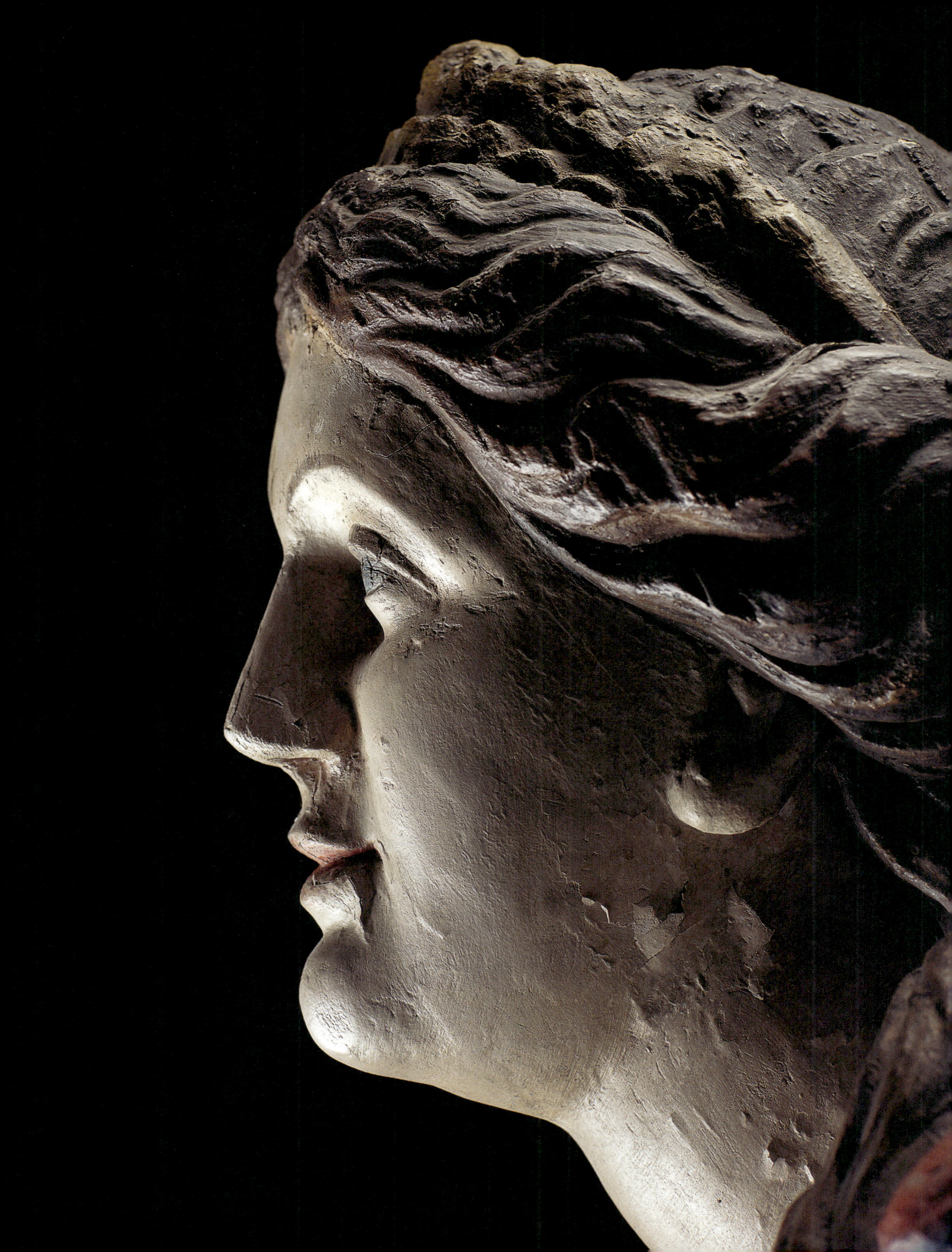

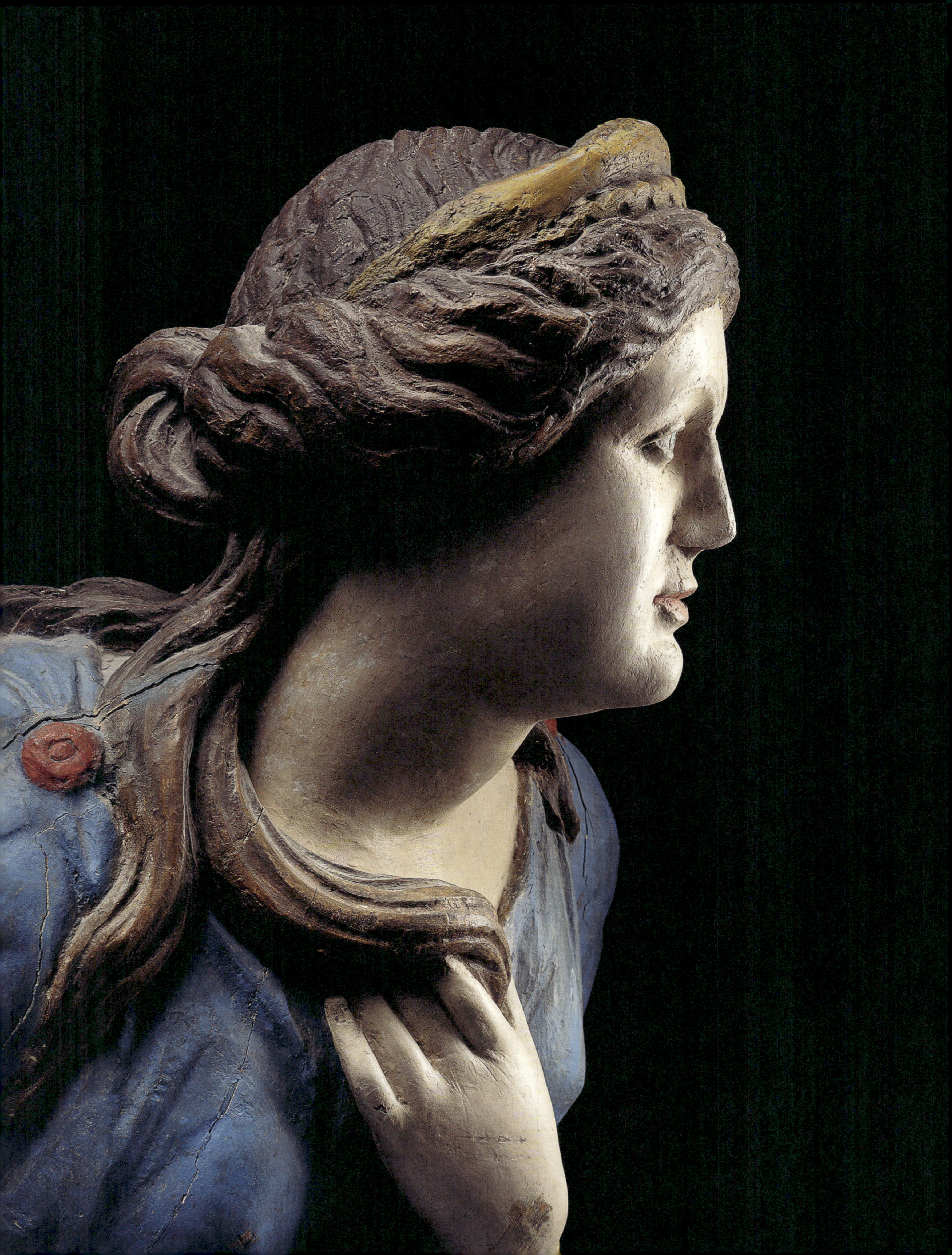

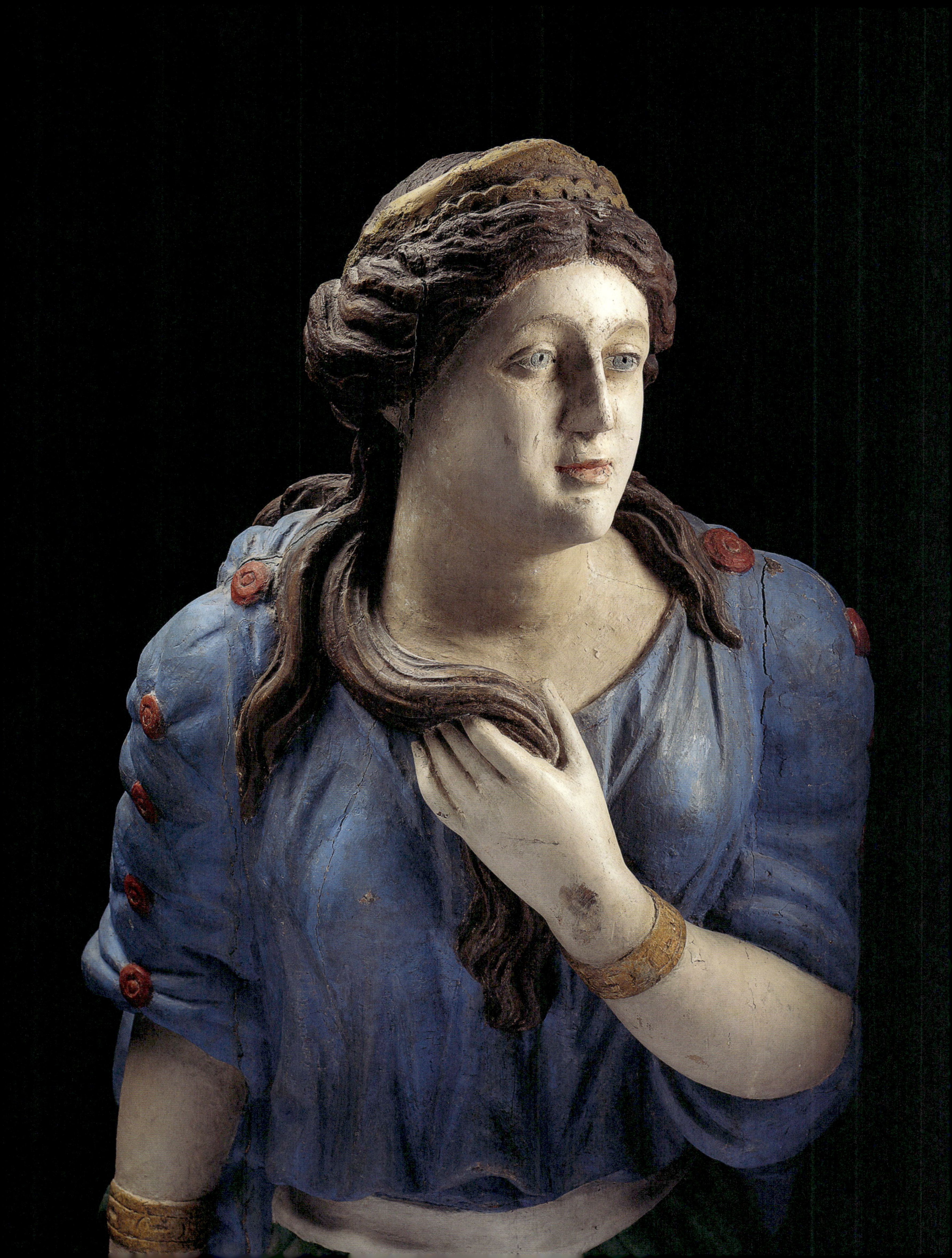

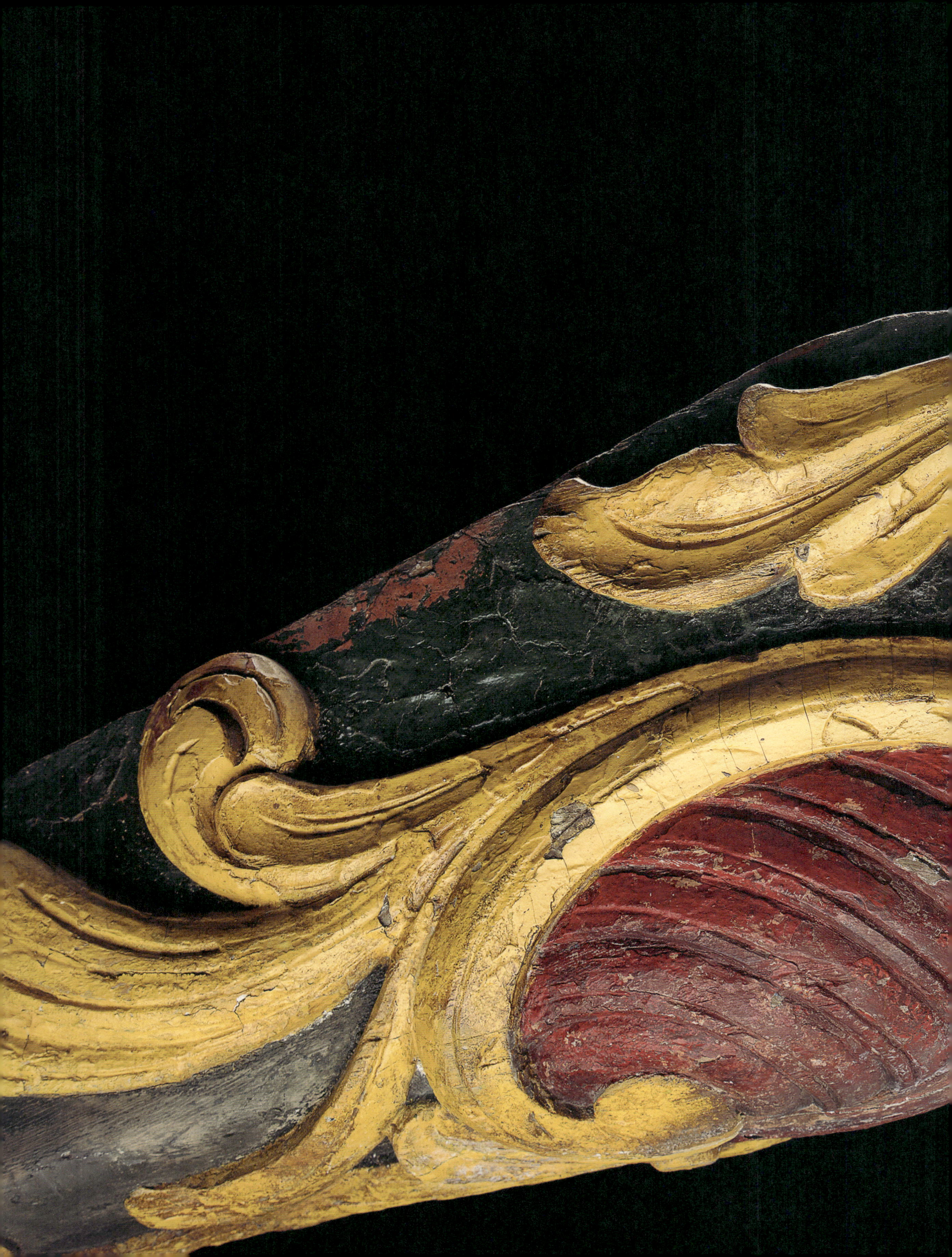

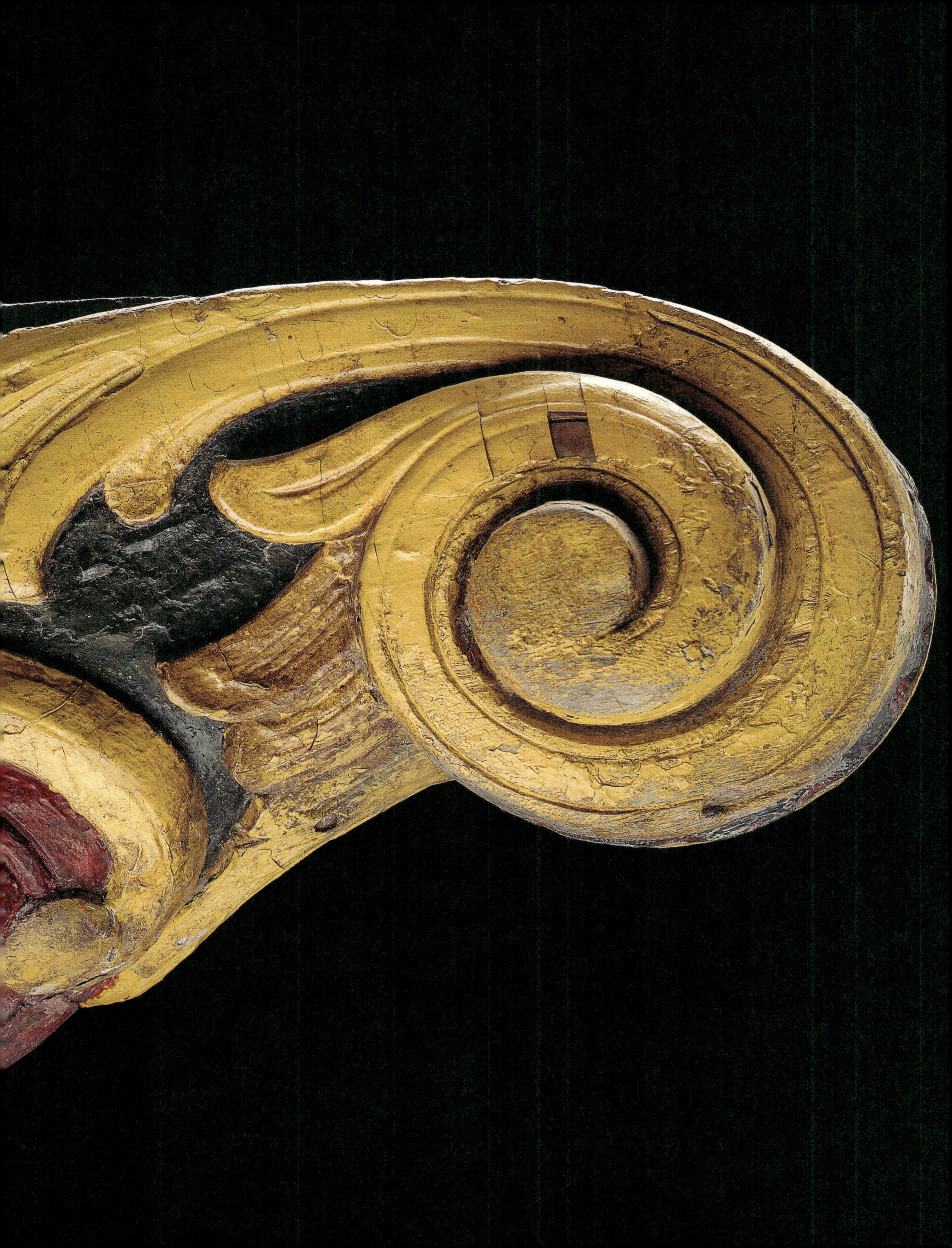

Preceding pages: Galleon ornament in the form of a so-called figurehead curl with red shield surrounded by yellow vegetal ornamentation. Origin unknown.

Full-length figure of woman in blue flounced dress, holding a rose and hymn book. Of the shipowner's wife type. Origin unknown.

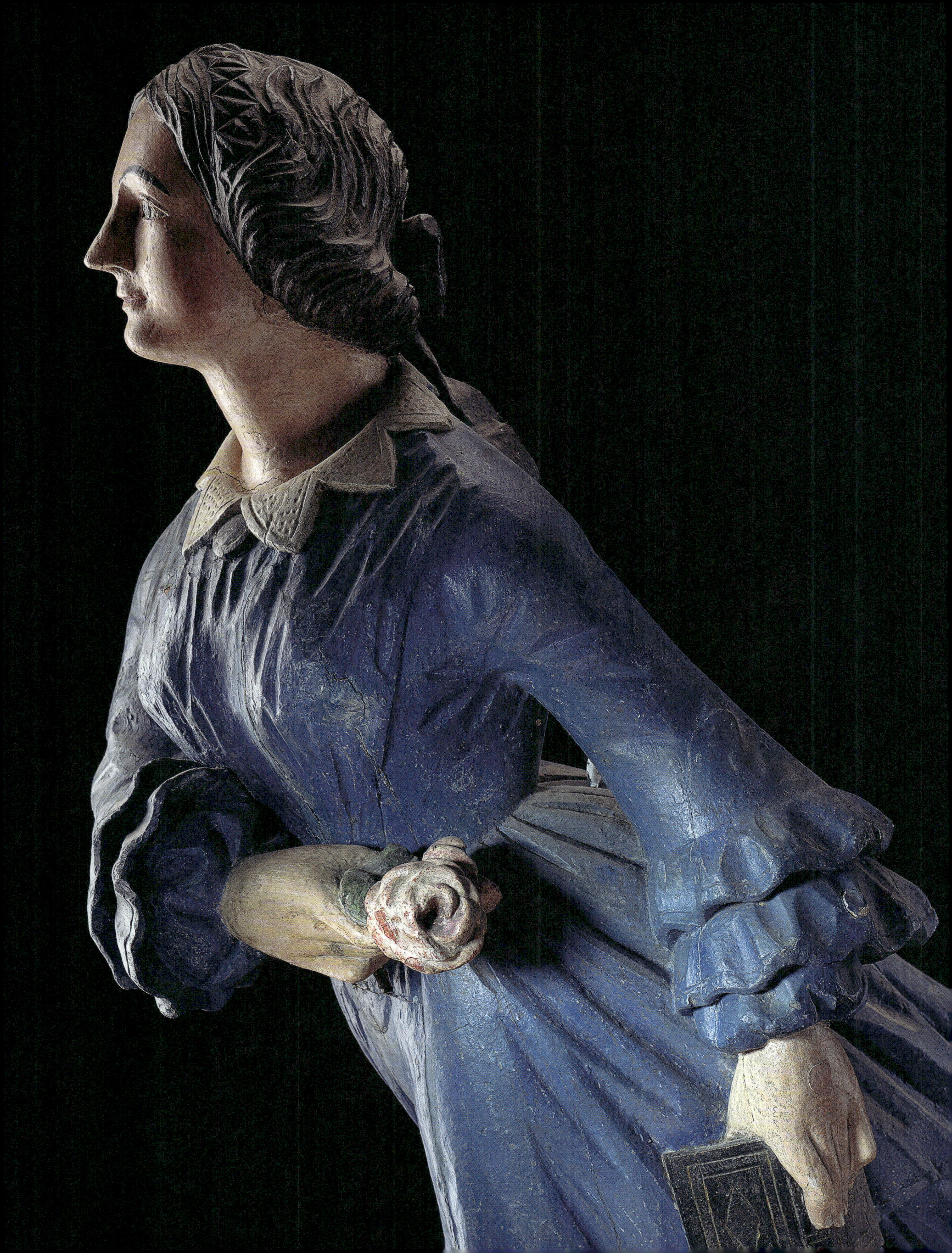

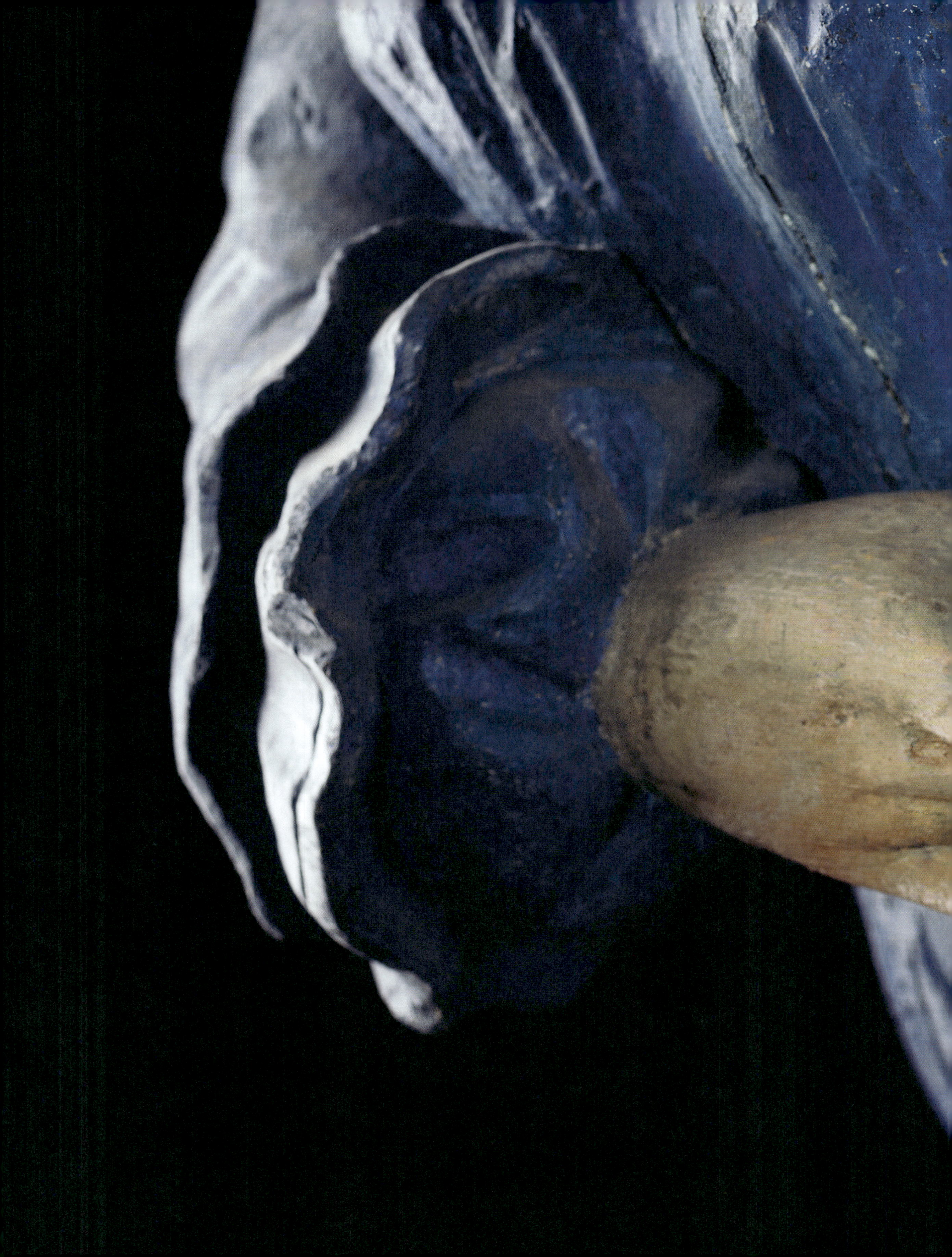

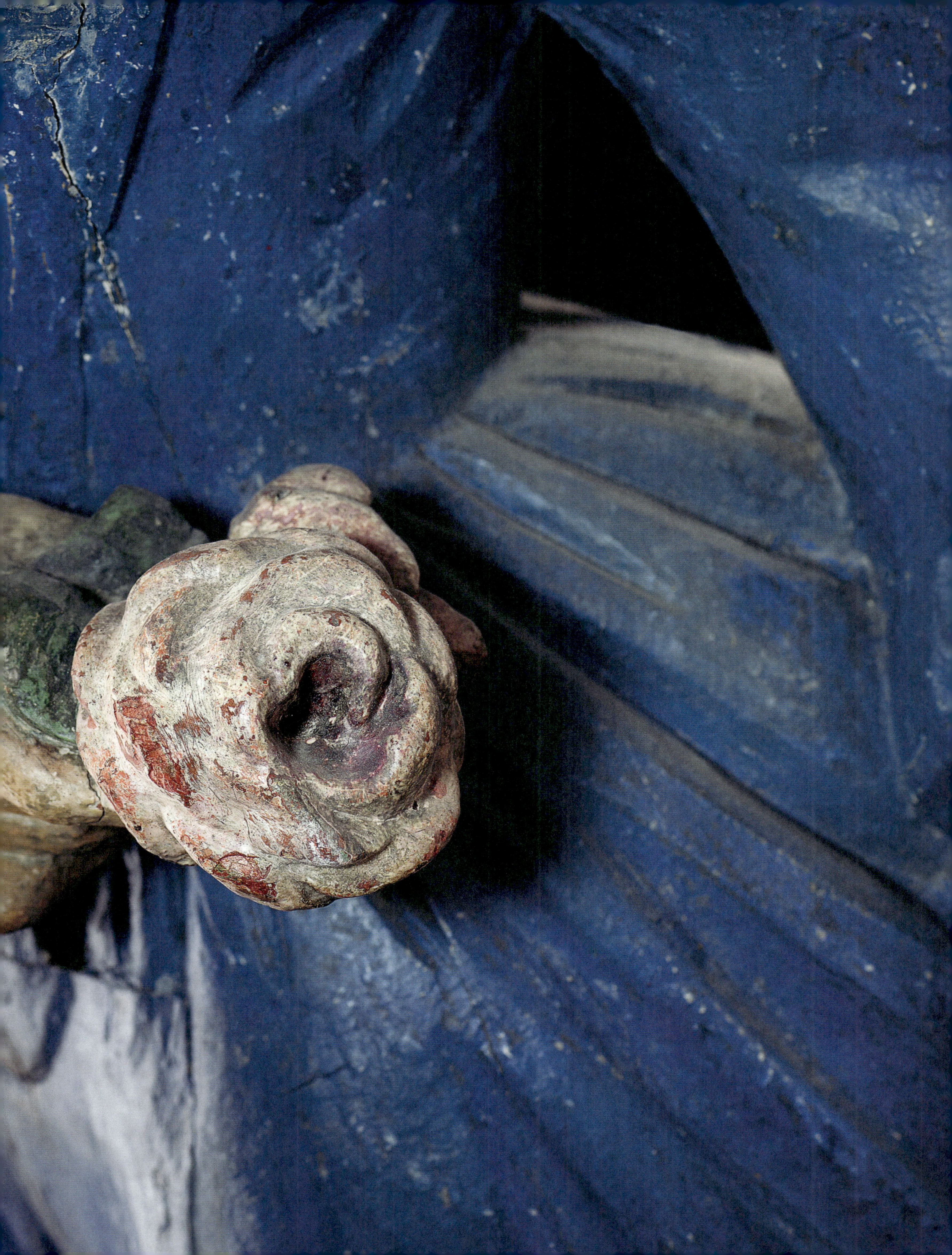

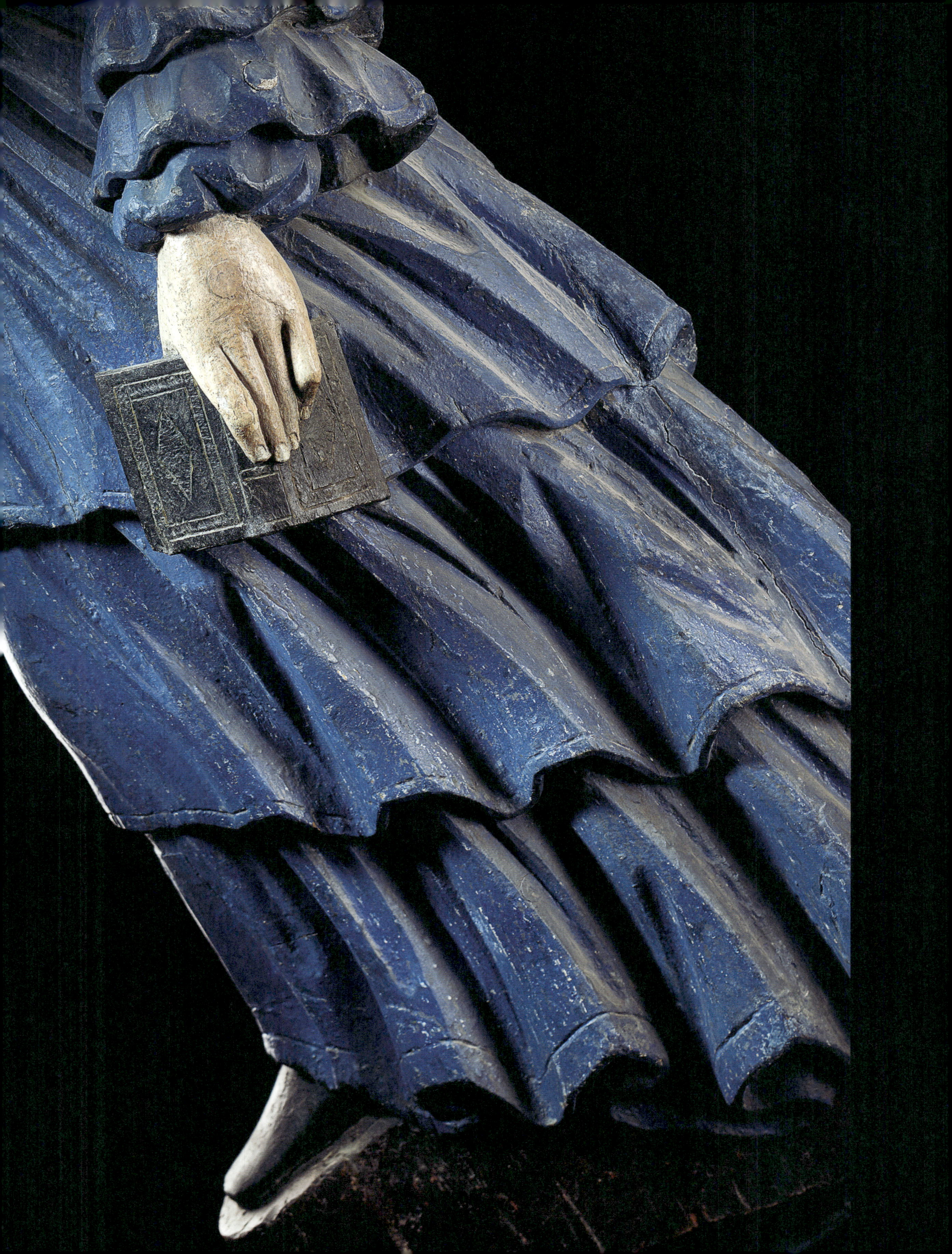

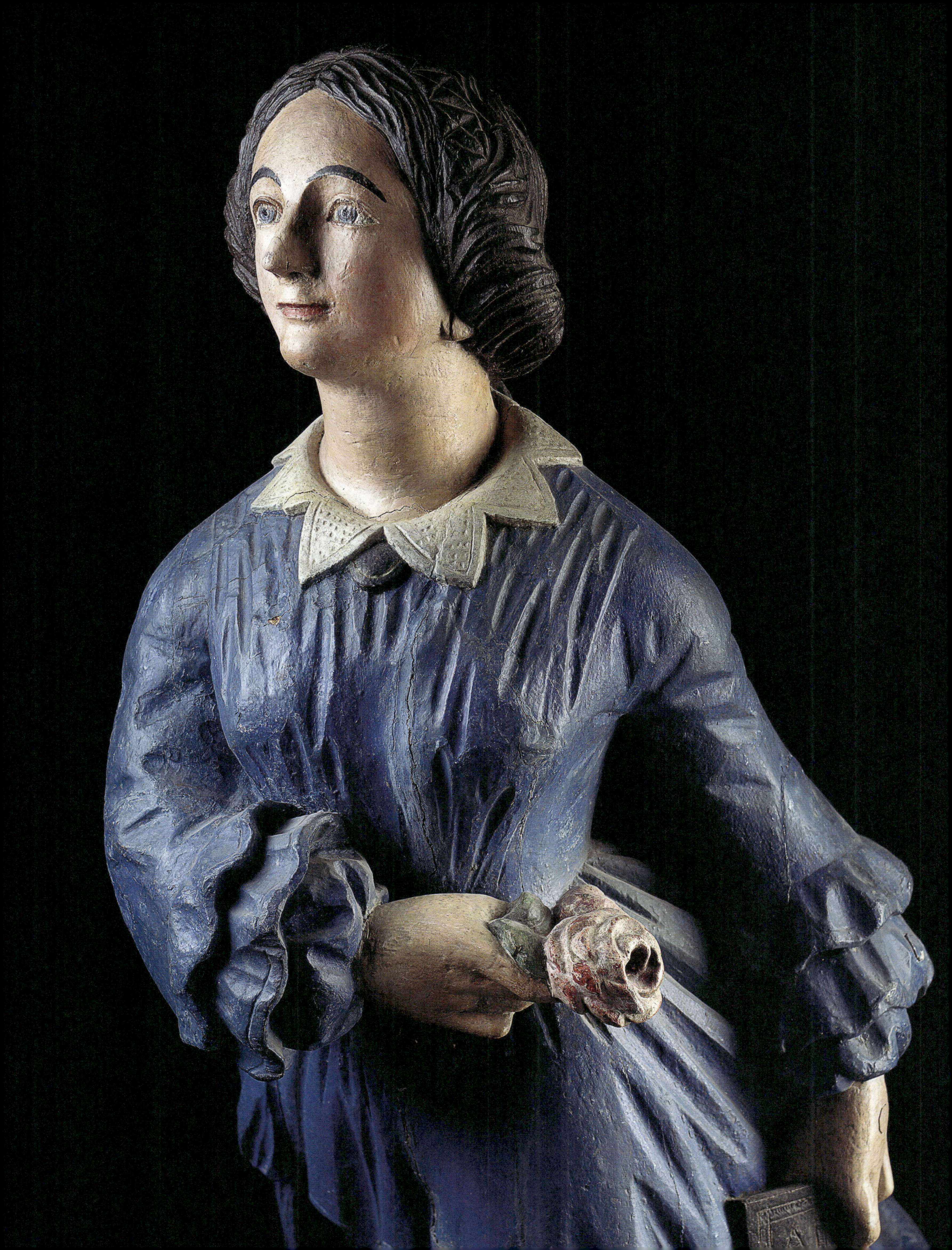

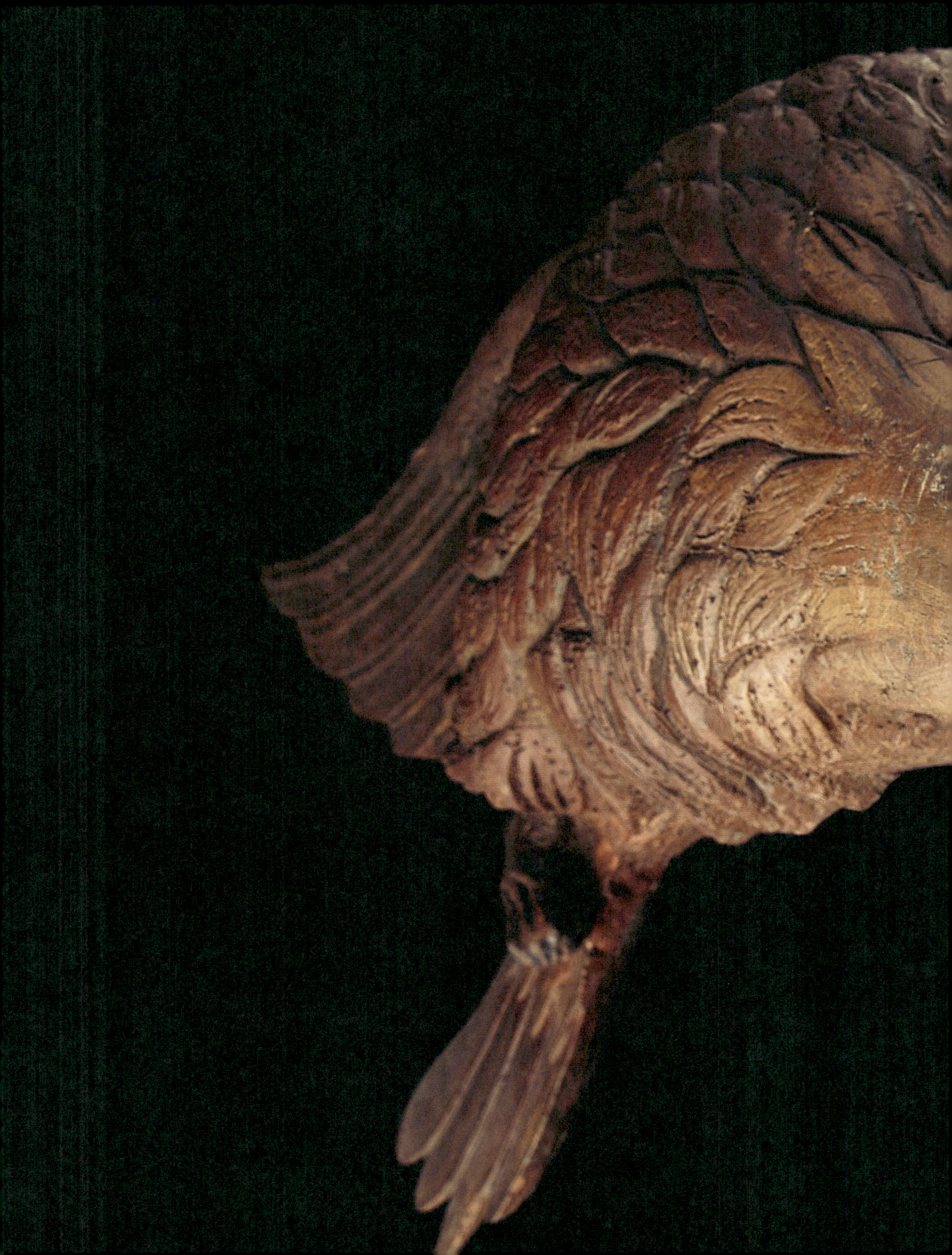

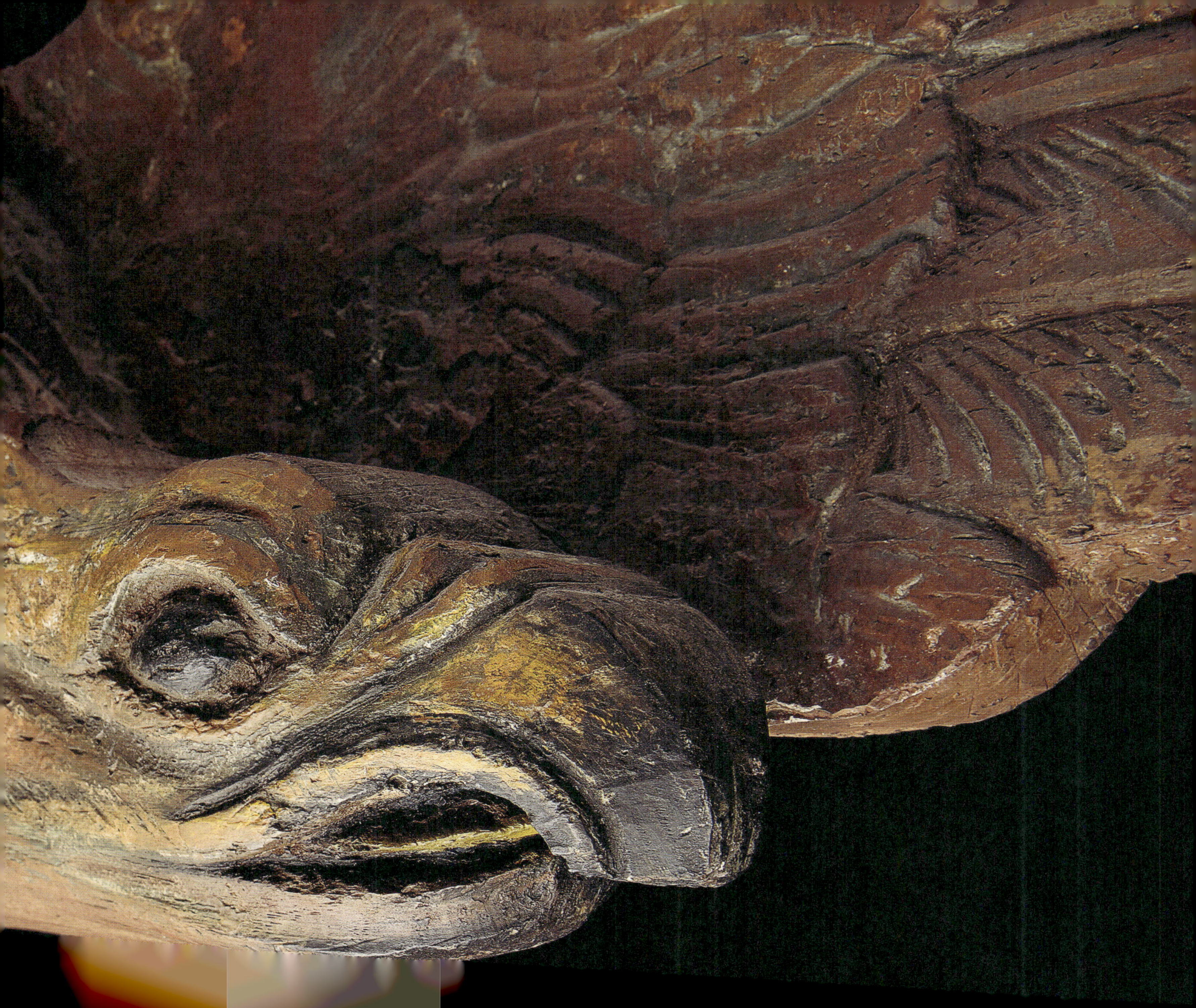

Preceding pages: Detail of eagle relief, the same found on pages 38–39.

Full-length figure of boy with dark hair, blue shirt and white collar. Origin unknown.

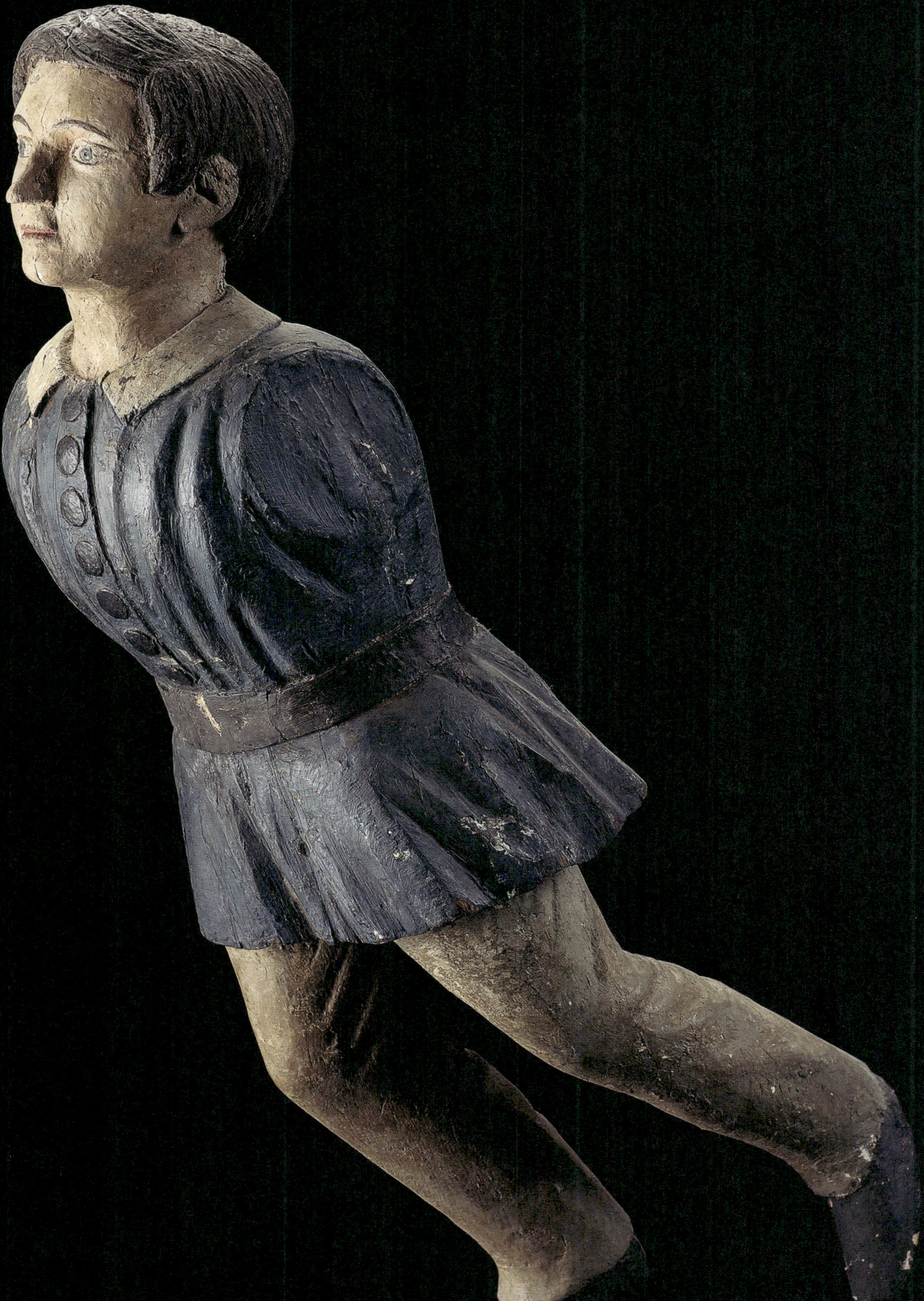

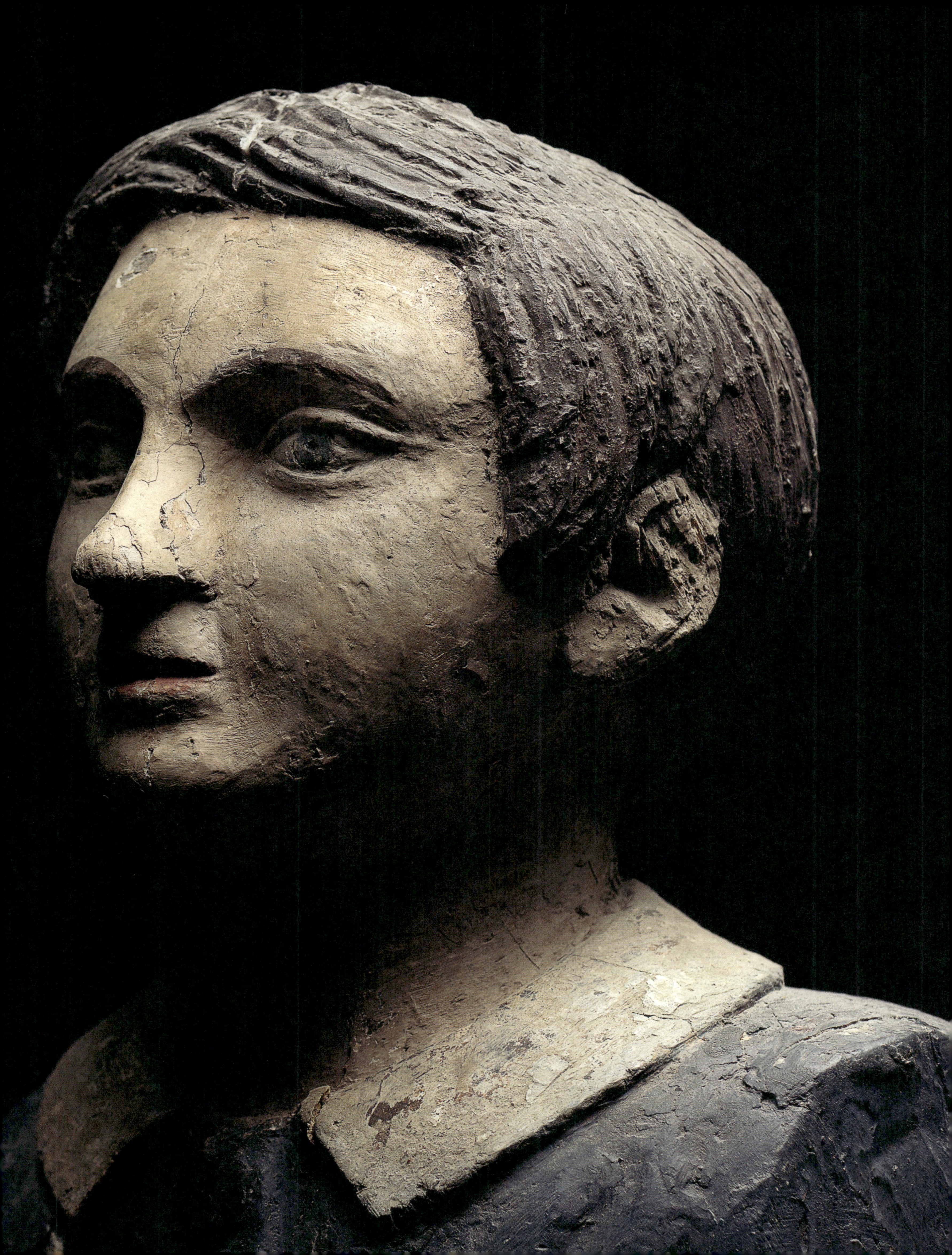

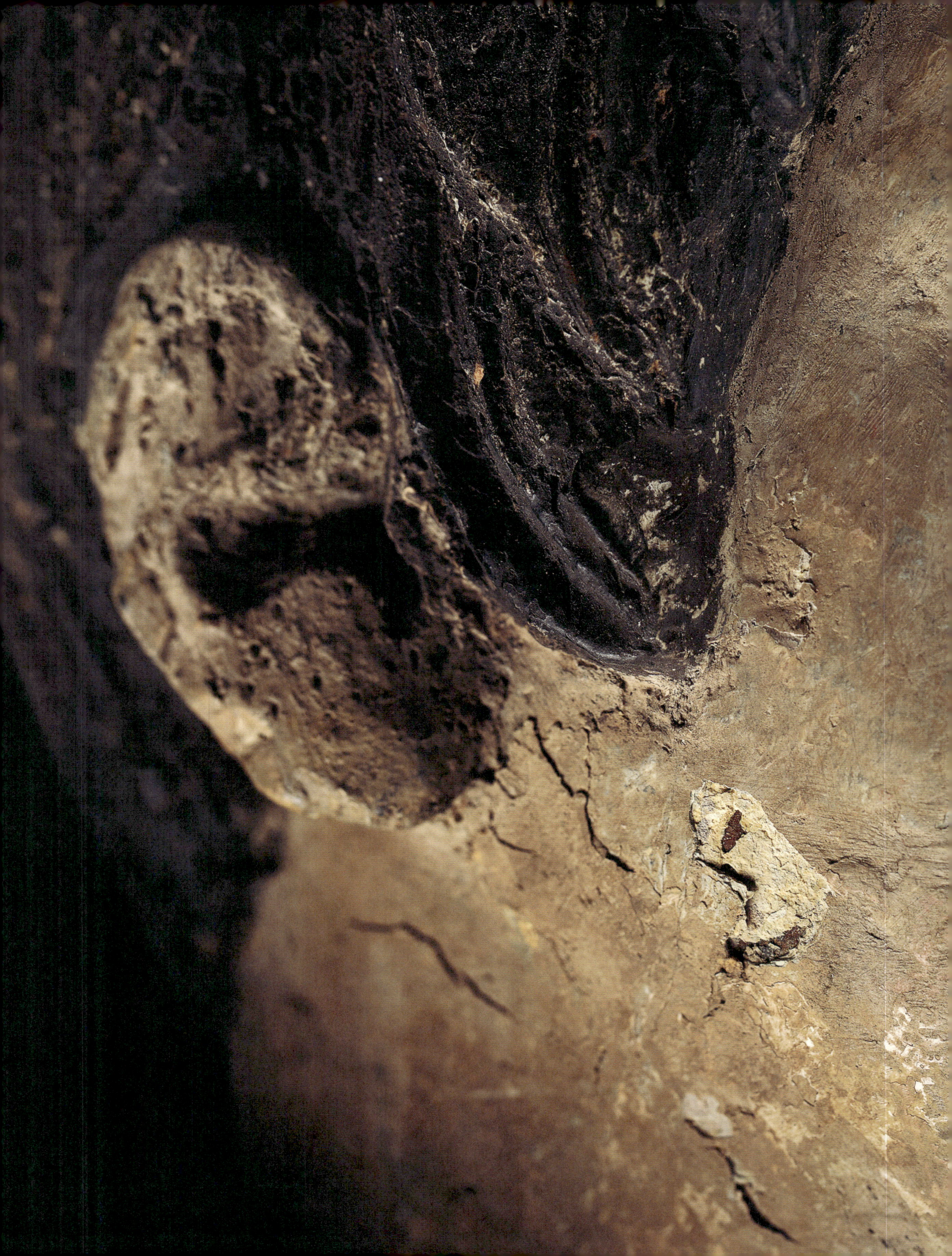

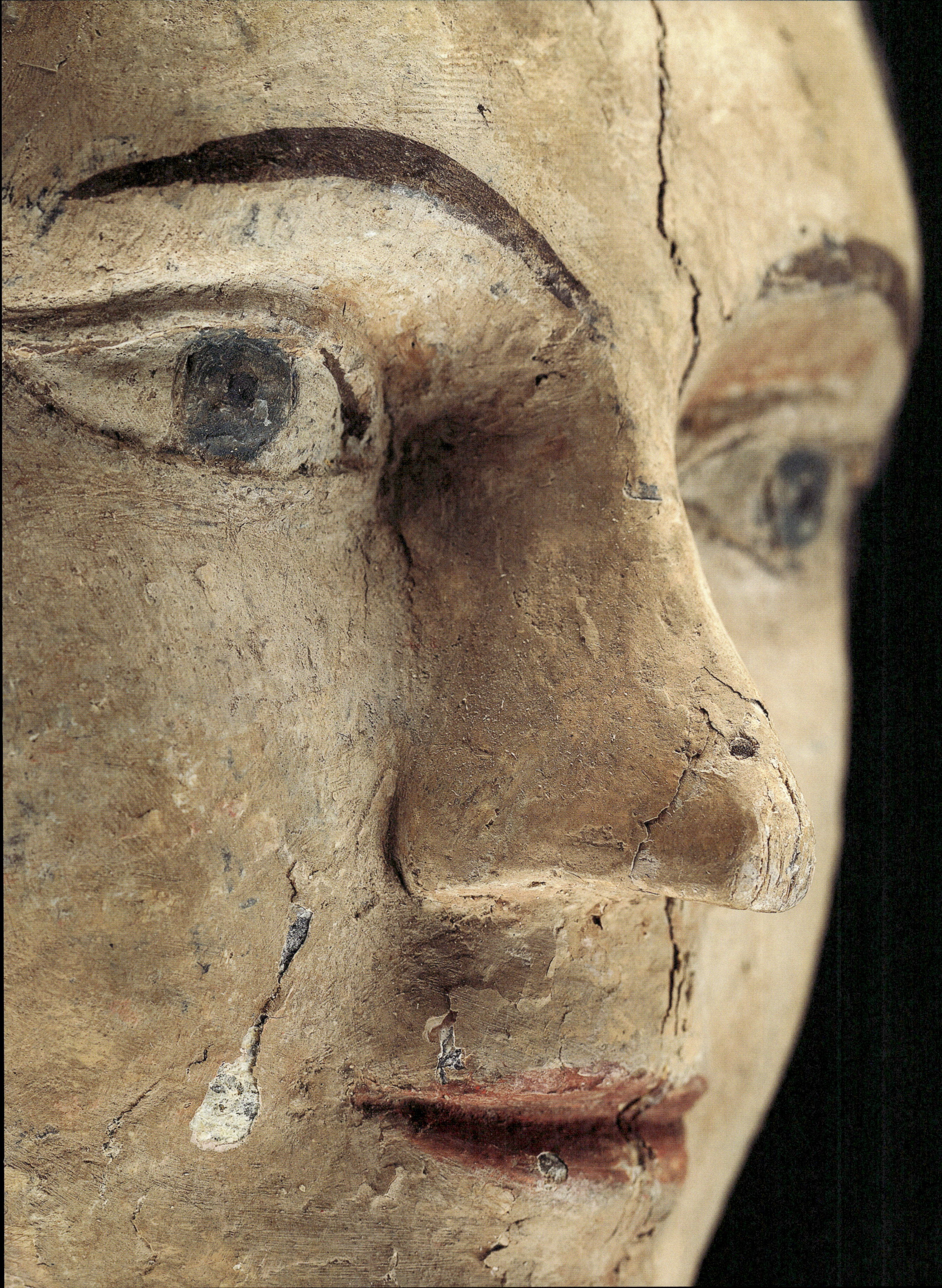

Full-length figure of medieval soldier in chainmail, with axe, helmet and shield. On the back is the inscription "J. BROOKER. M. PORT". Origin unknown.

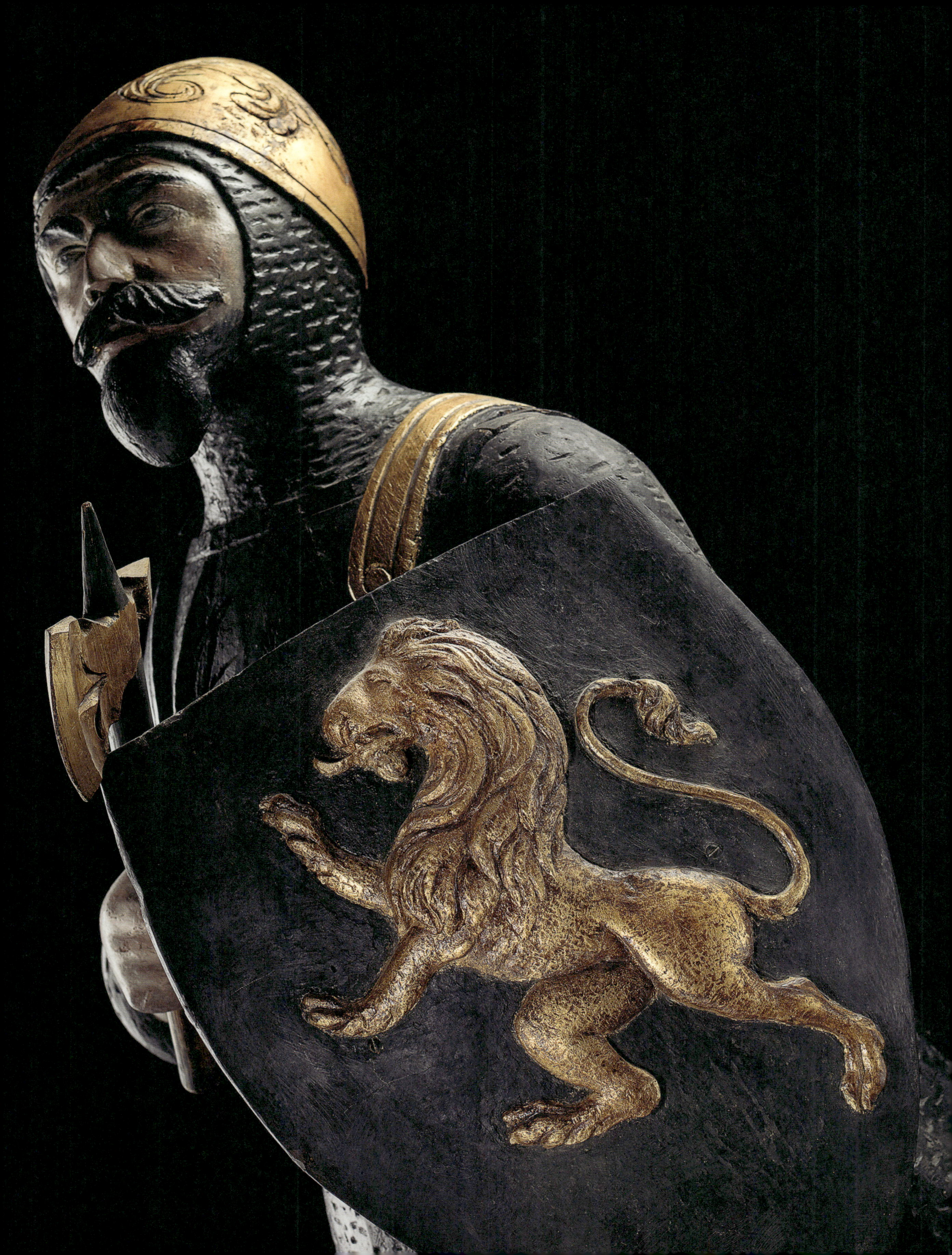

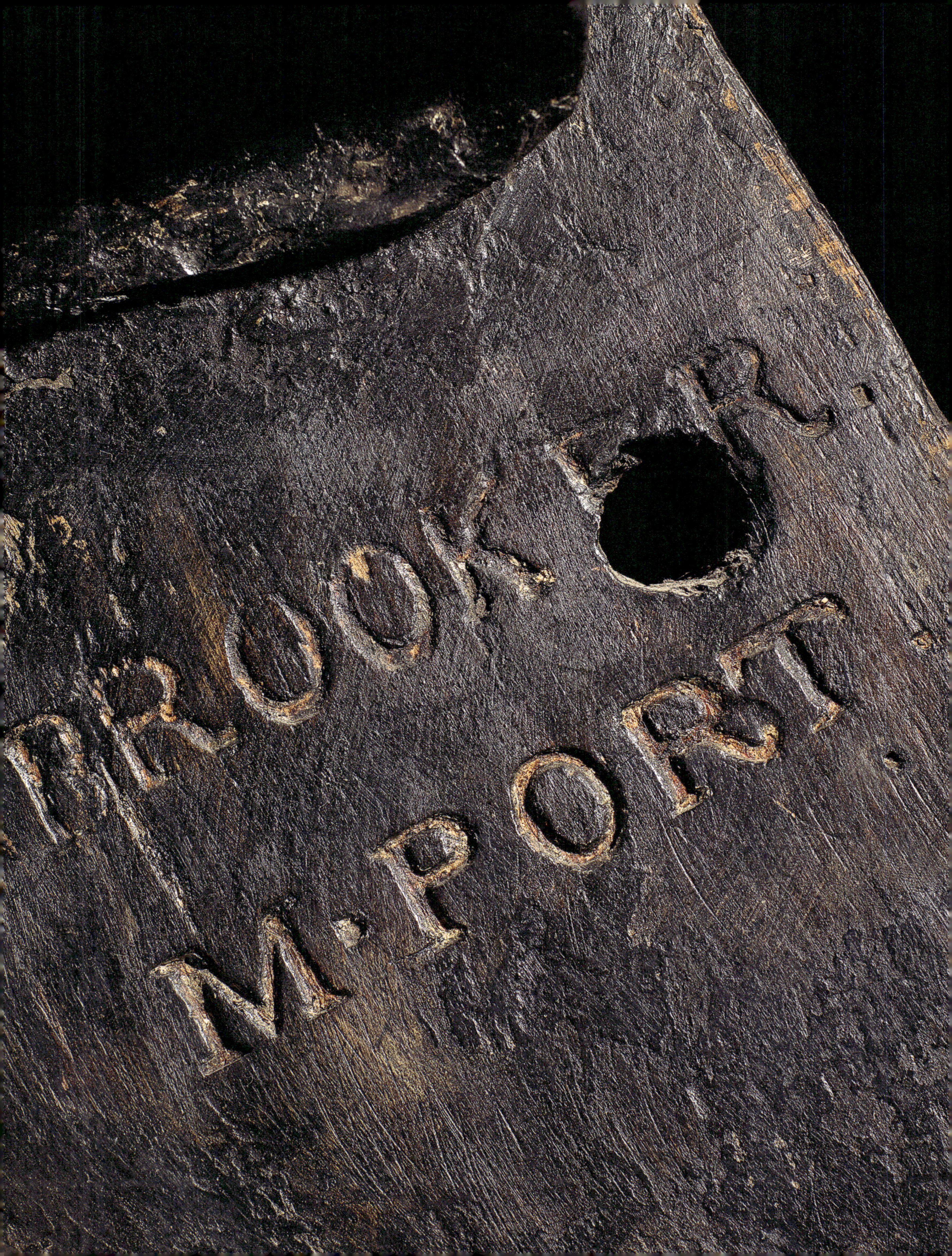
M·PORT

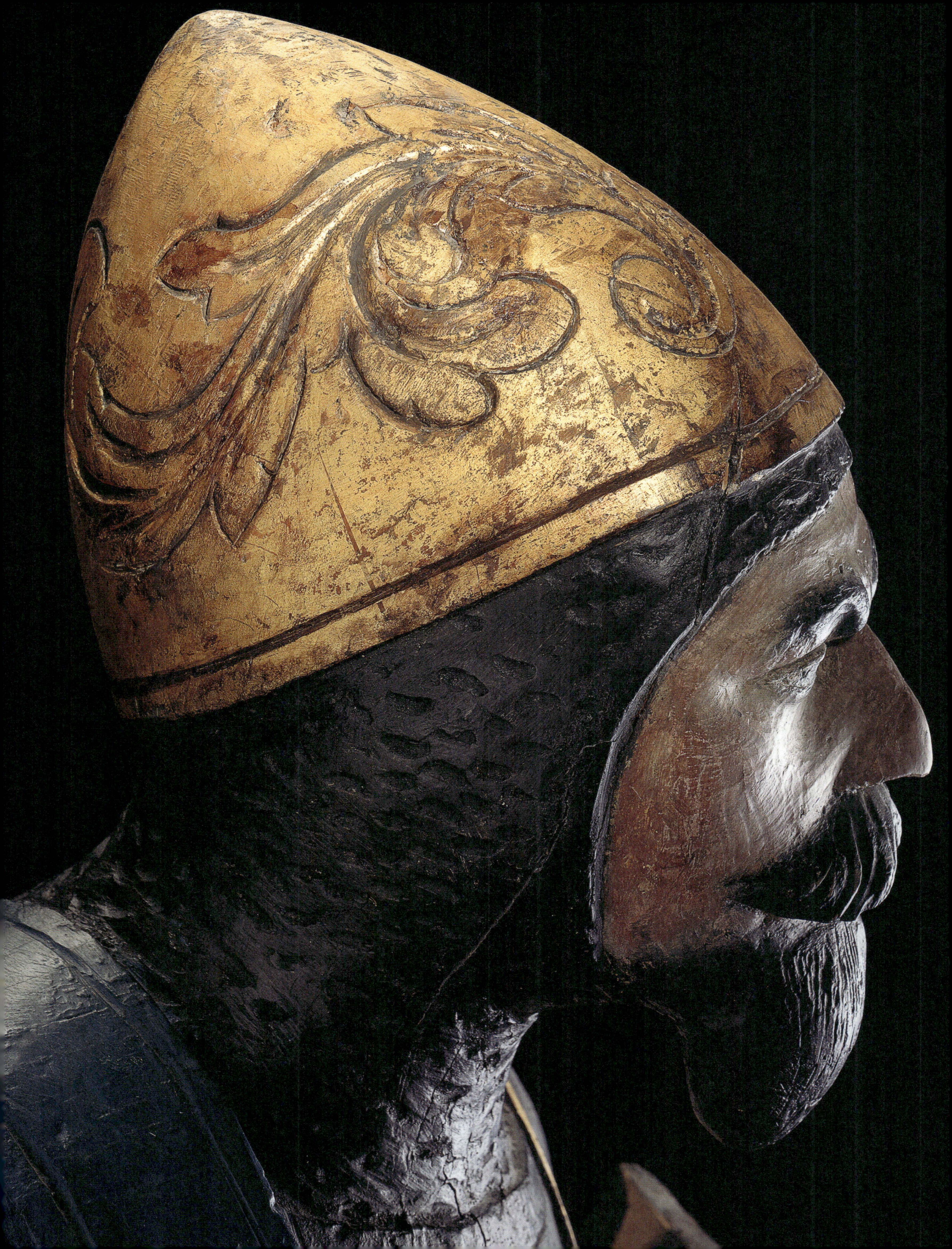

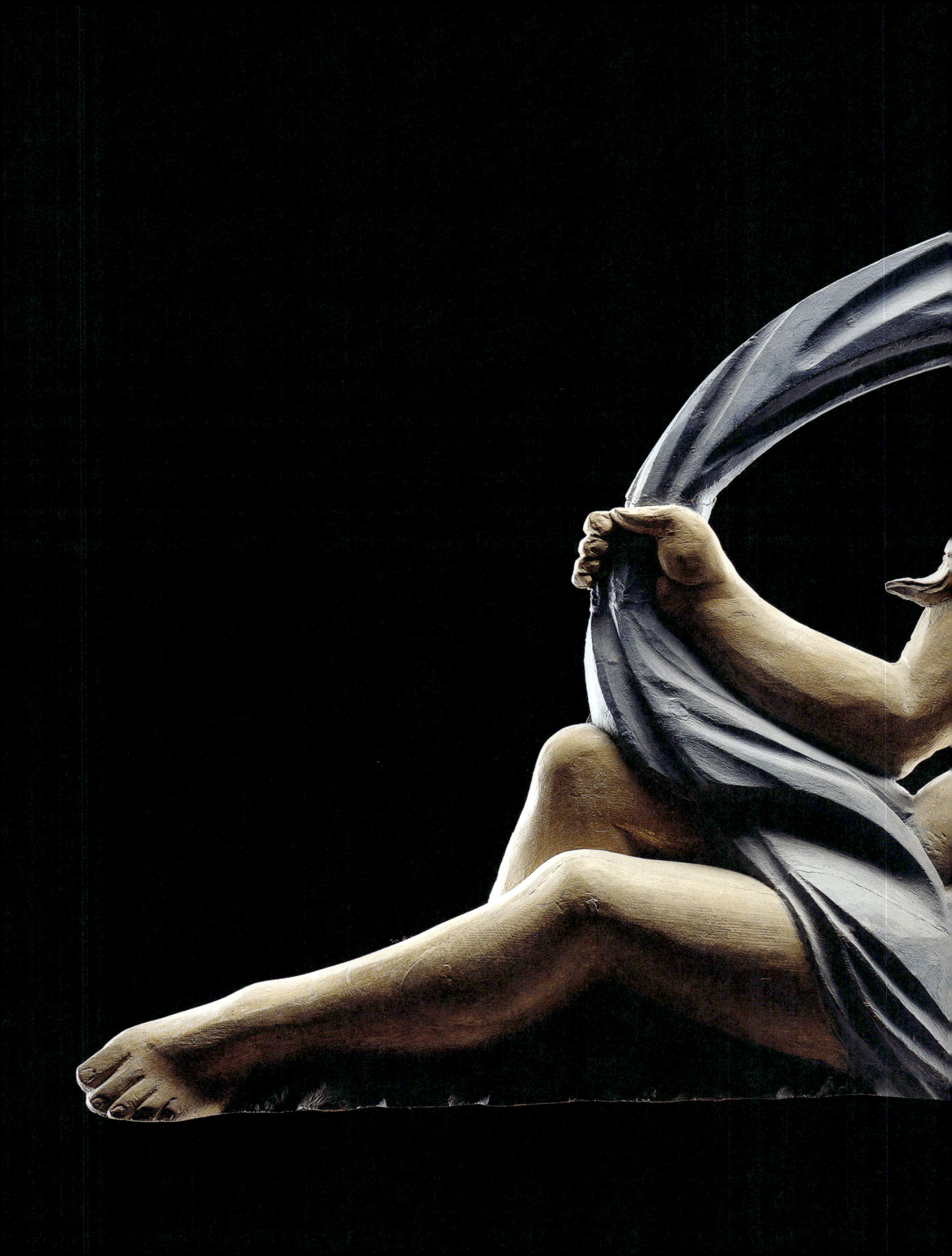

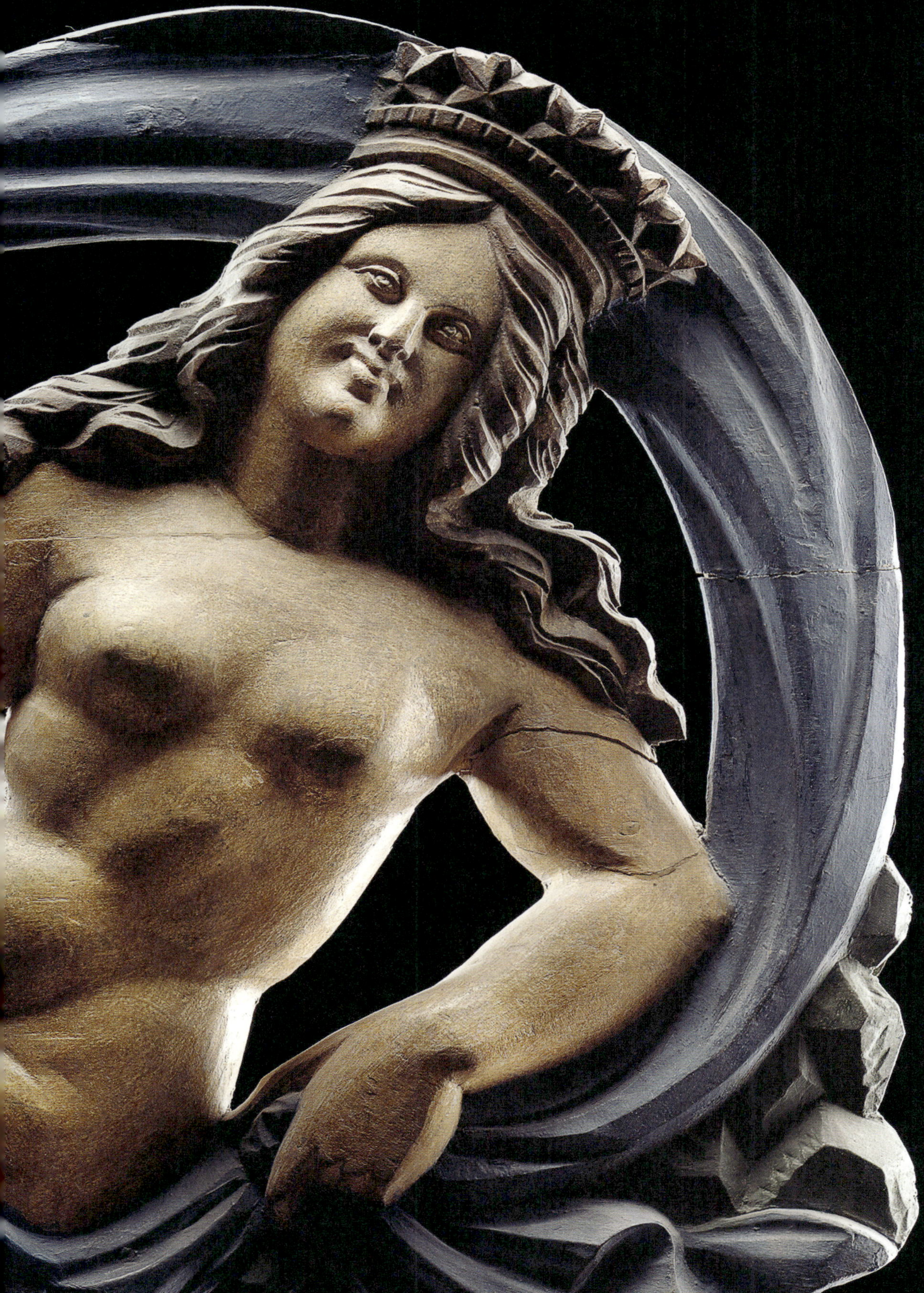

Preceding pages: Sternpost ornament with framed relief portrait of dark-haired woman with red hat and gown. Probably from the brig *Lady Stanley* of West Hartlepool, which ran ashore near Tyborøn on the west coast of Jutland in 1888.

Three-quarter-length figure of boy in dark-blue costume with pink bow, holding a flower basket. Origin unknown.

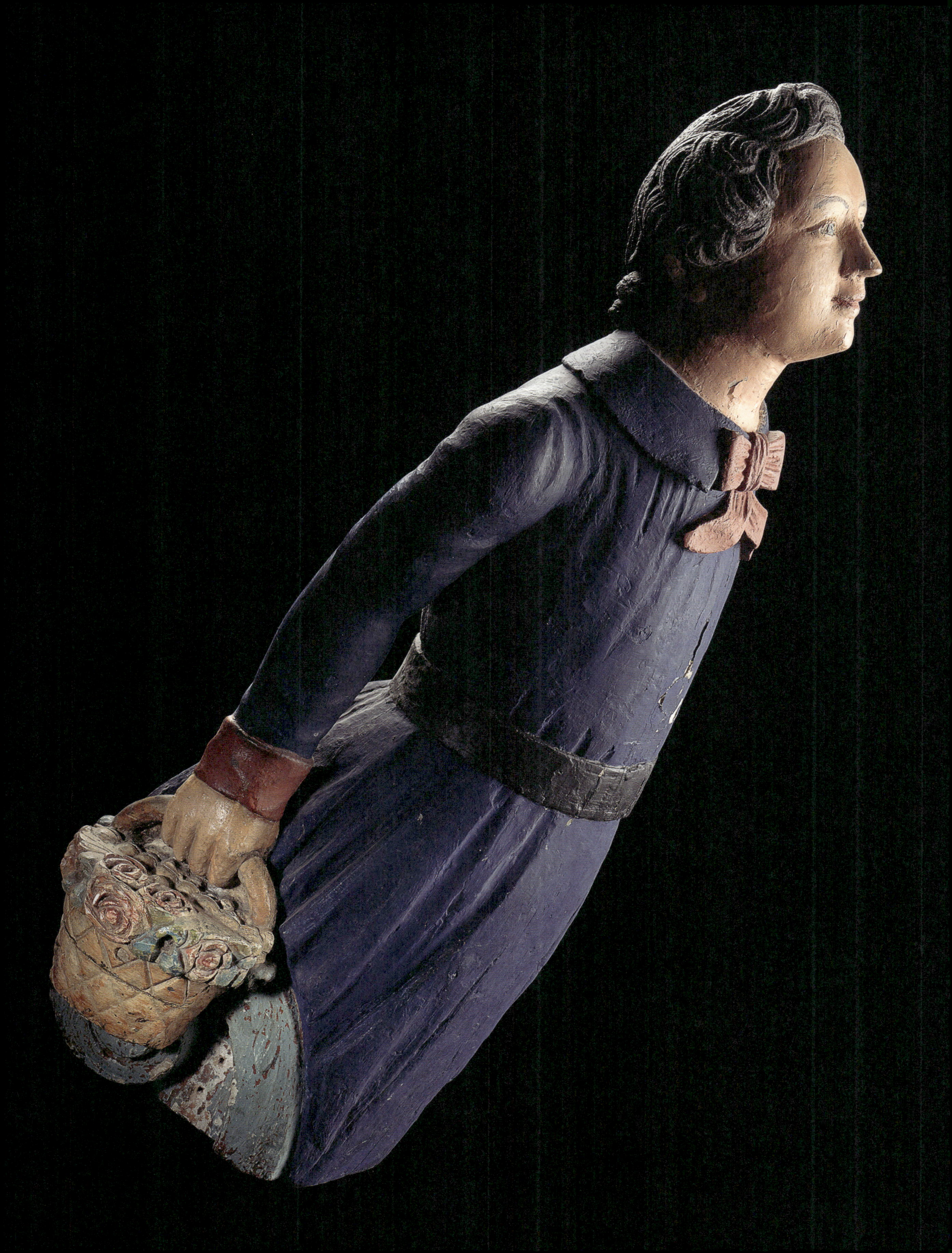

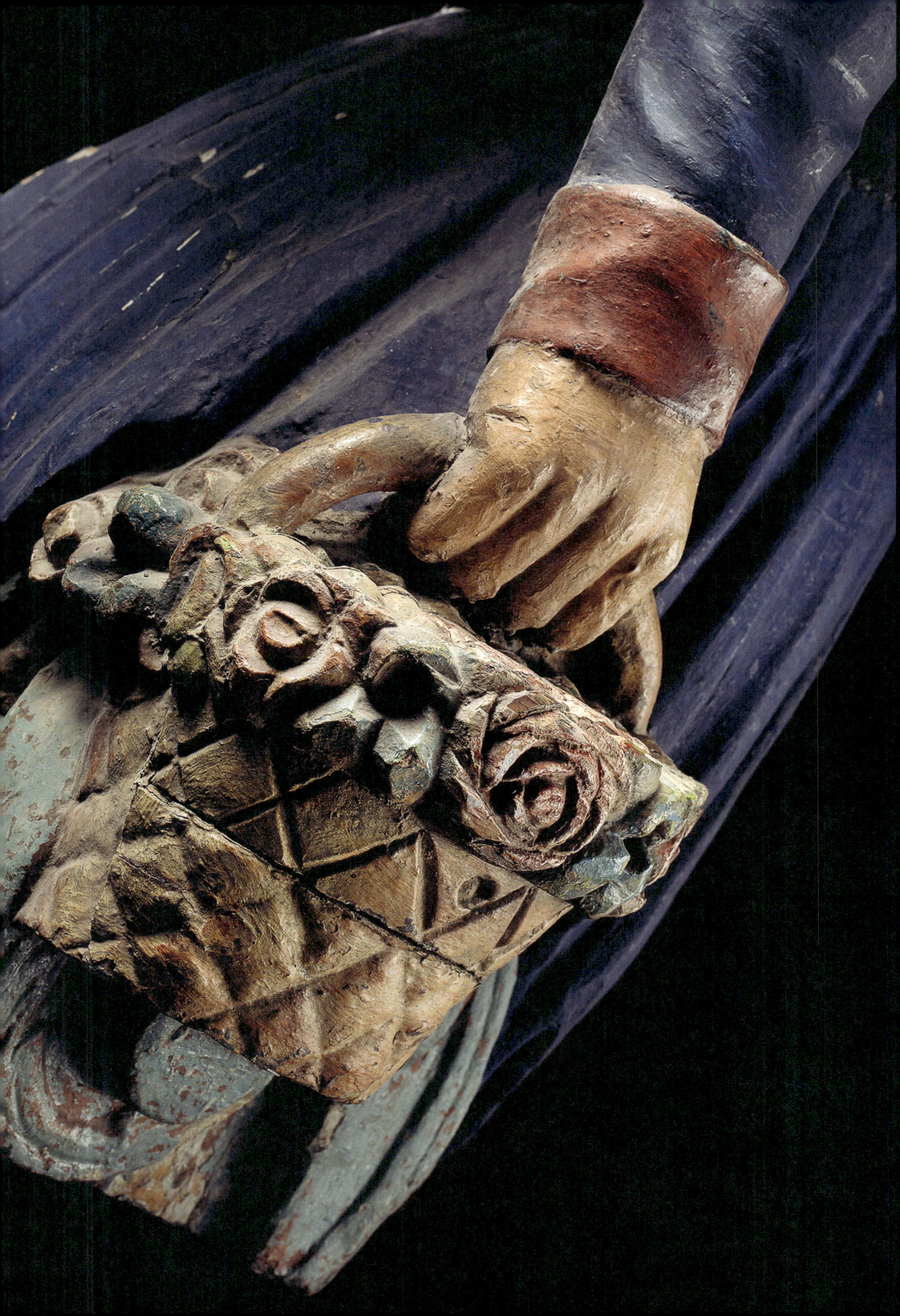

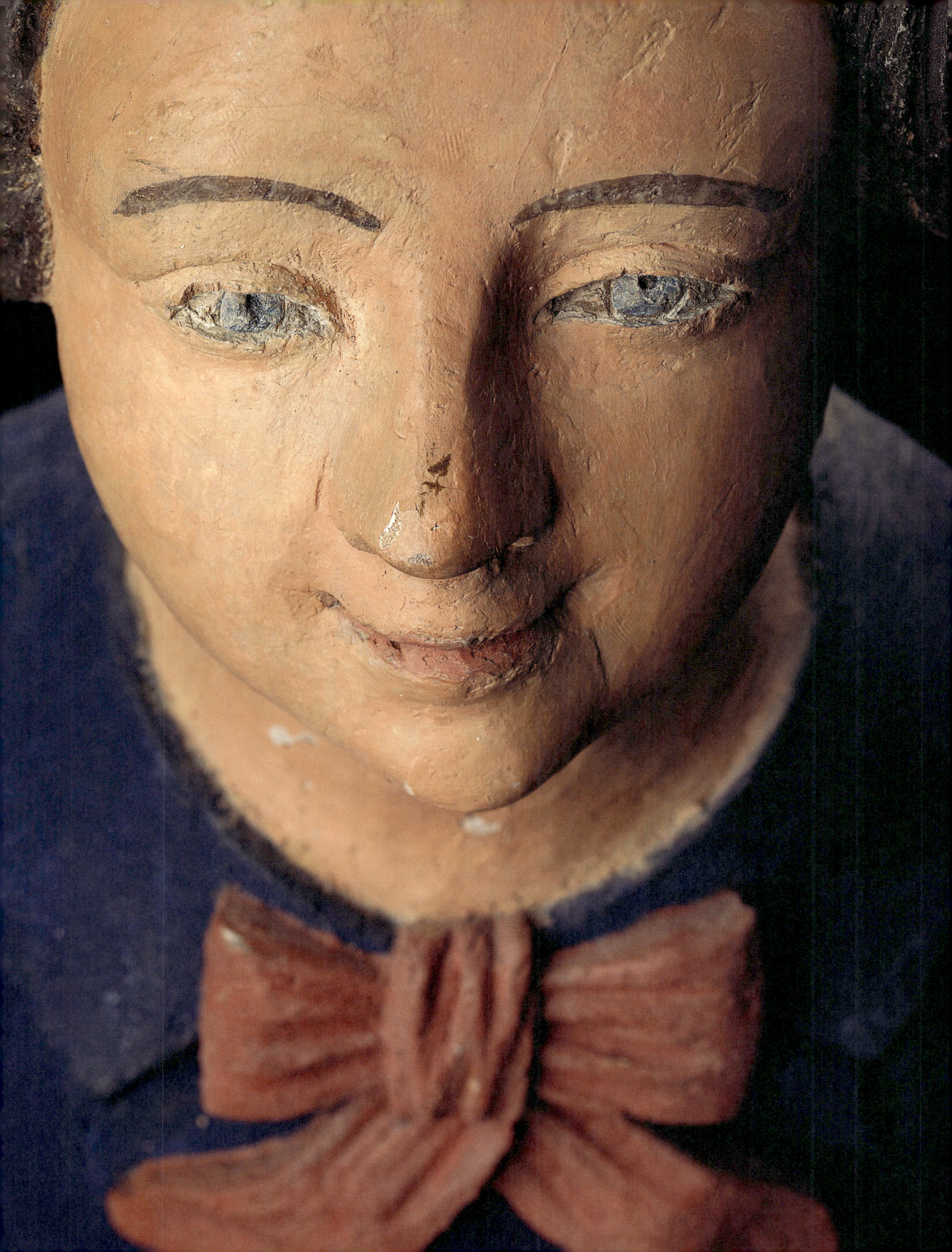

Female figure of shipowner's wife type from the barque *Elina* of Åland, which ran ashore at Thorup Strand in North Jutland in 1882.

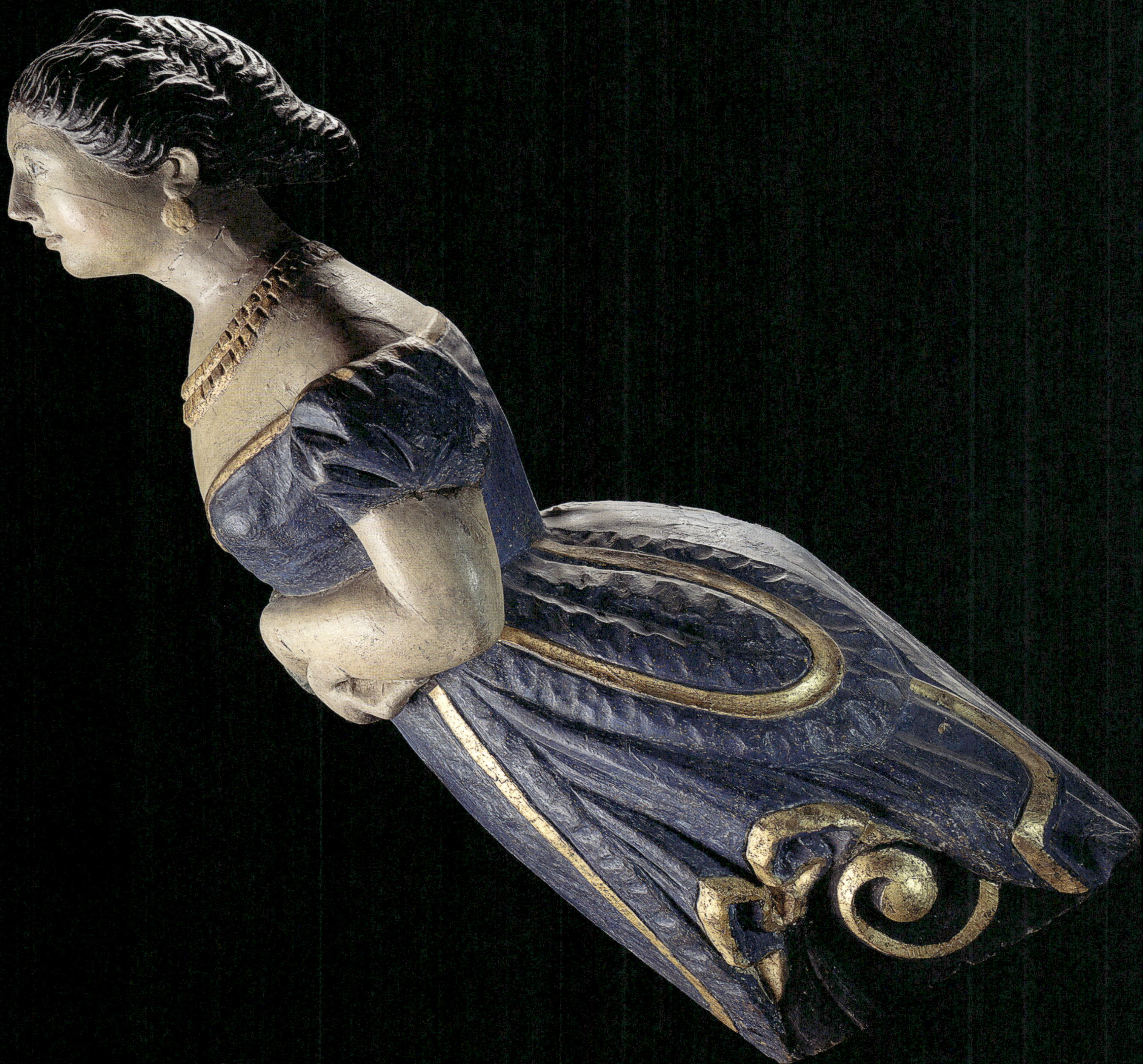

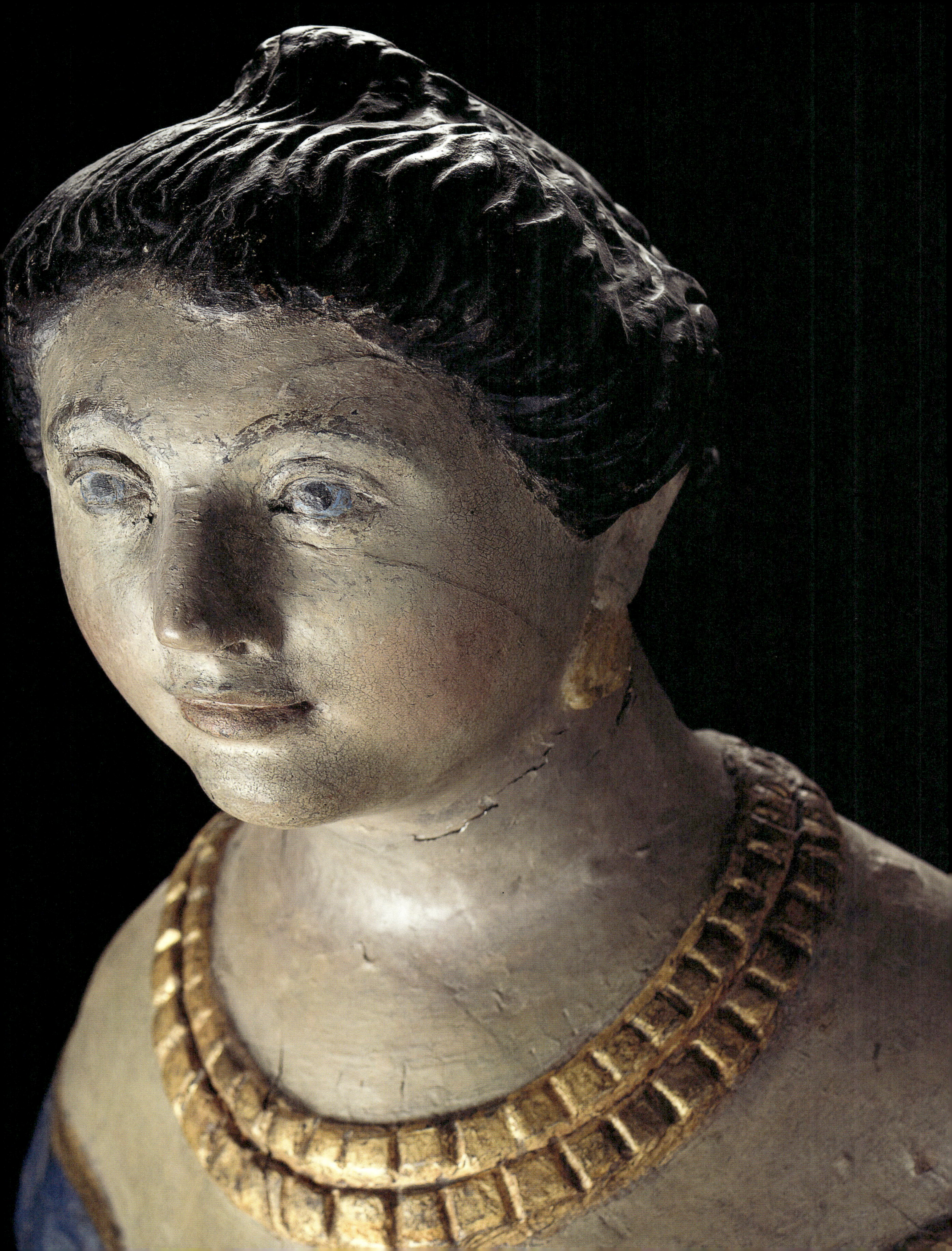

Demi figurehead in black frock coat of the shipowner type, from the frigate *Friedrich Wilhelm IV* of Gdánsk, which ran ashore off Løkken in 1864.

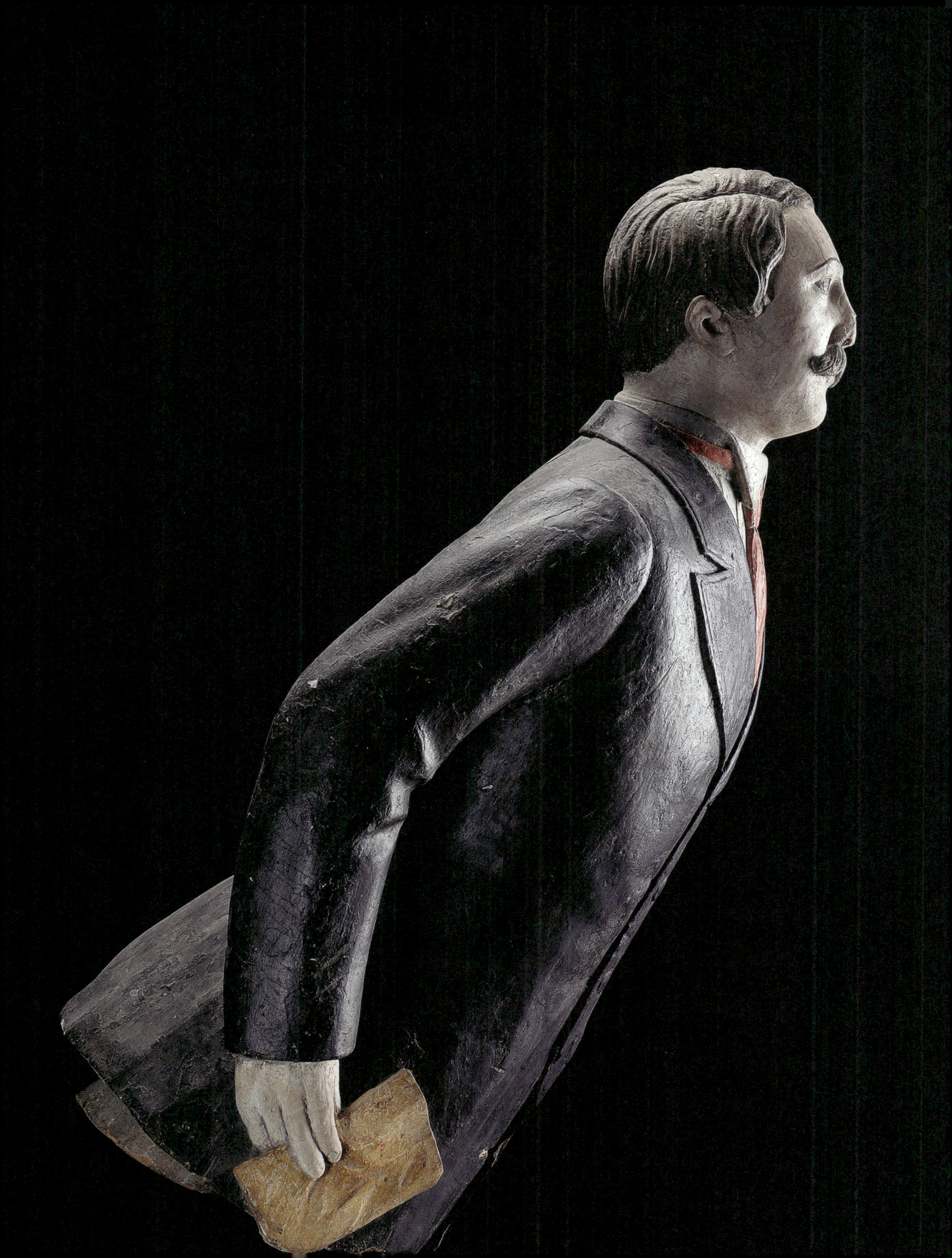

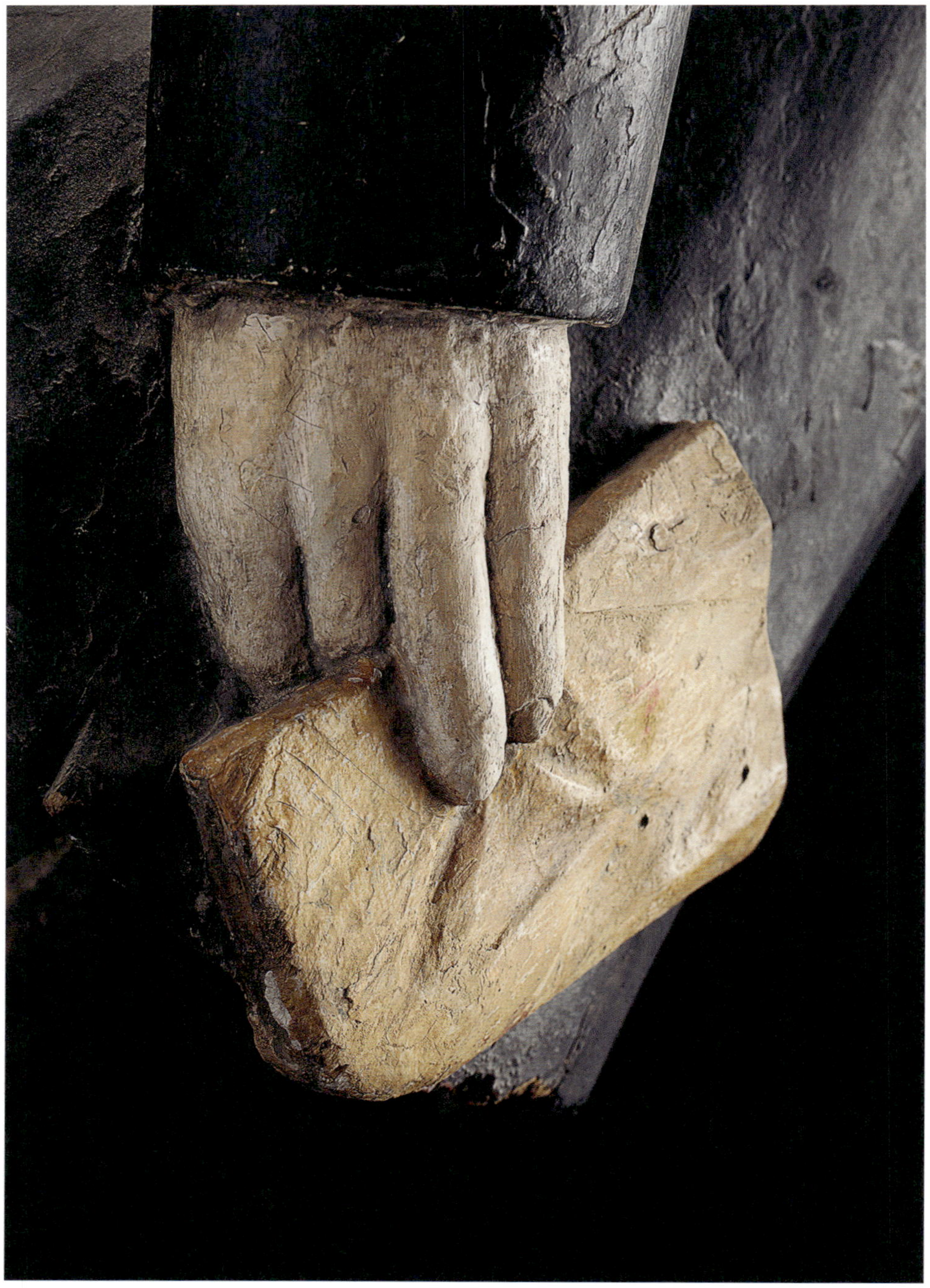

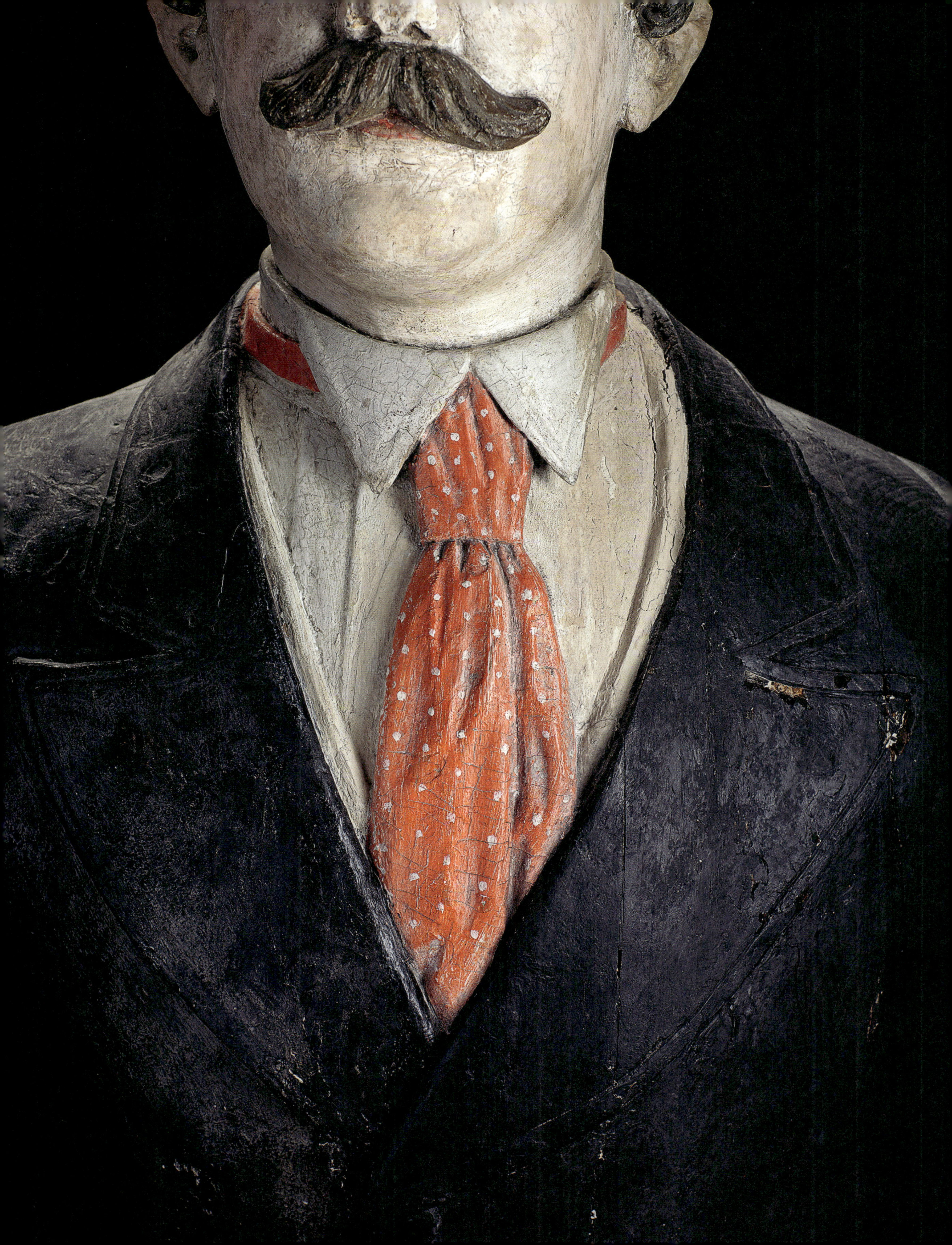

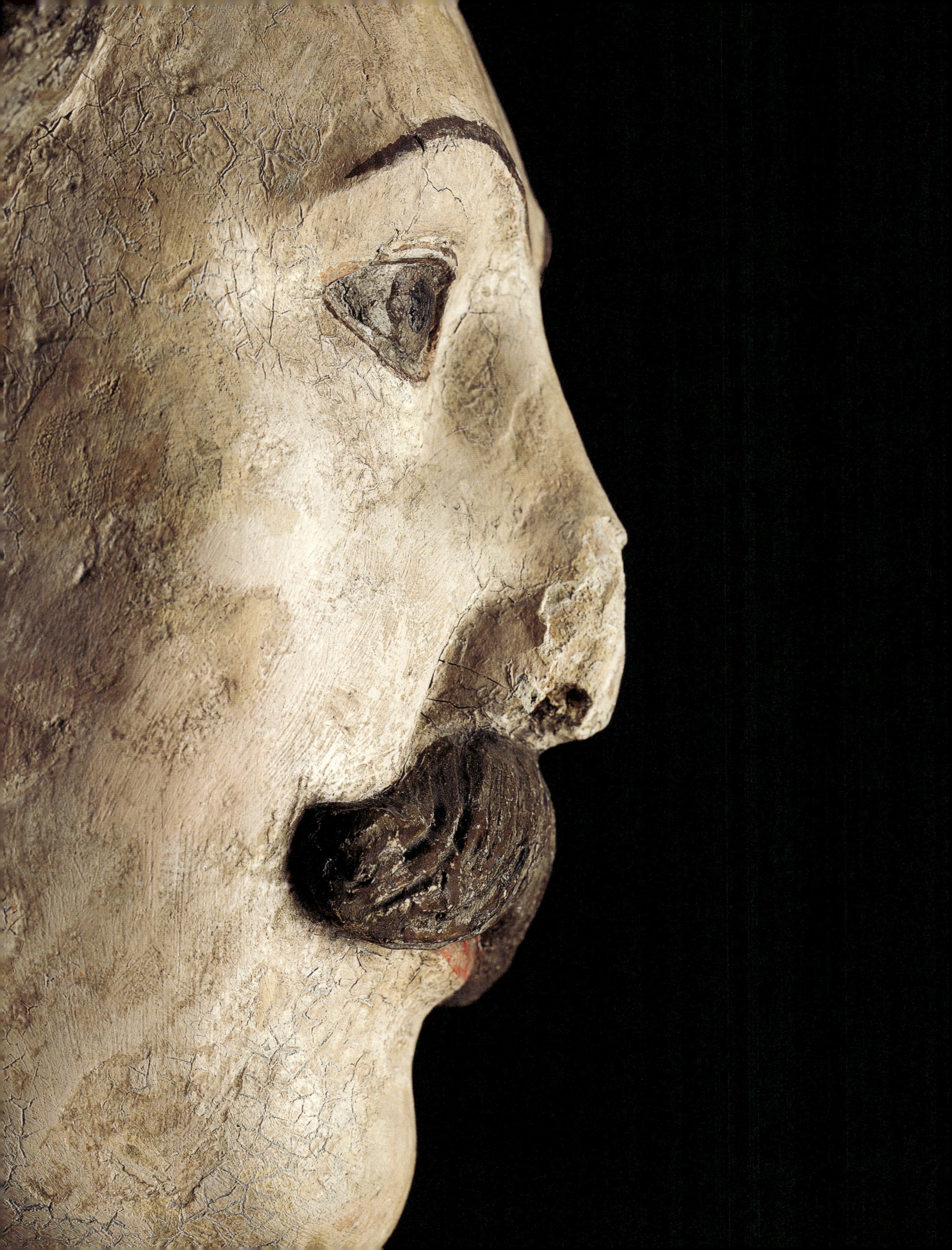

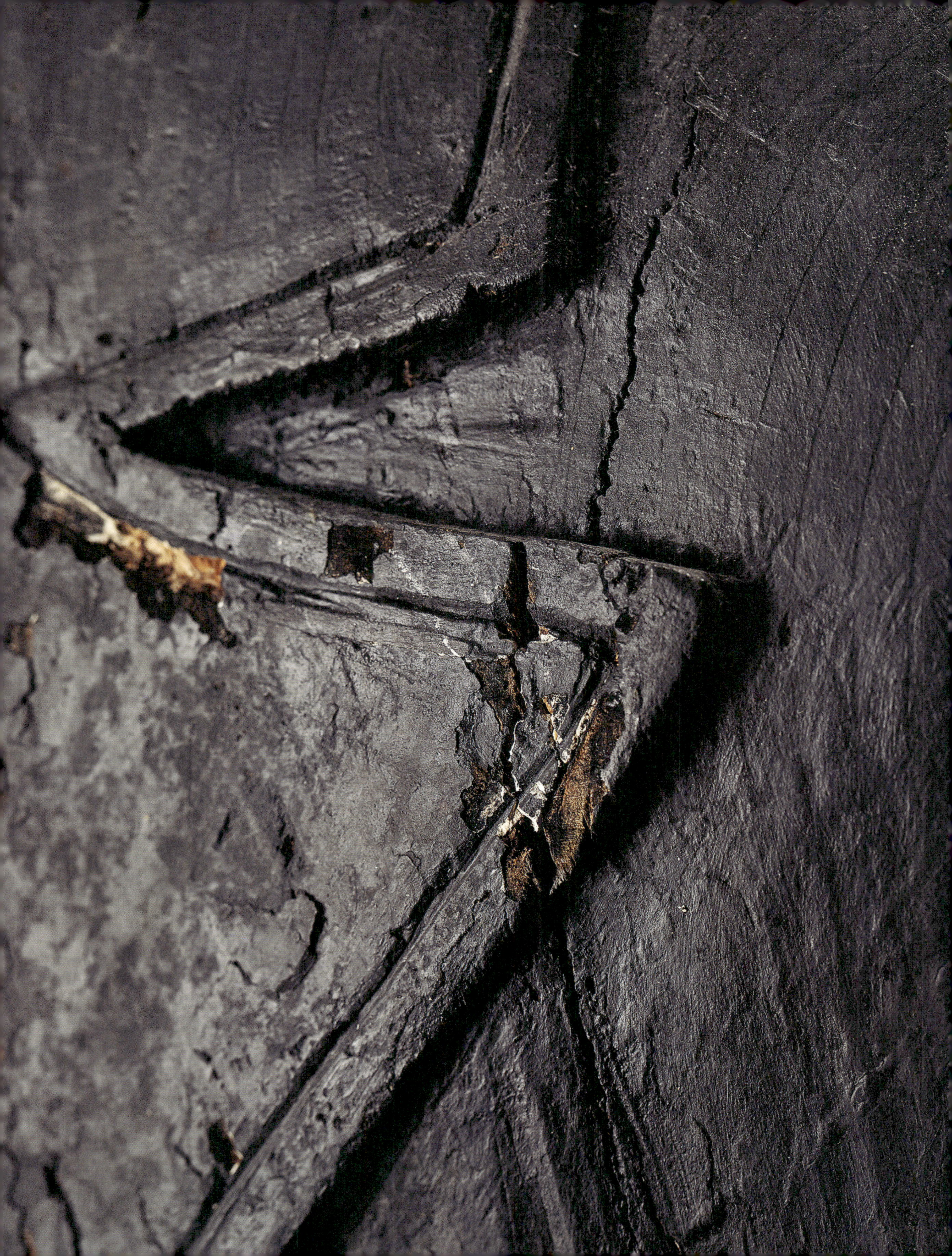

Full-length figure of standing lion painted yellow, possible 18th century. Origin unknown.

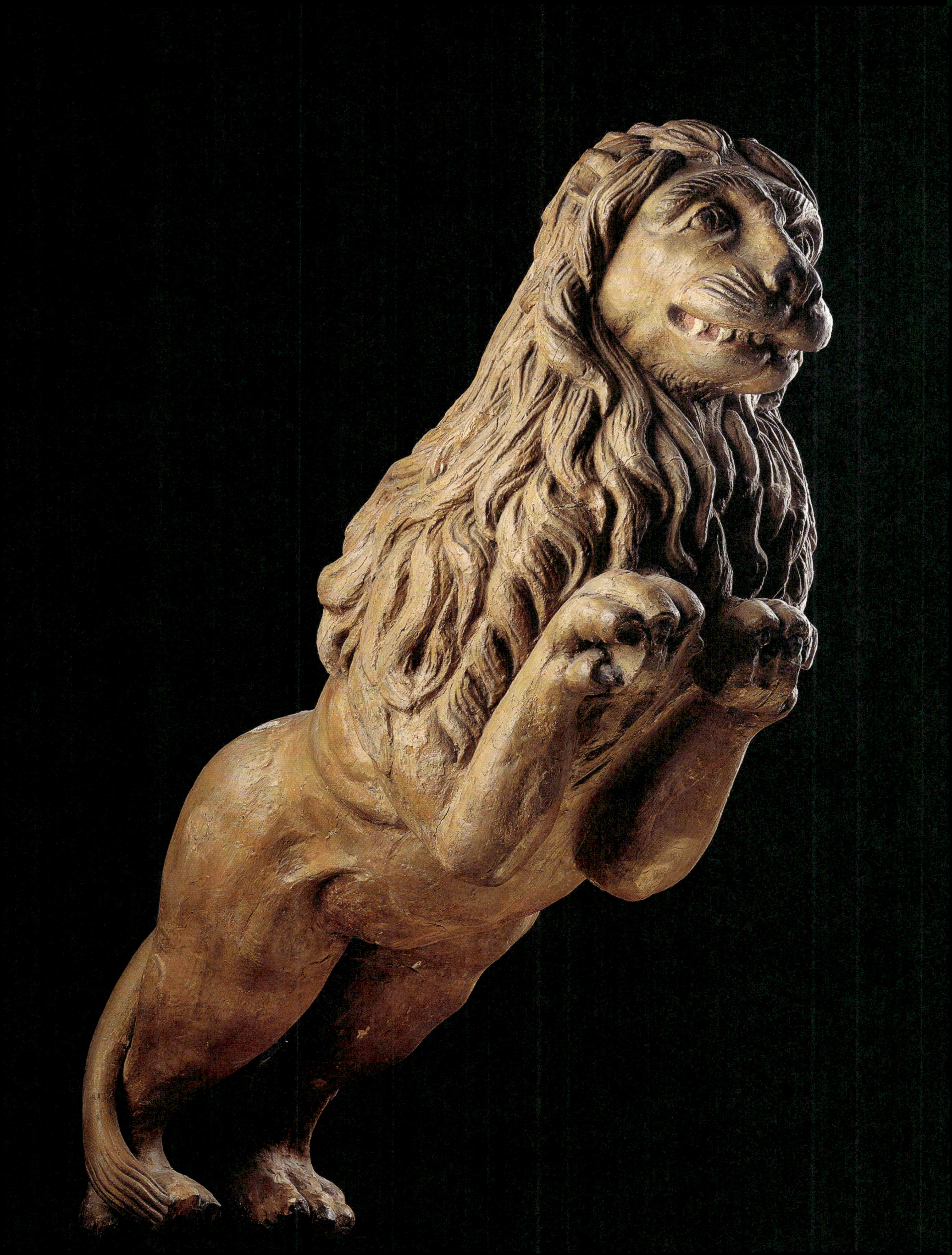

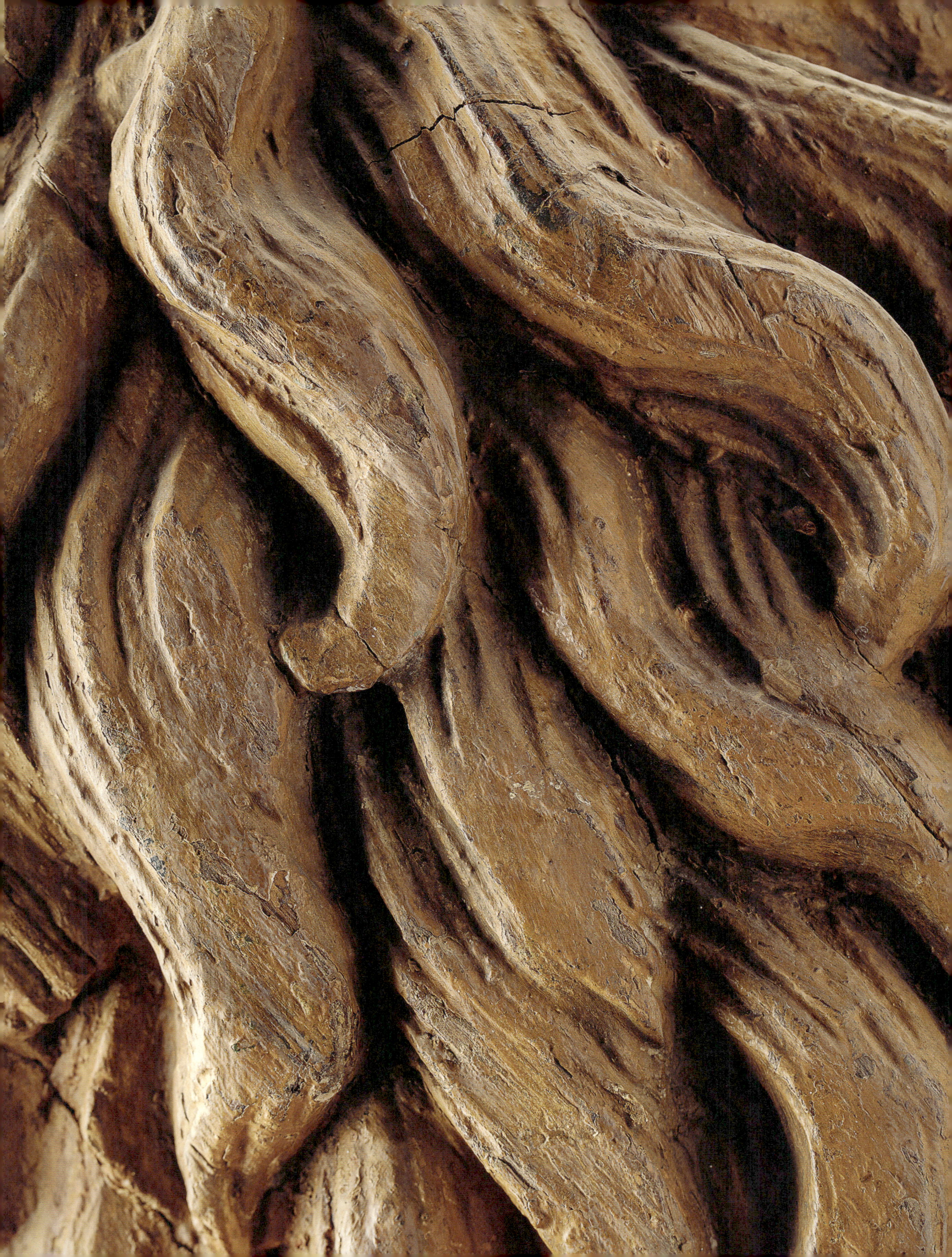

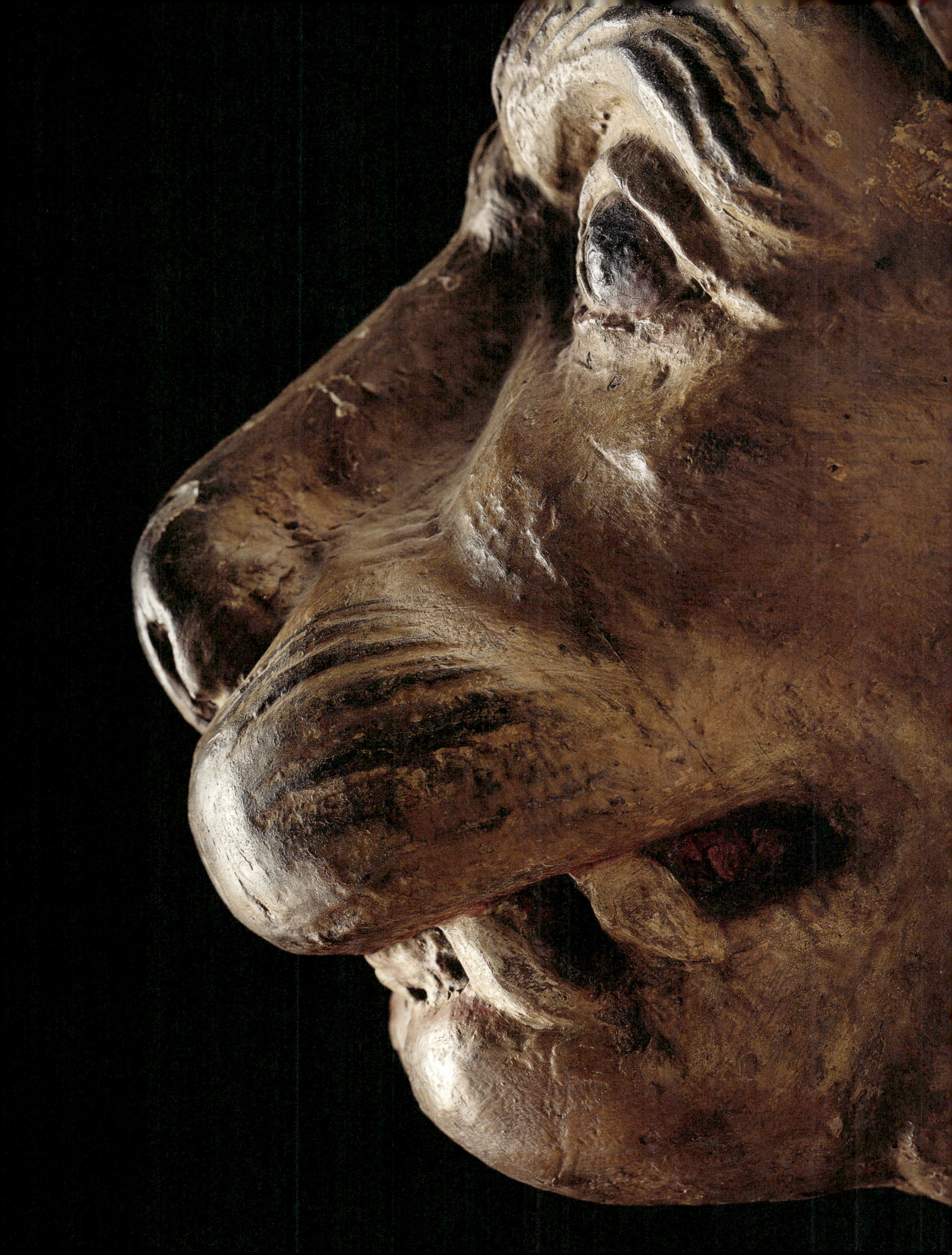

Demi figure of woman, painted in white and gold wearing necklace and earrings. Origin unknown.

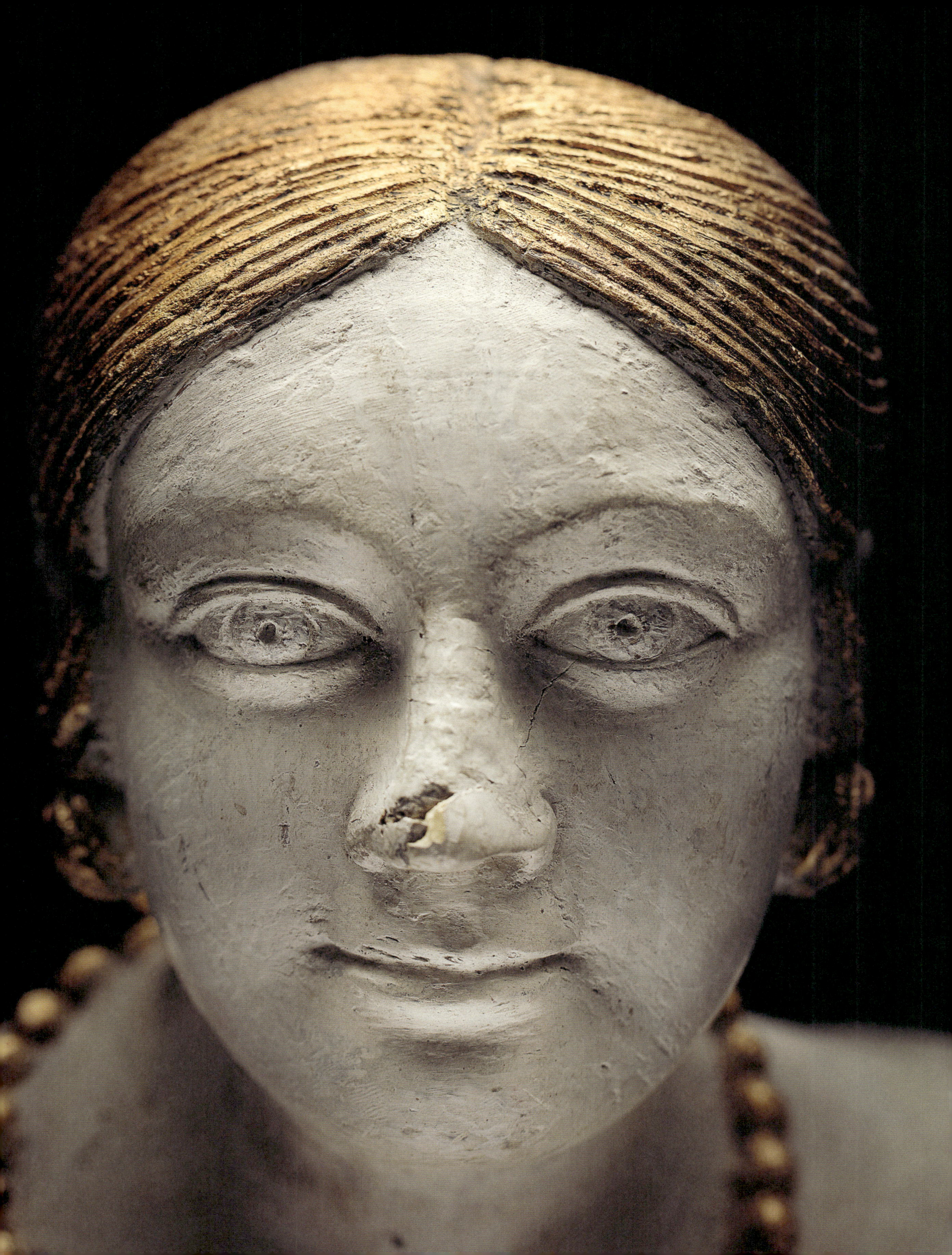

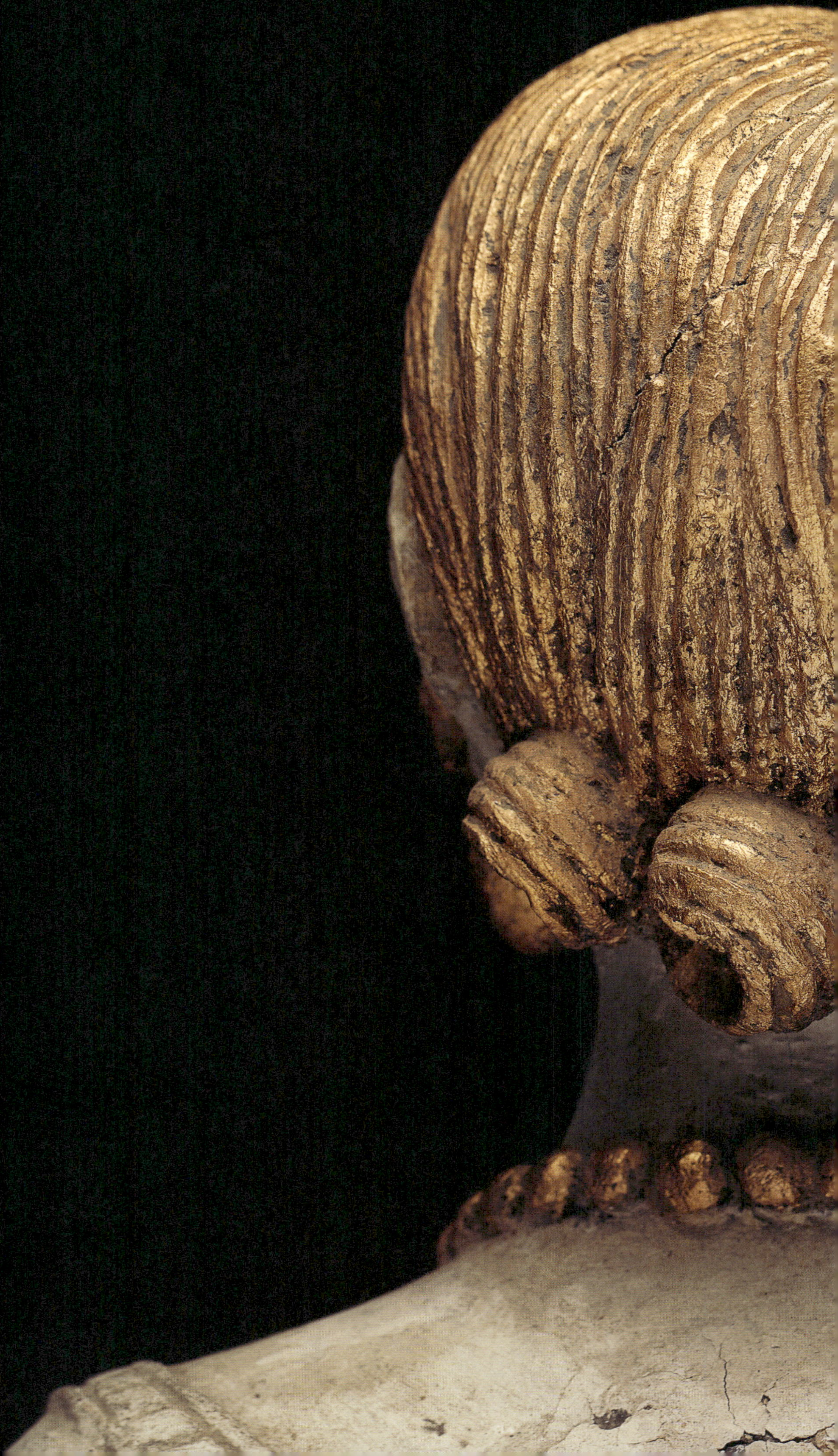

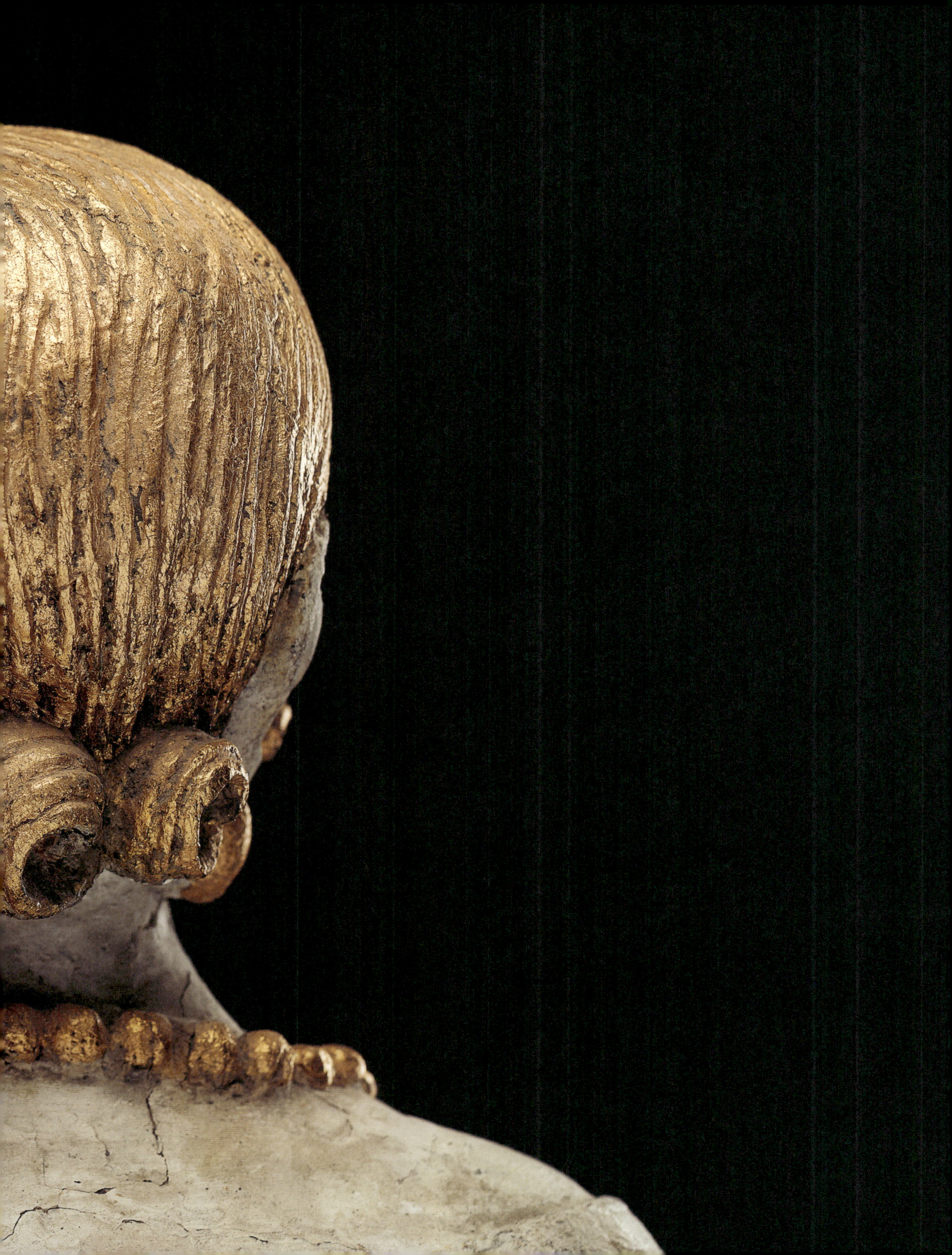

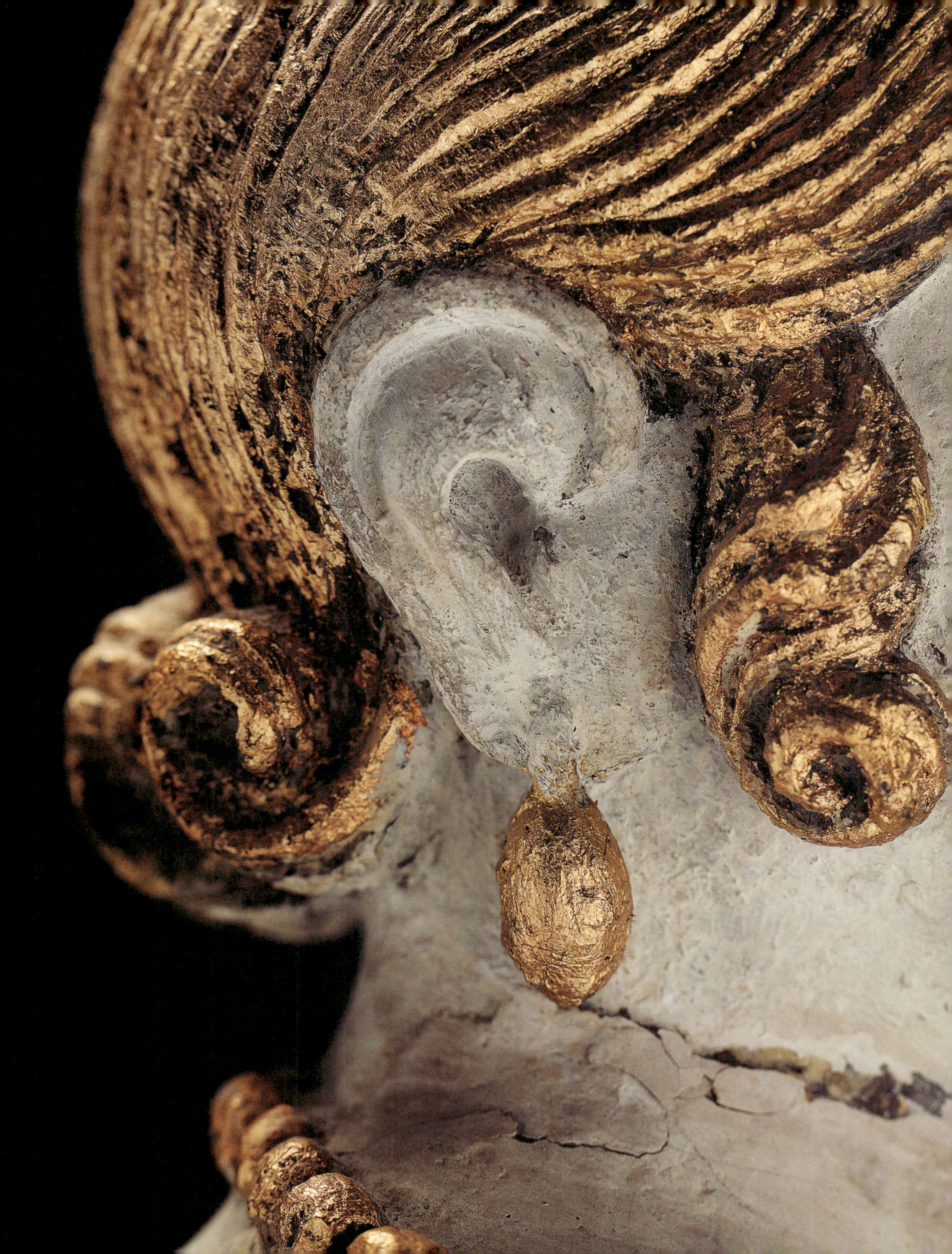

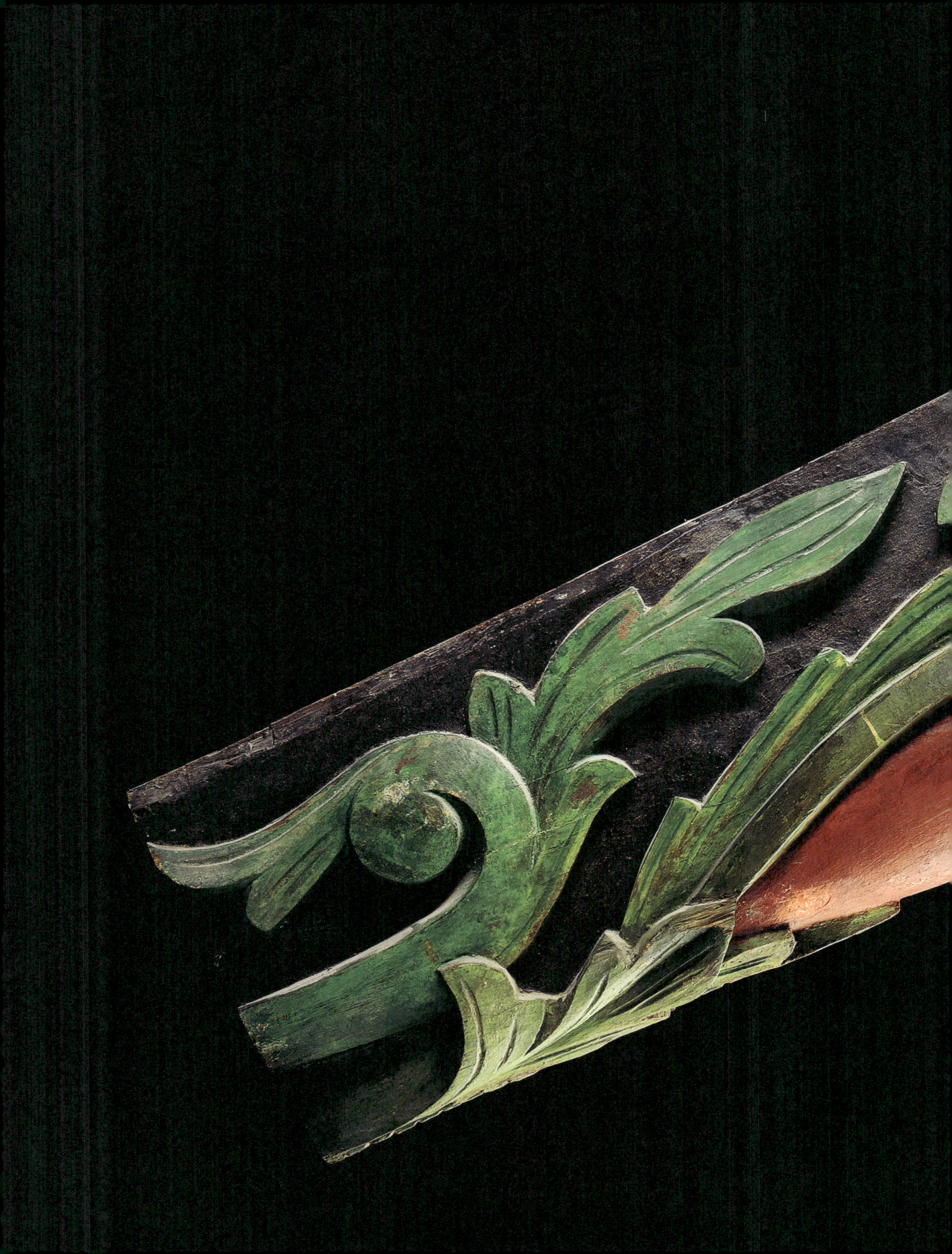

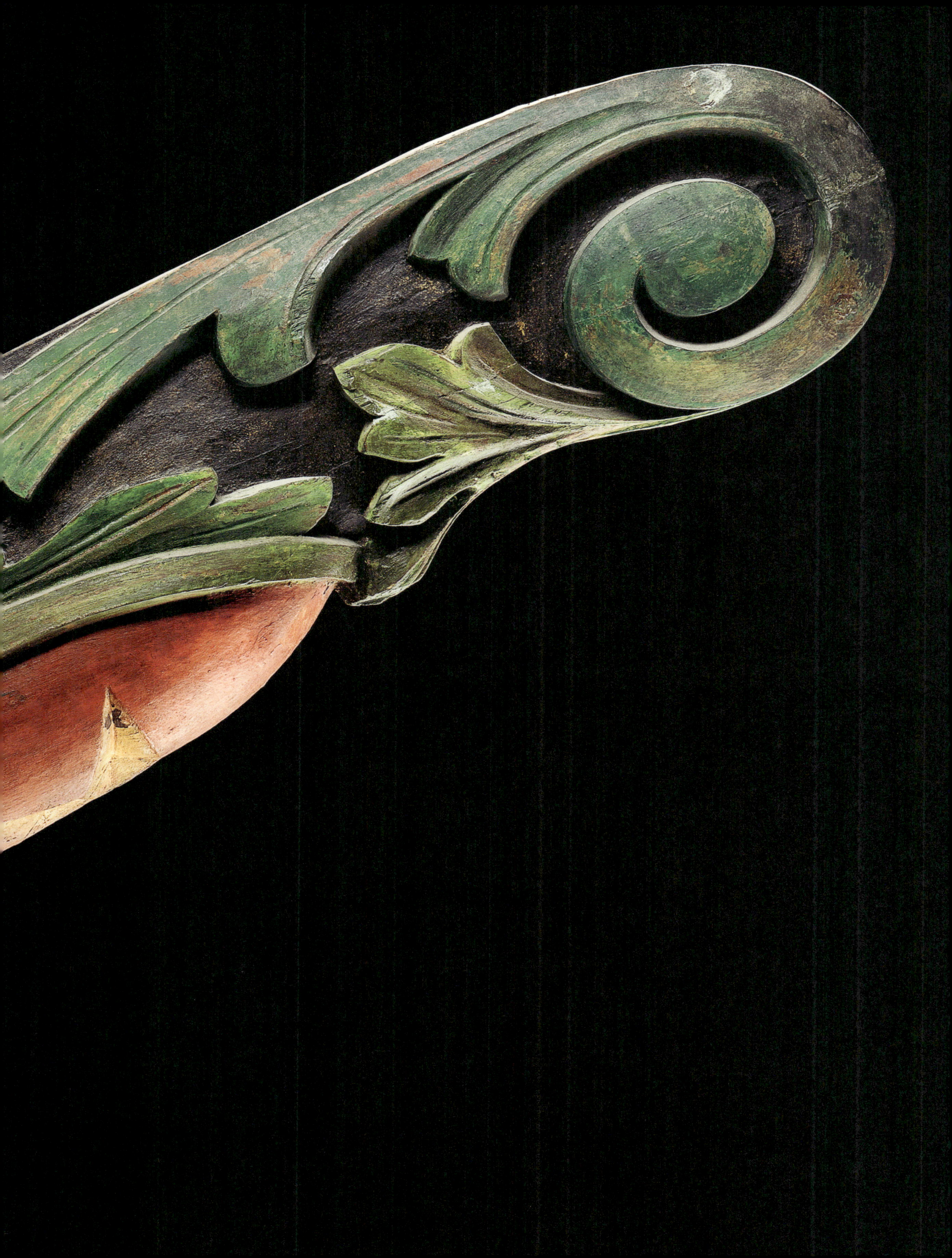

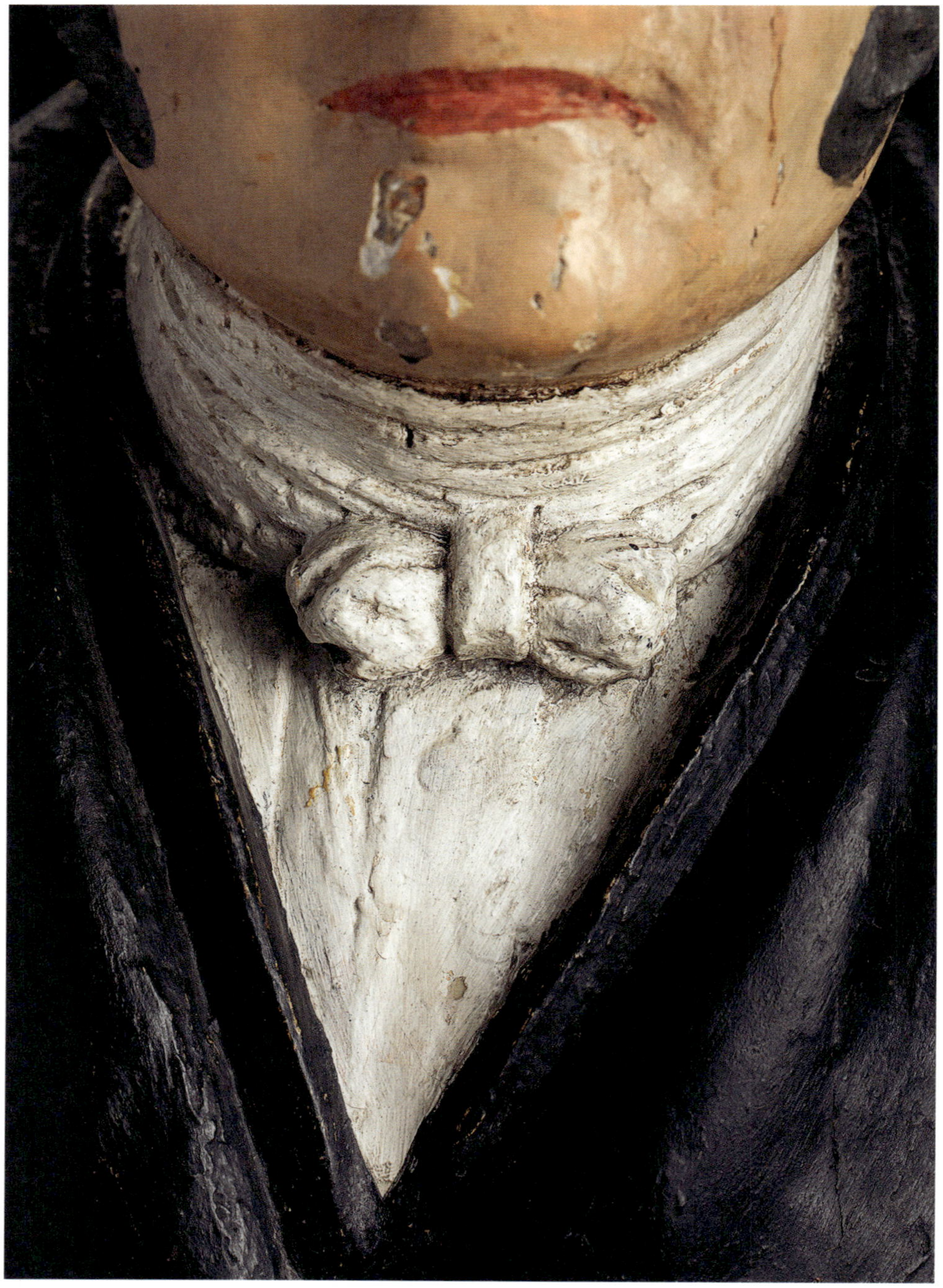

Preceding pages: Figurehead curl with shield formation in red with yellow star, surrounded by vegetal ornamentation in green. Origin unknown.

Bust of the shipowner type, wearing sideburns and dressed in black jacket and white linen neckband. Origin unknown.

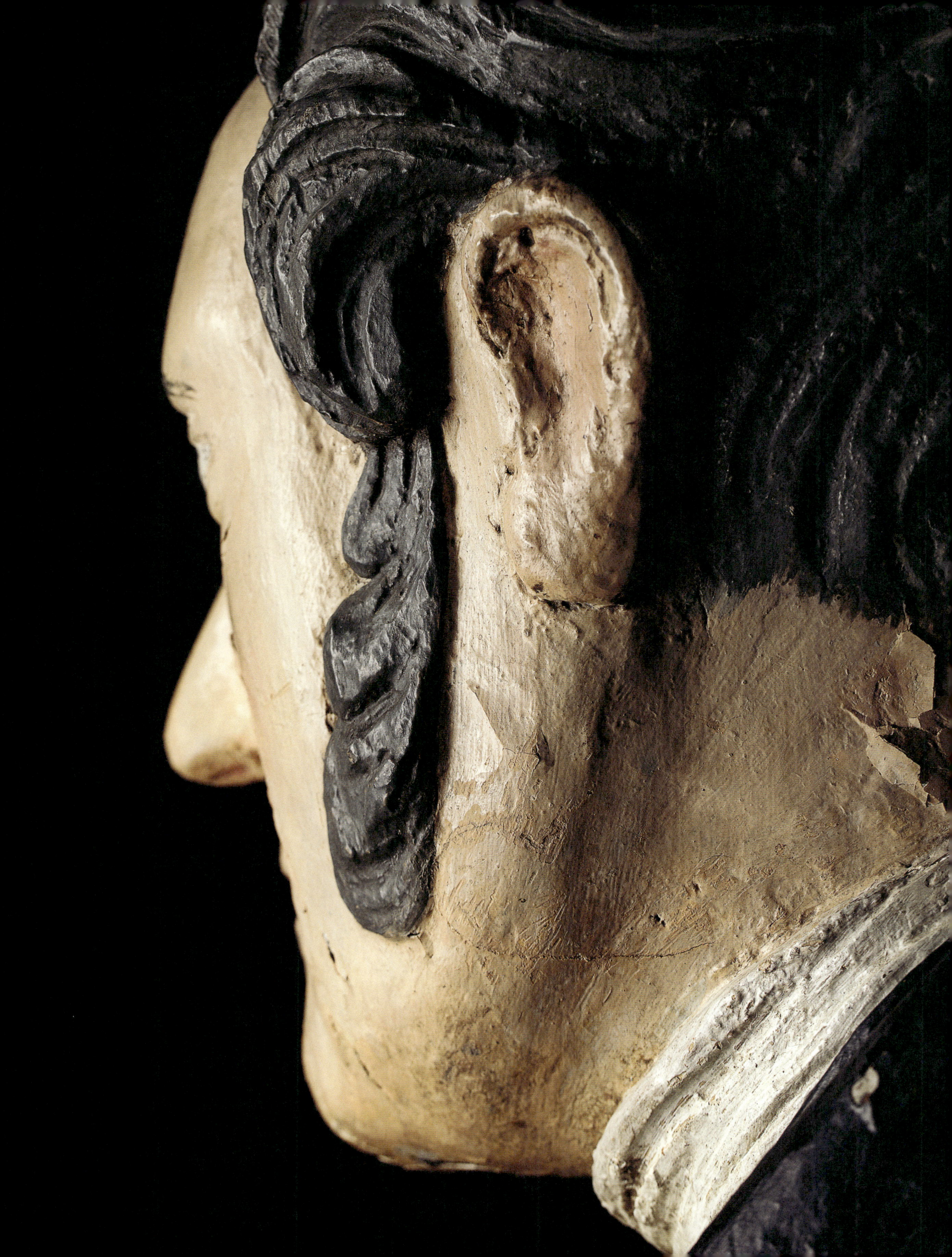

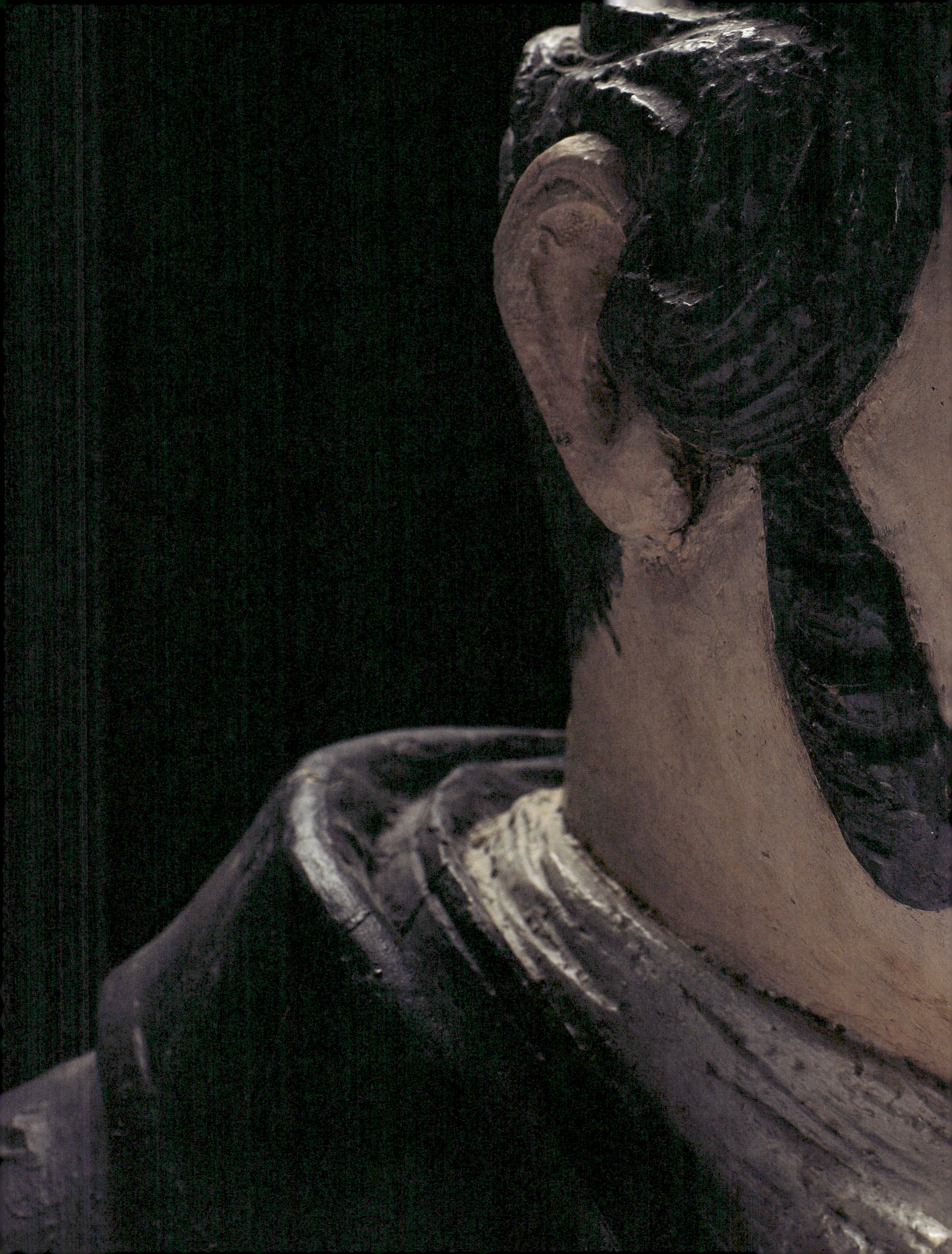

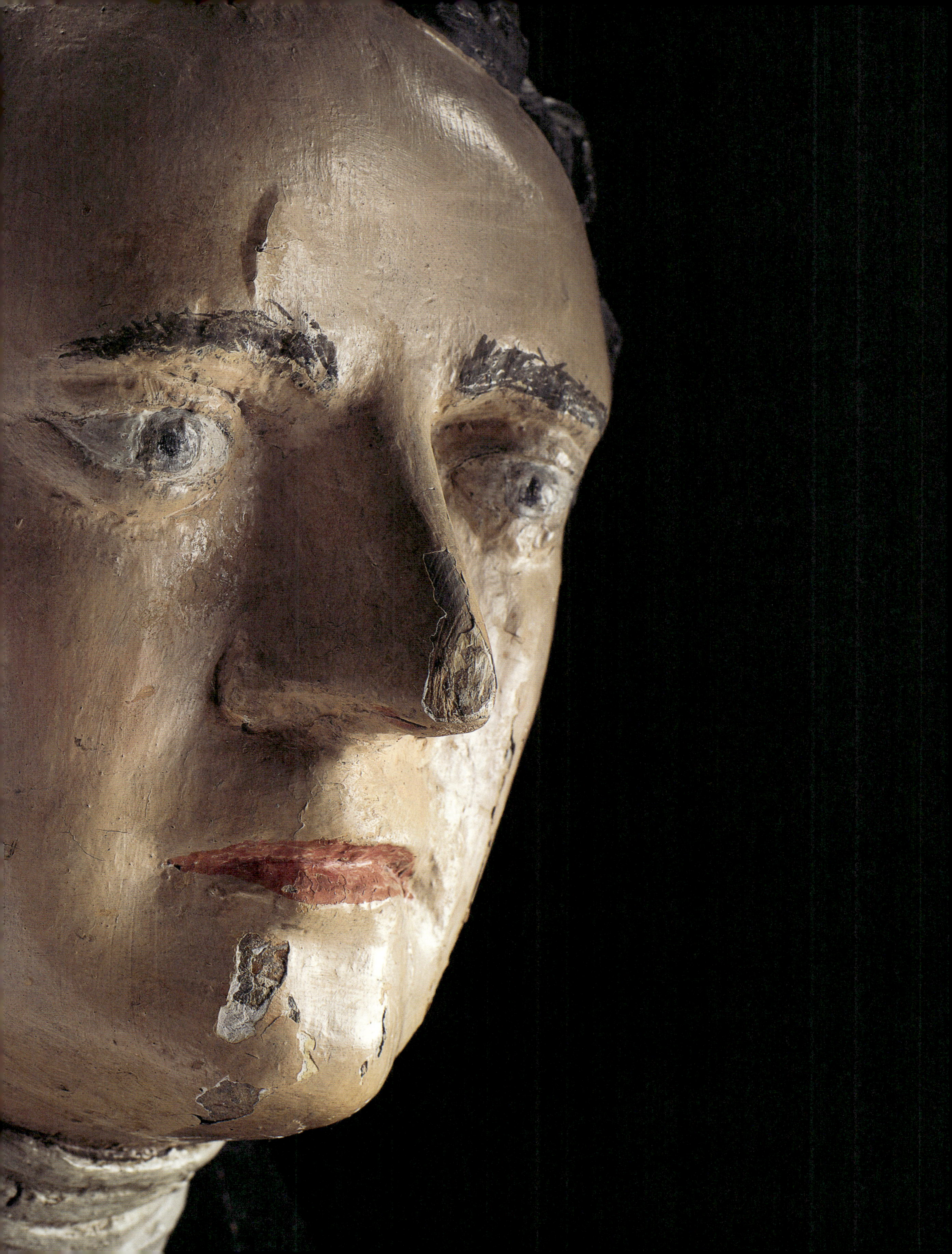

Relief of woman and dog in full-length figure, possibly part of a sternpost ornament. Origin unknown.

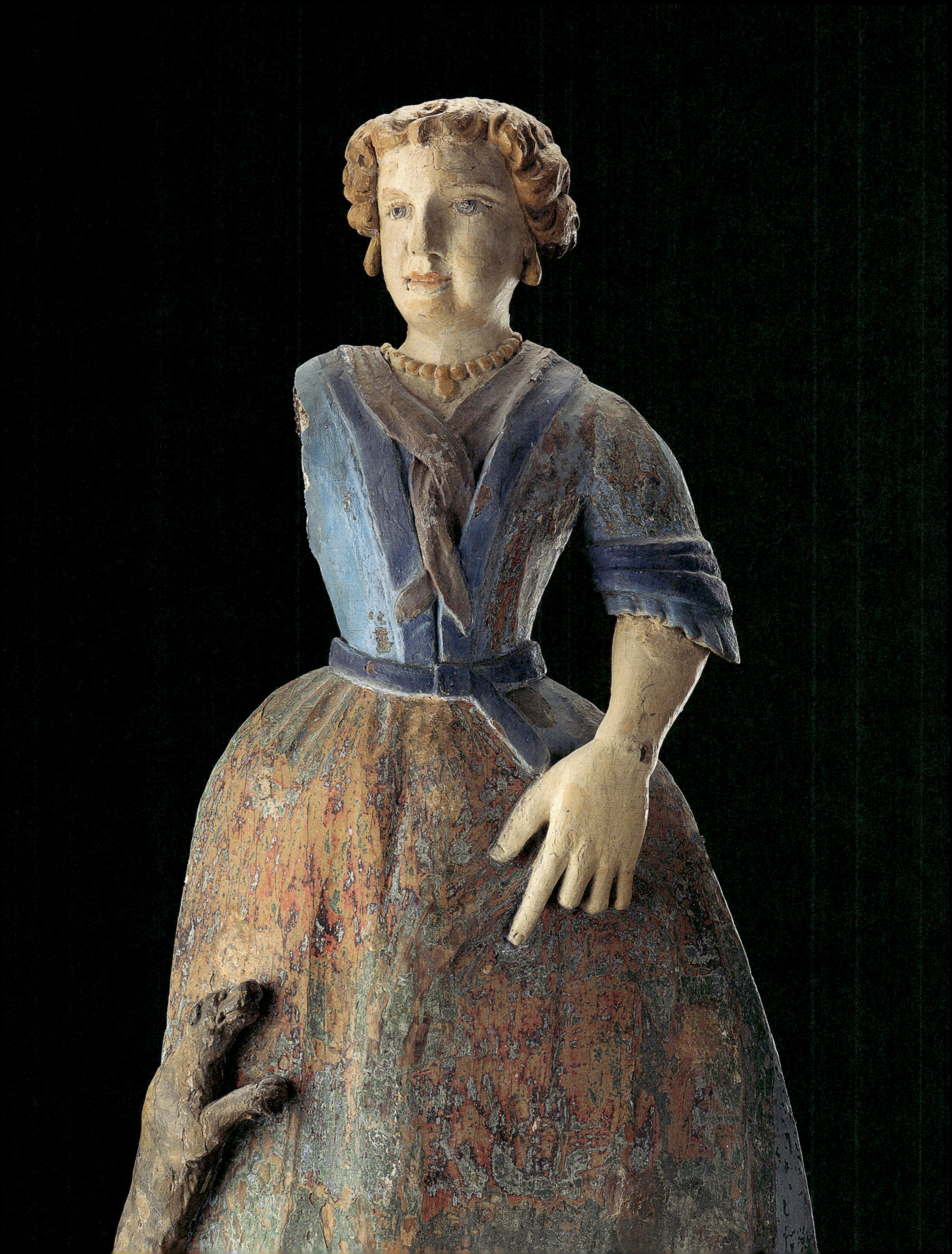

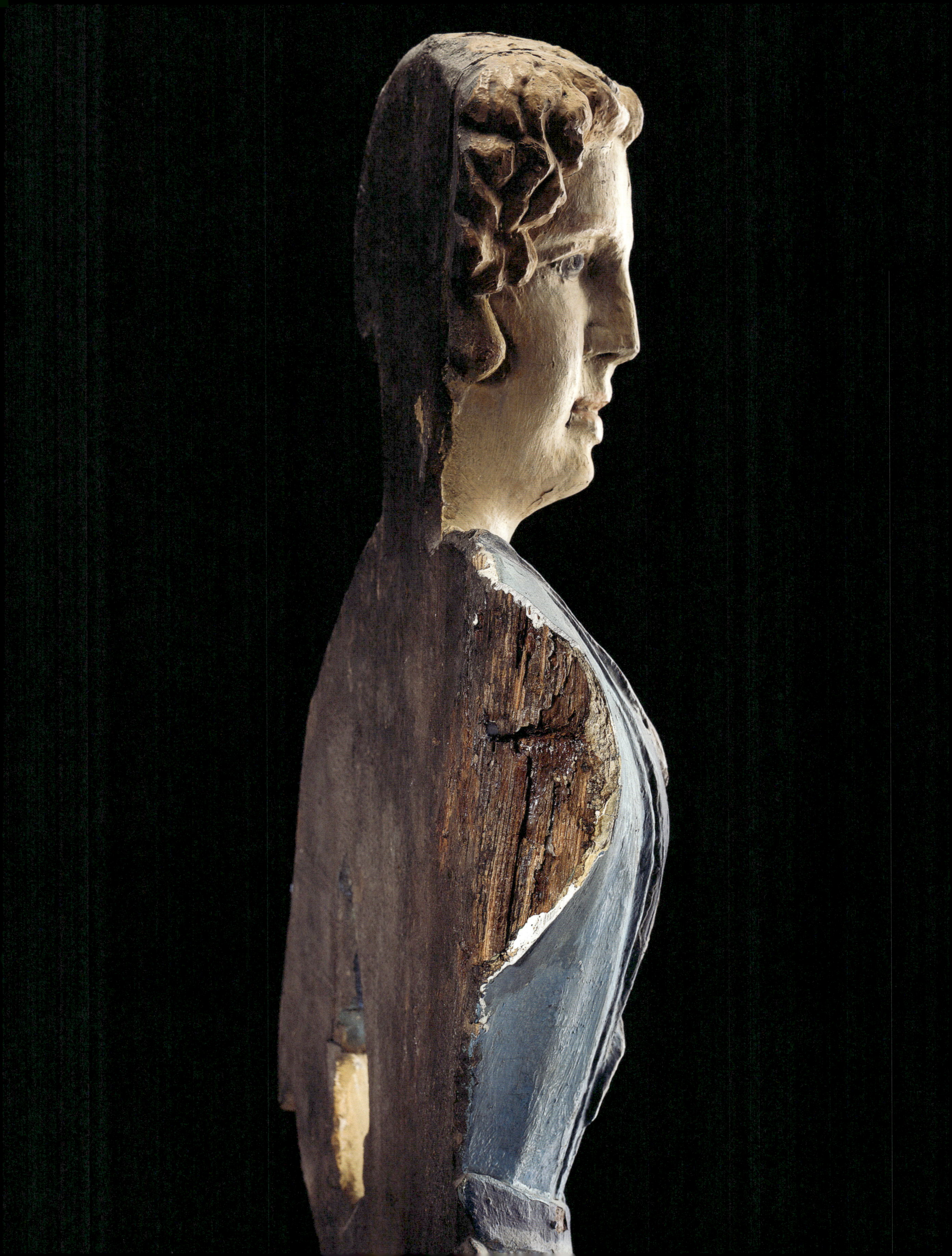

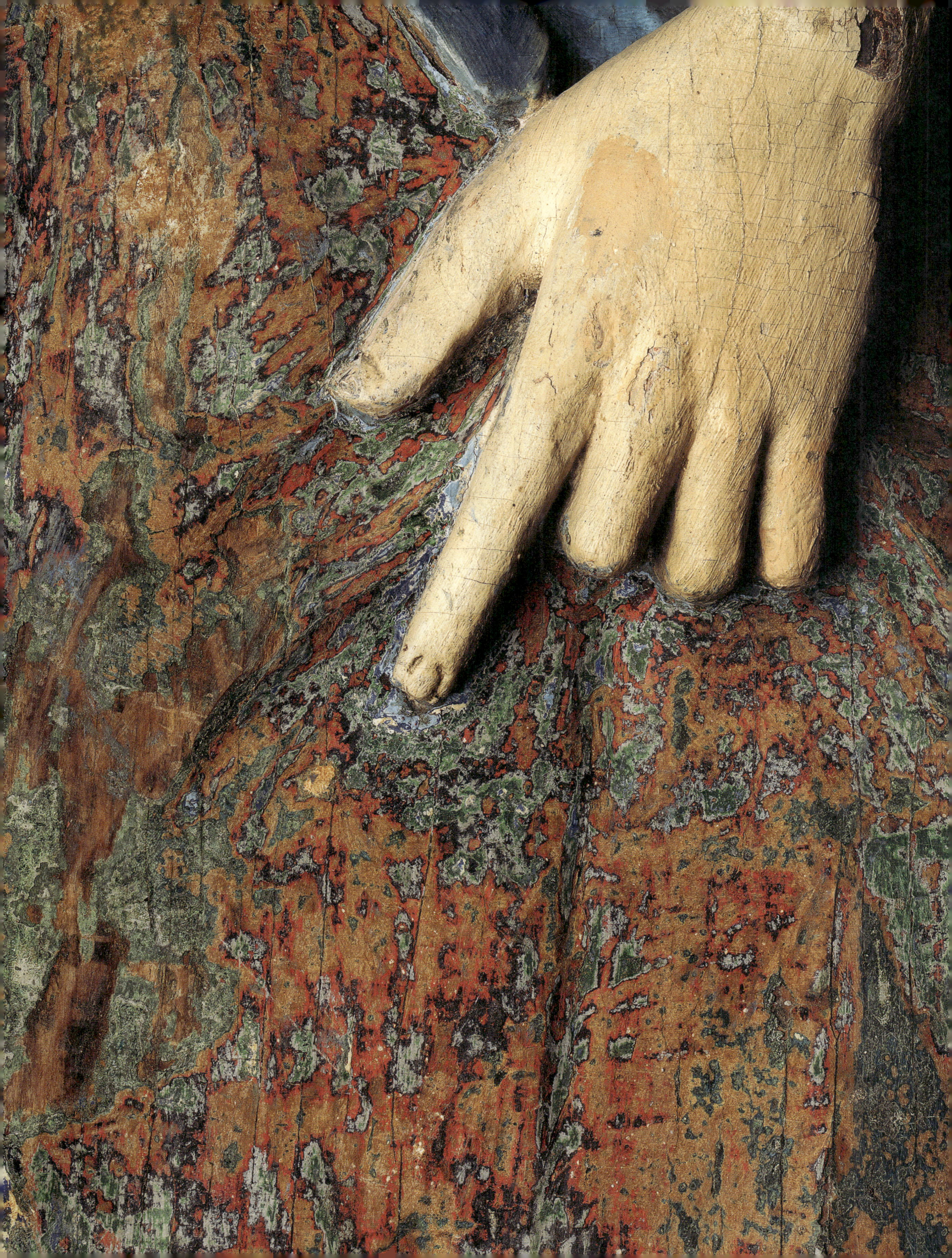

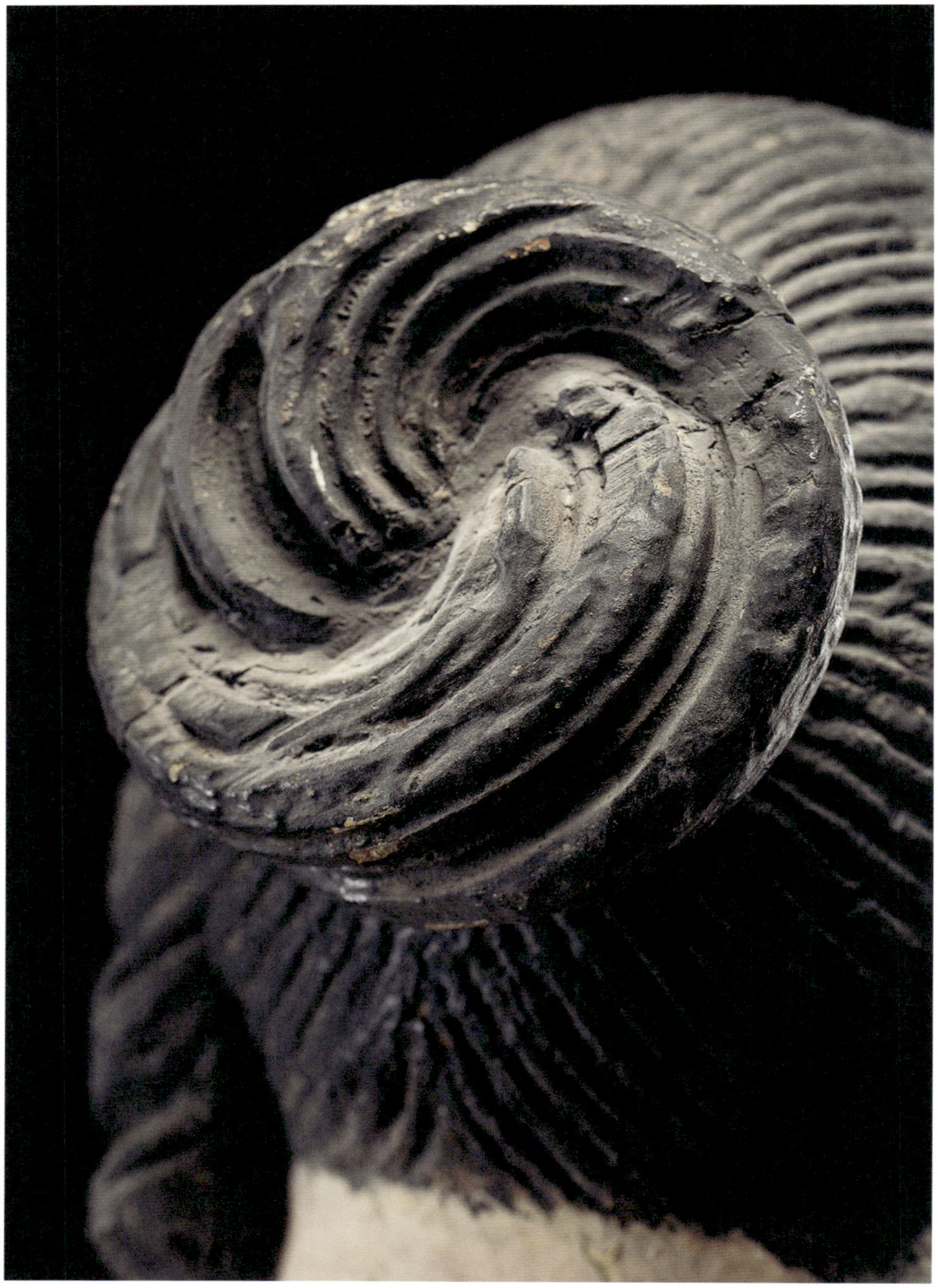

Woman in quarter figure with dark hair, blue décolletage blouse trimmed with lace and red frock. Origin unknown.

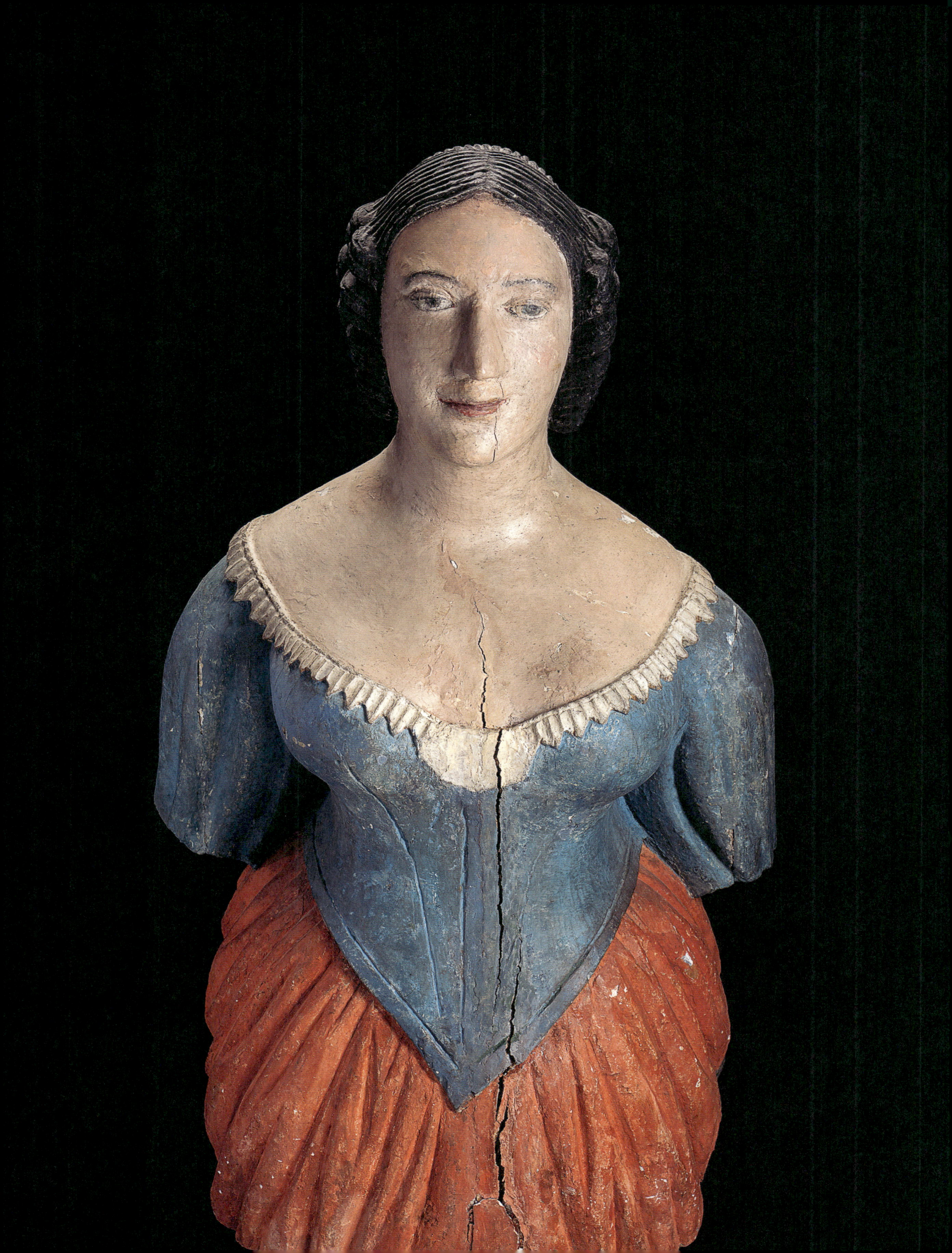

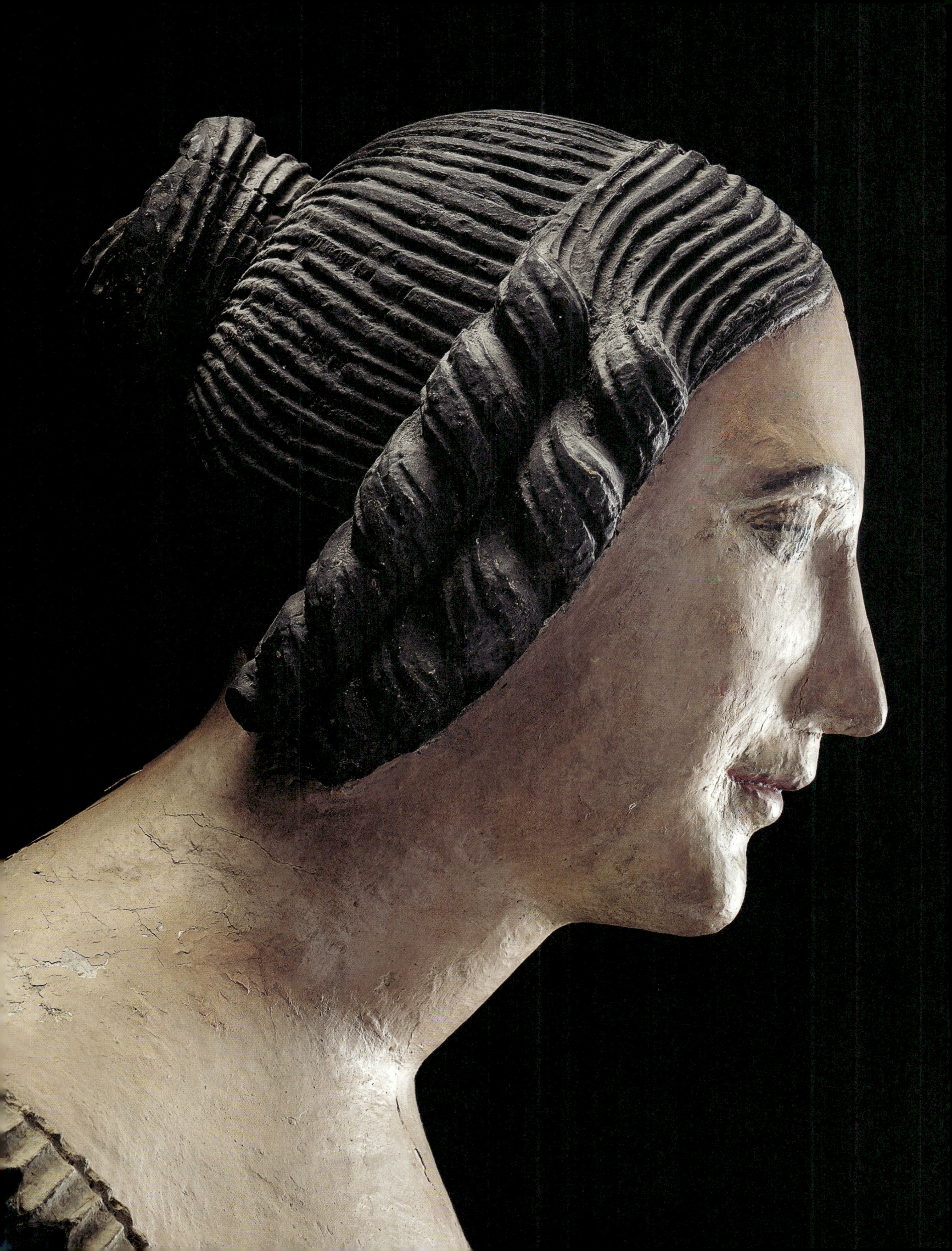

Three-quarter-length figure of woman dressed in Classical style, with dark hair, tiara and other jewellery. Origin unknown.

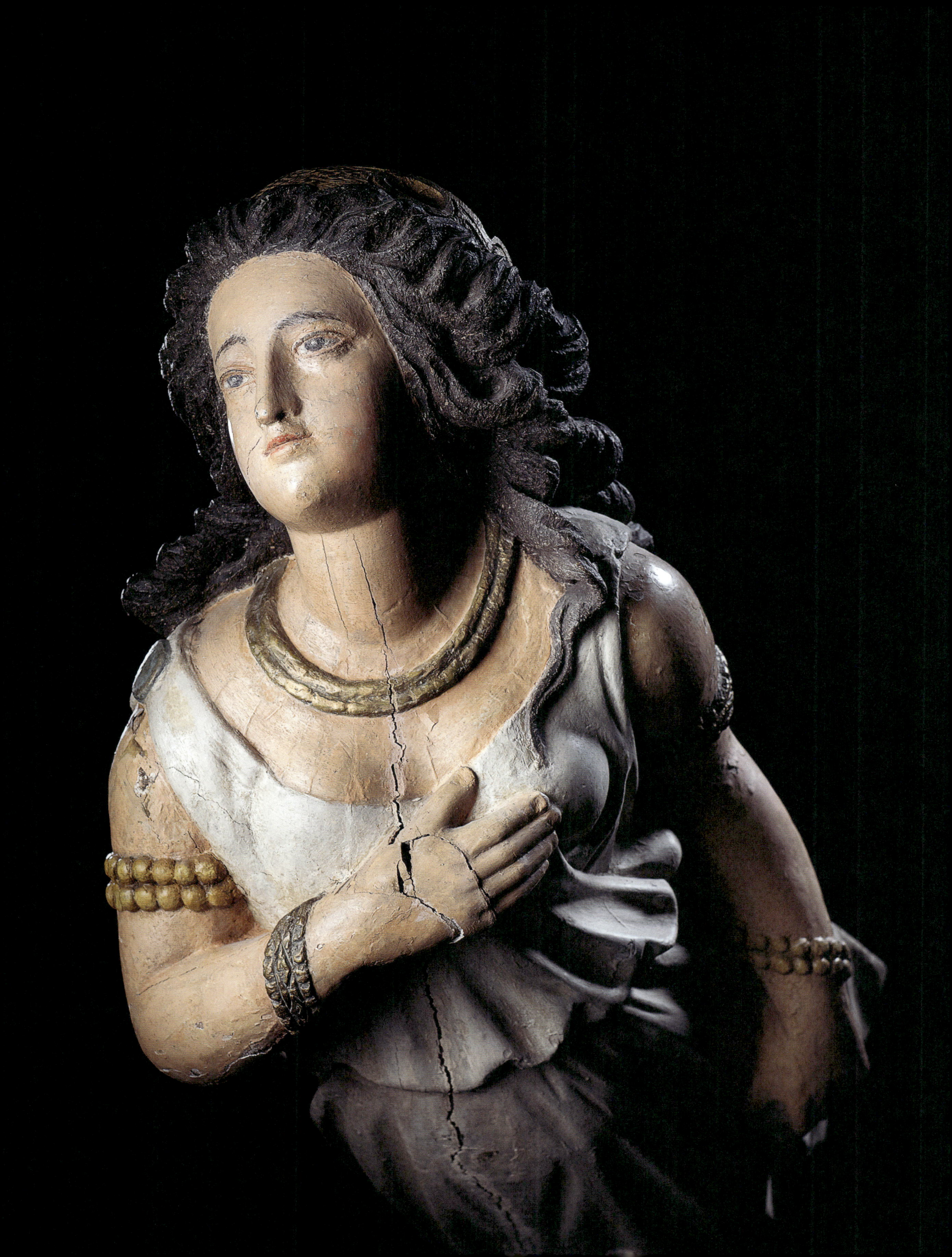

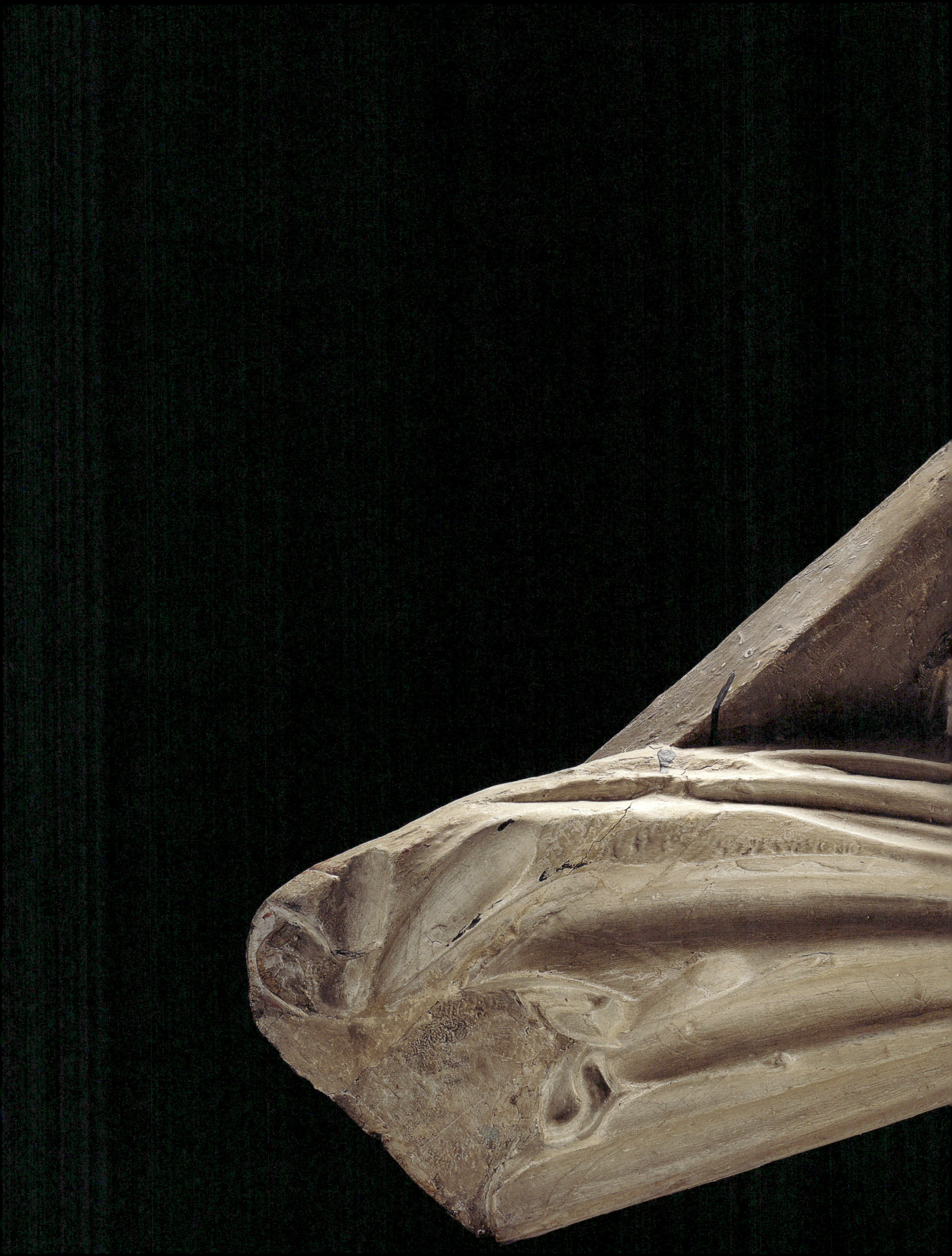

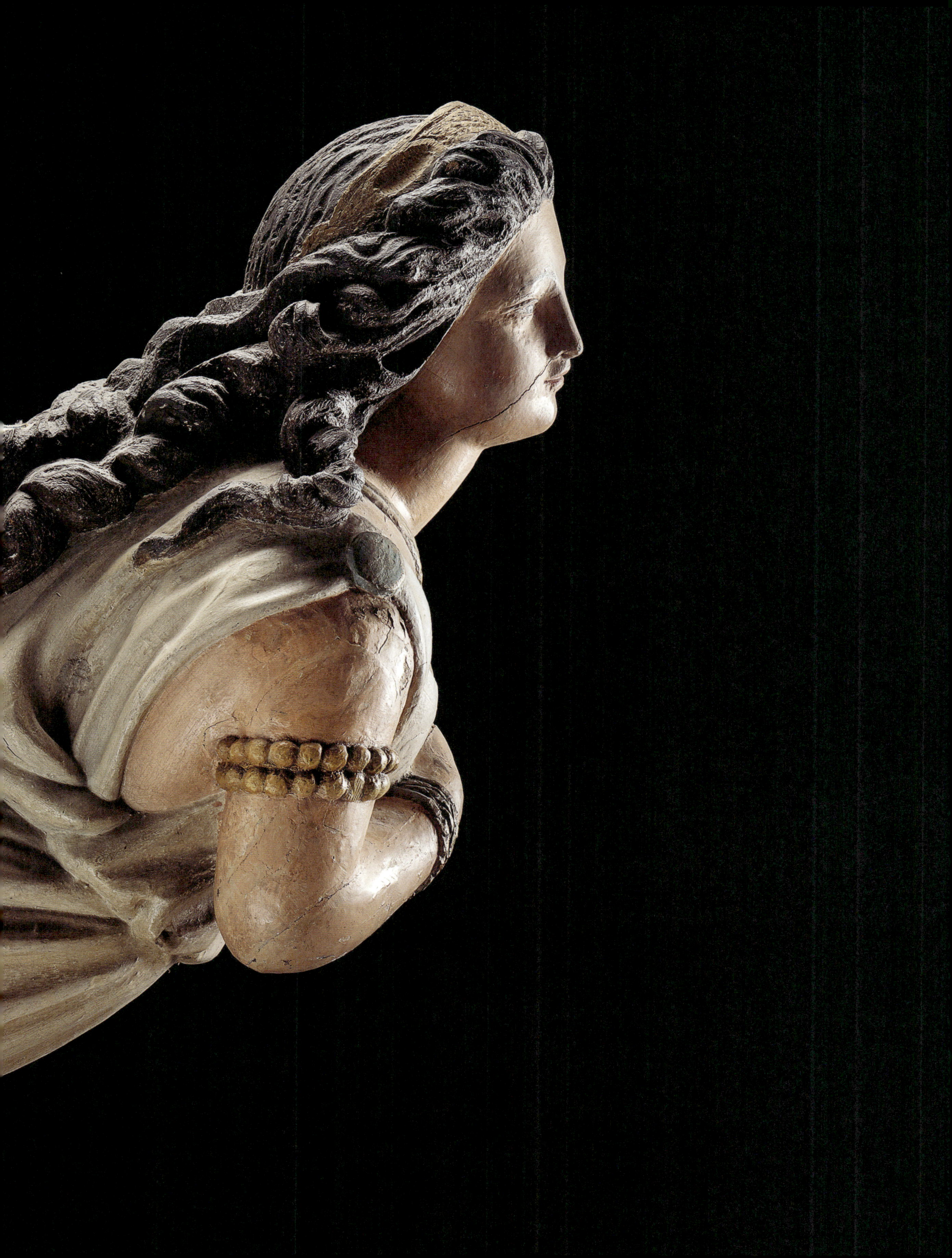

Head of bird painted white, black and yellow with half-open bill and richly detailed plumage. Origin unknown.

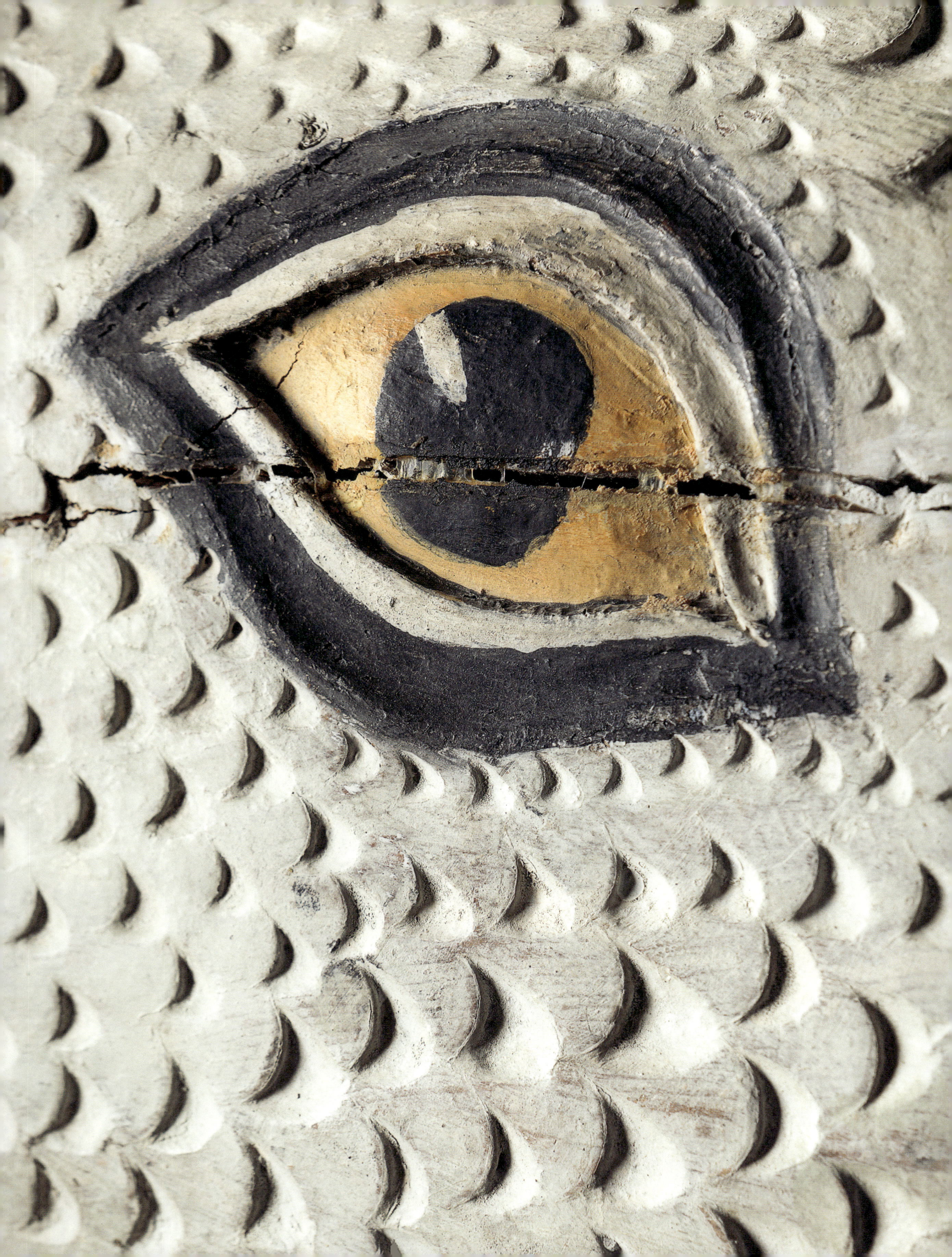

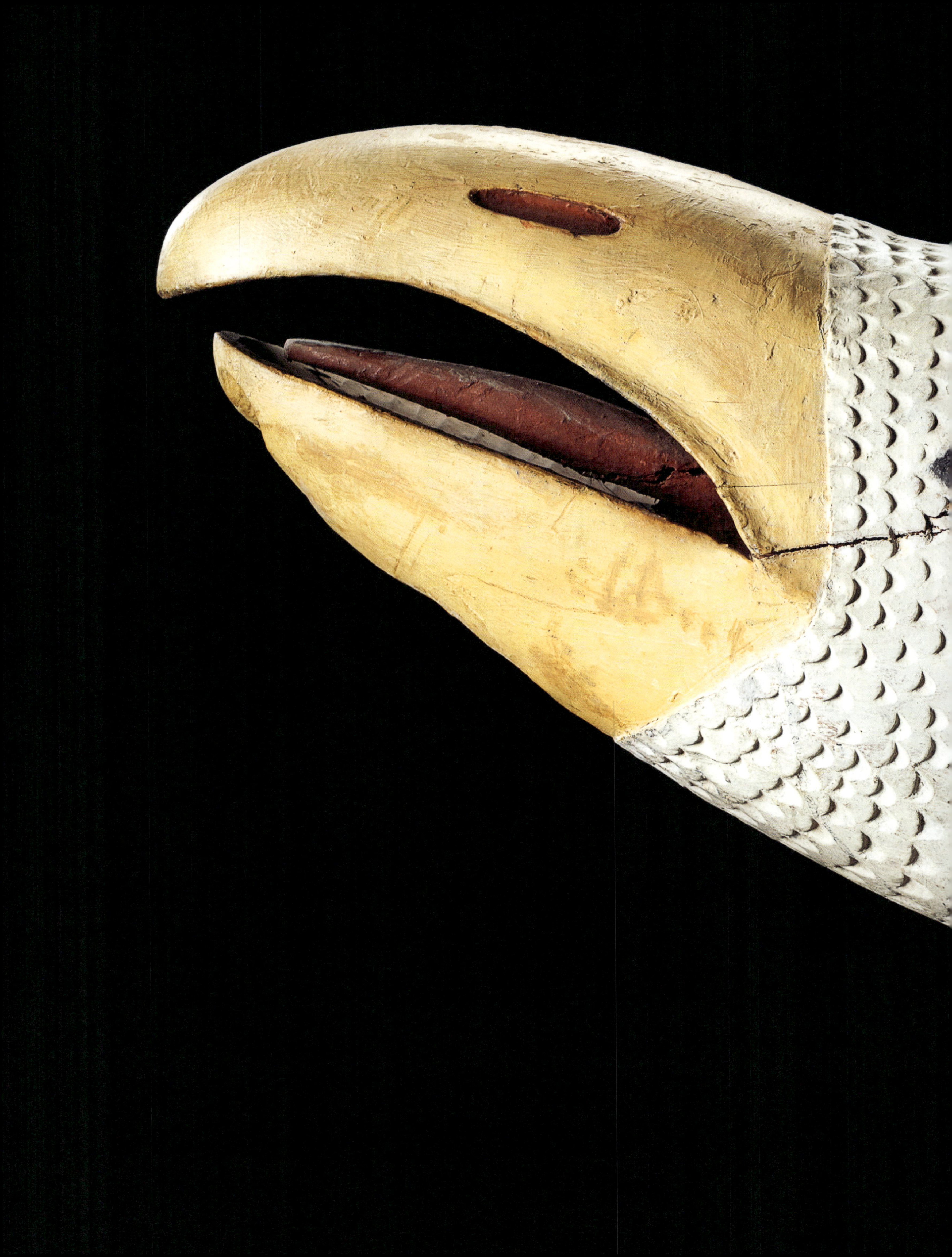

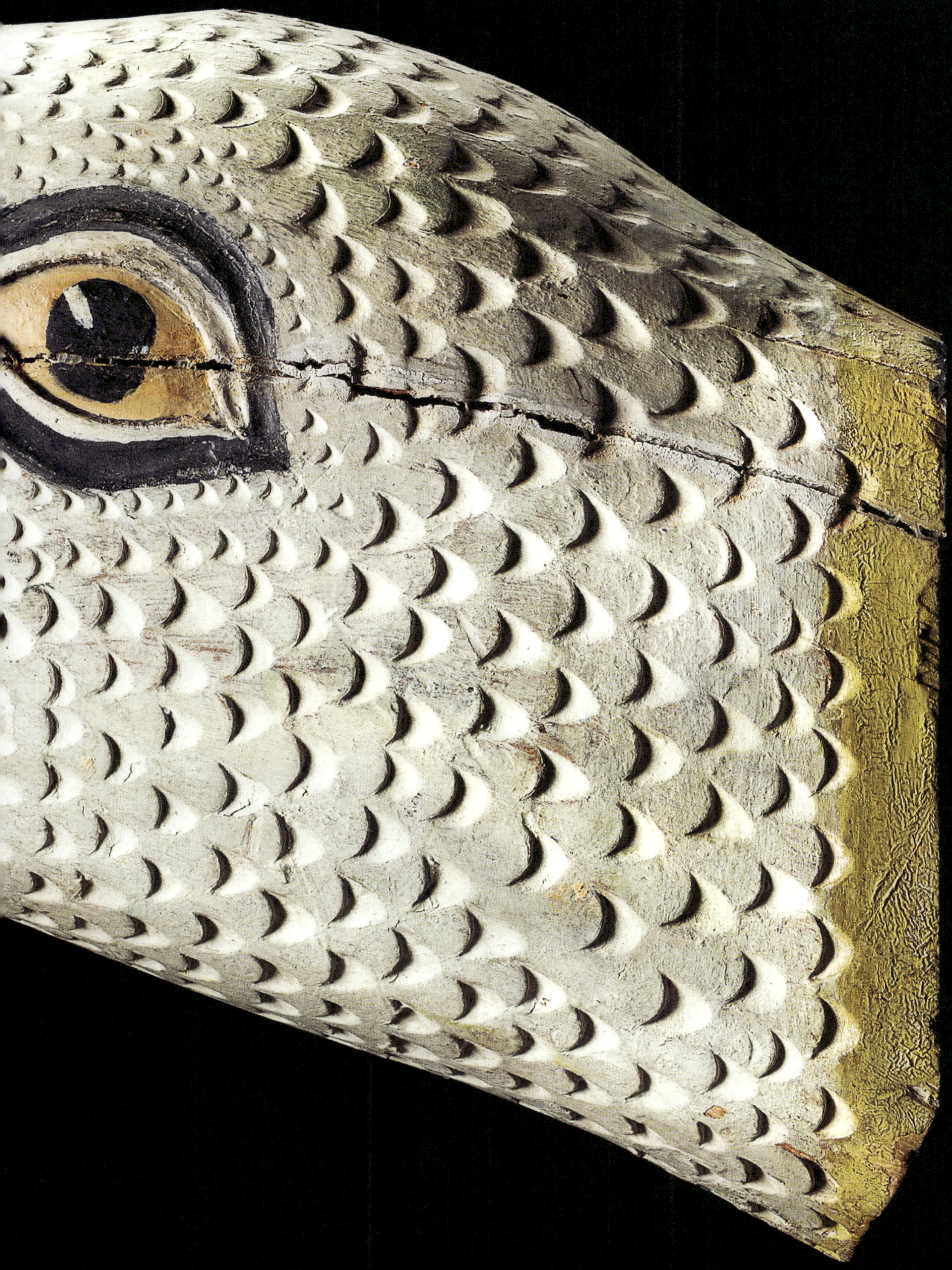

Woman in full-length figure with blonde hair and tiara, blue gown and blouse with details in red, gold and white. Origin unknown.

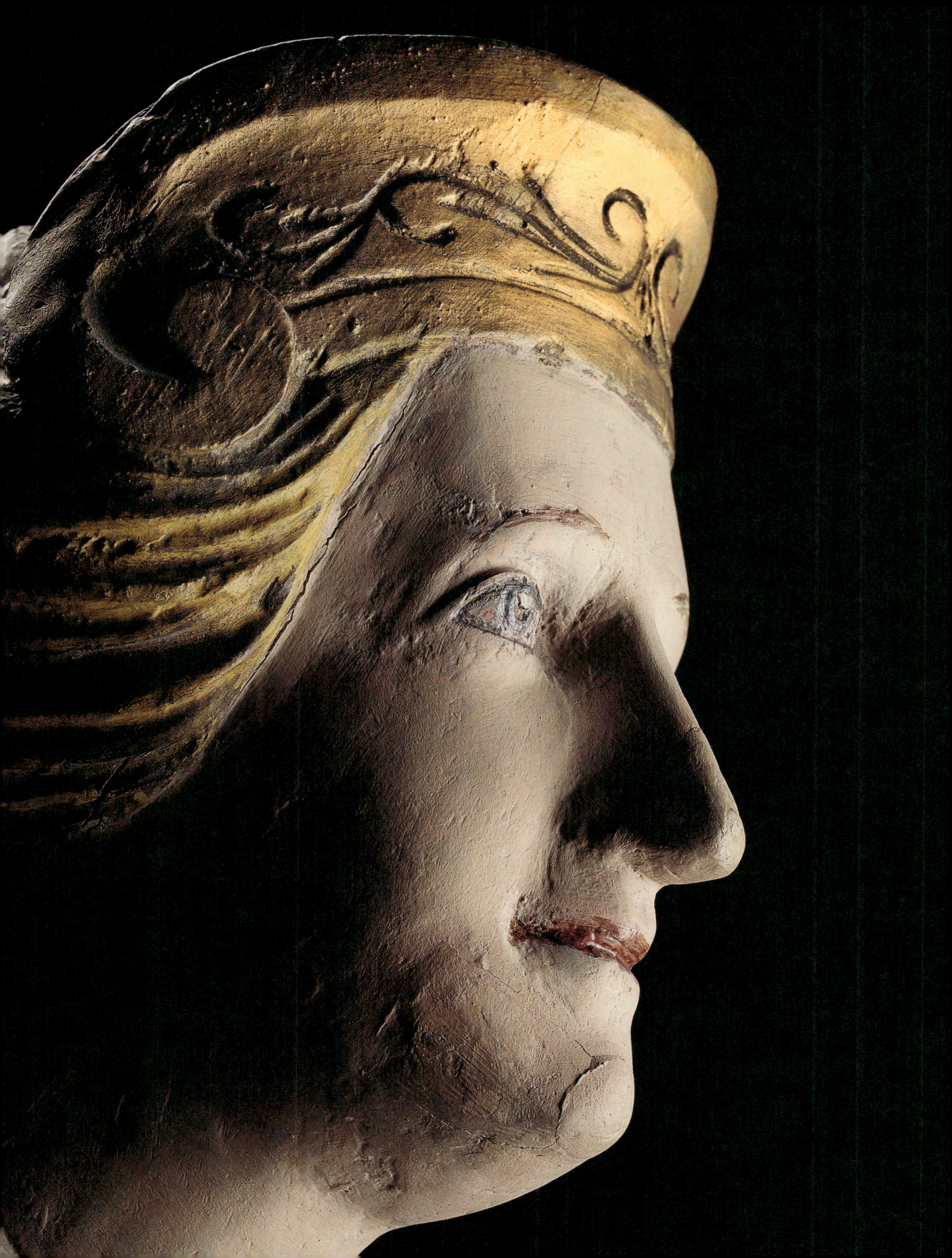

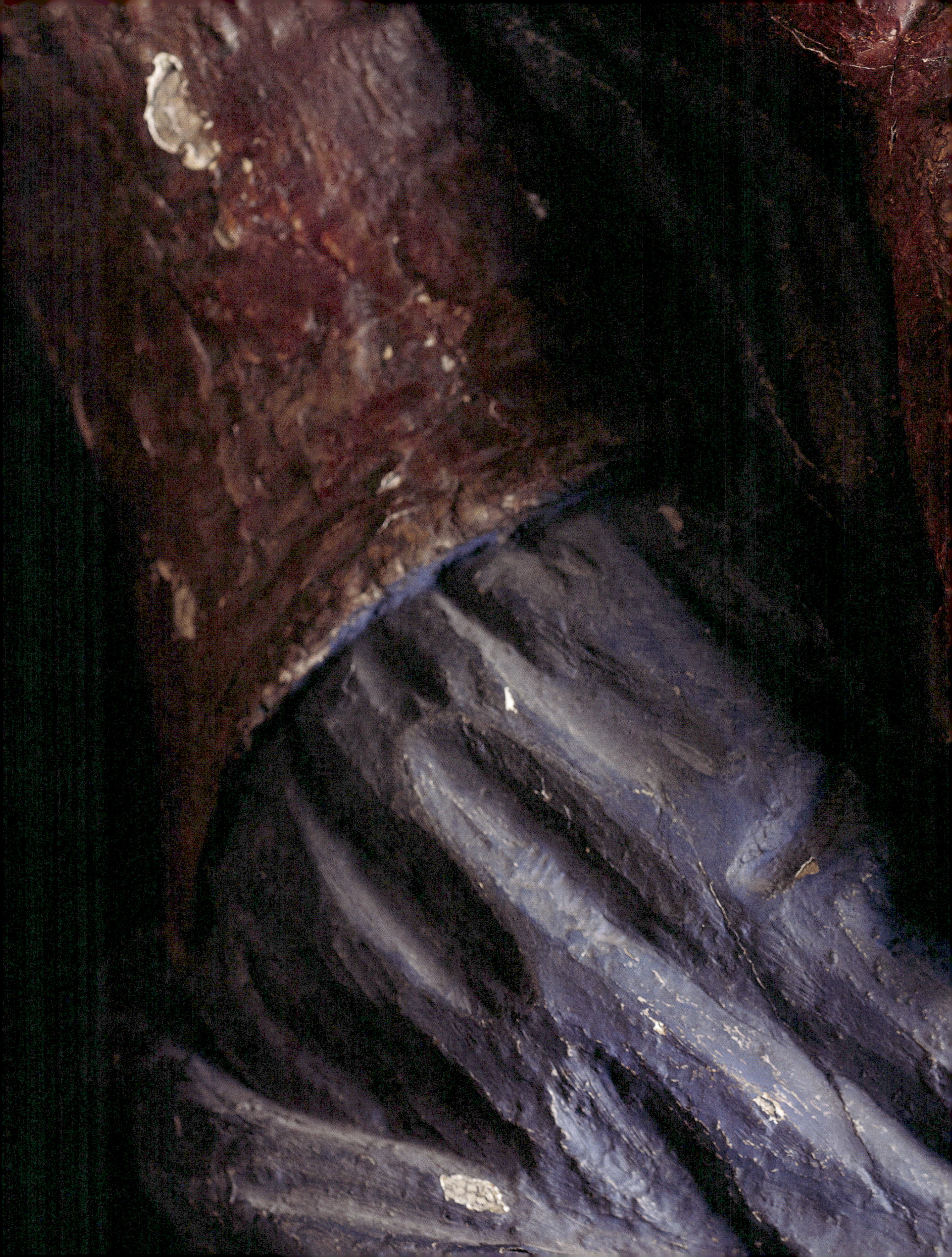

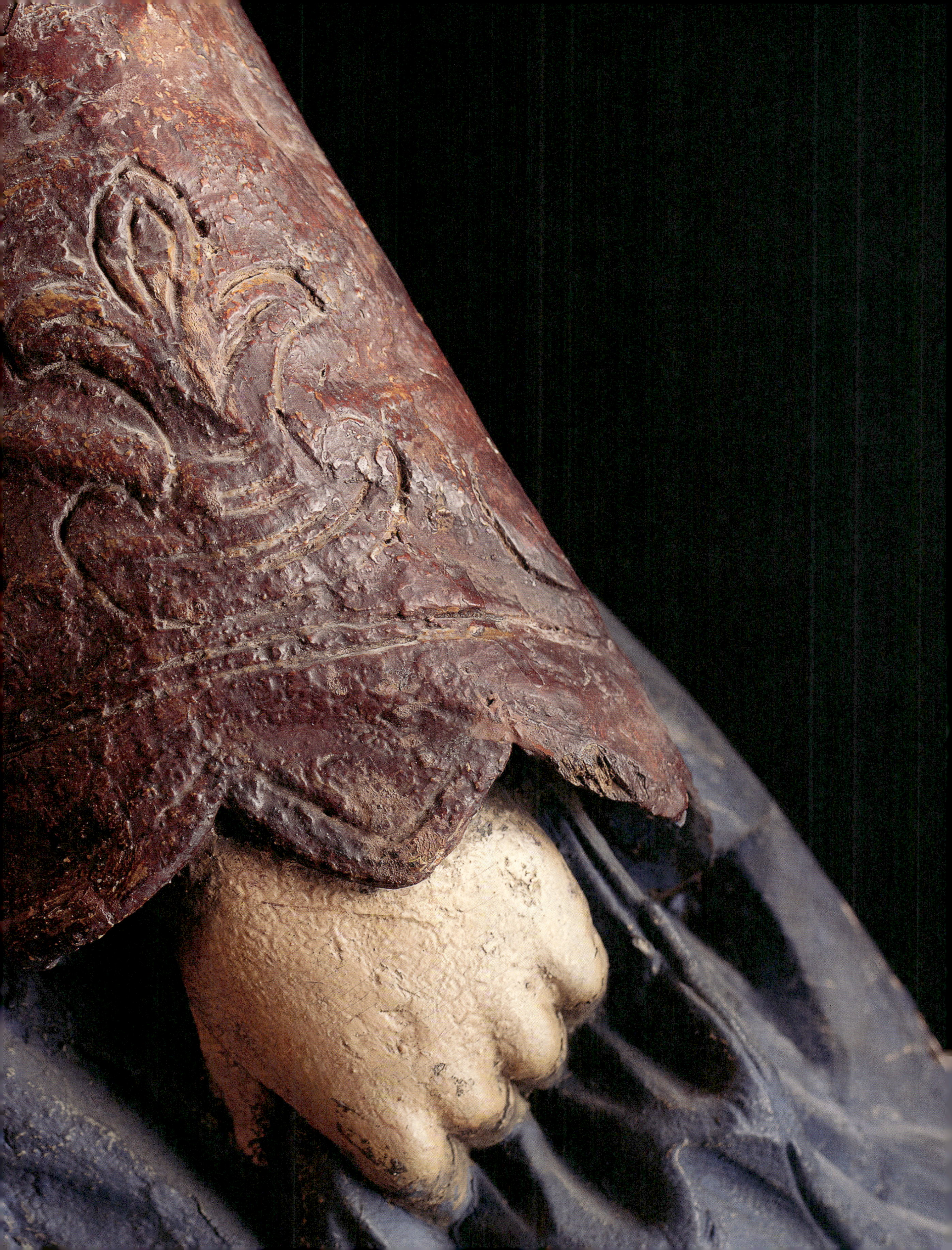

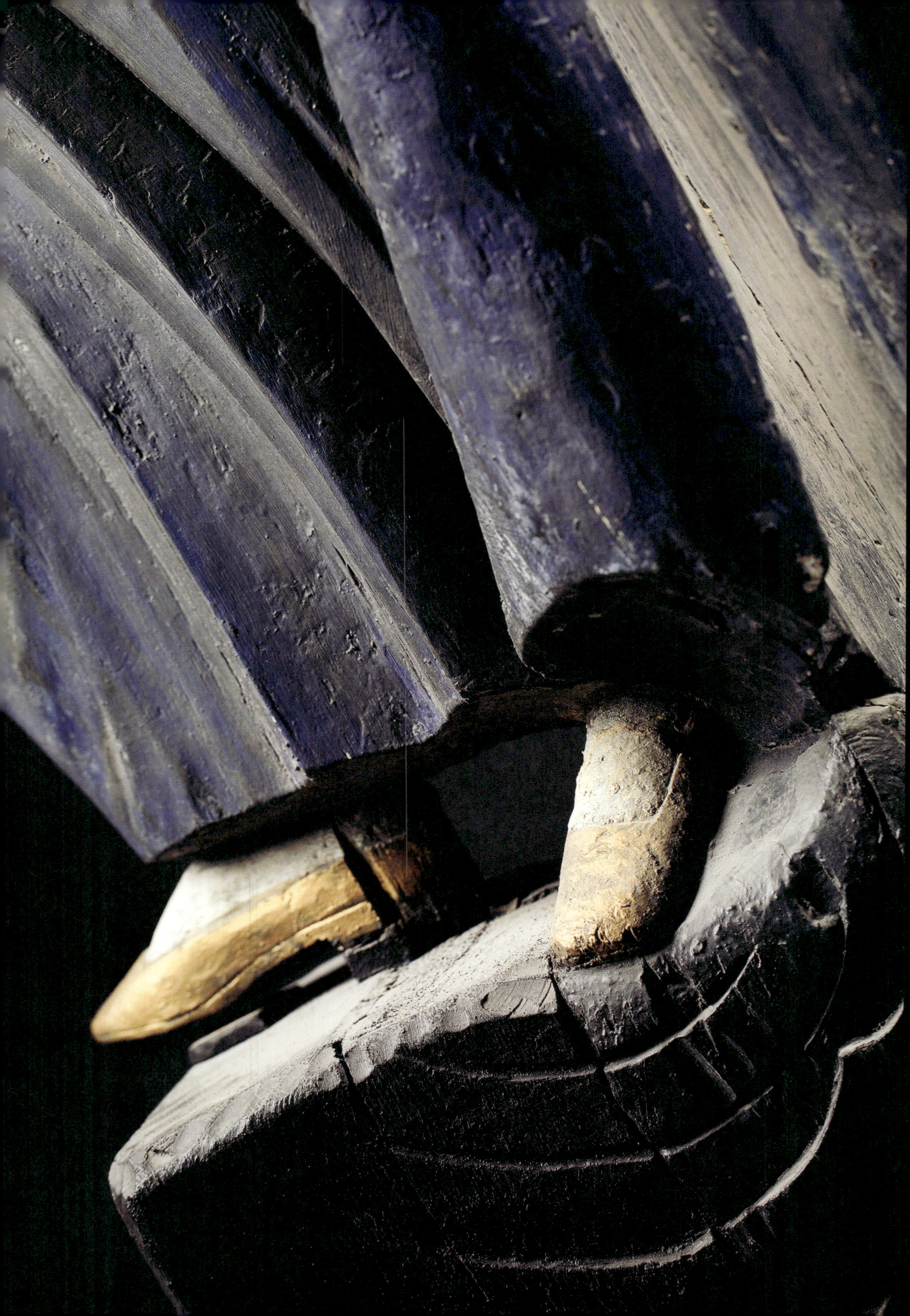

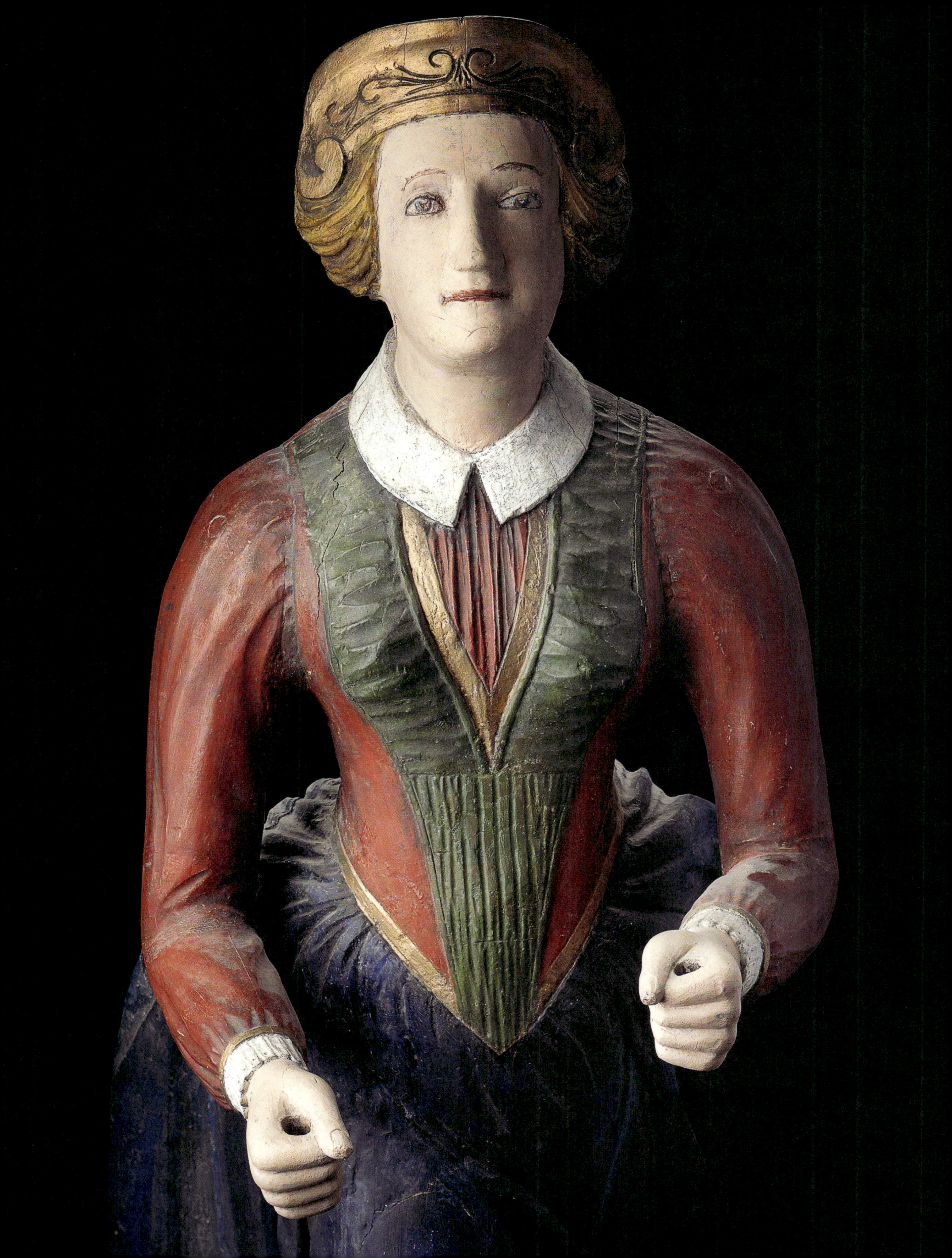

Full-length figure of woman in light-blue jacket and dark-blue skirt with details in red and gold. Probably belonged to the barque *Panama* of Blyth from Australia, which ran ashore off Skagen in 1872.

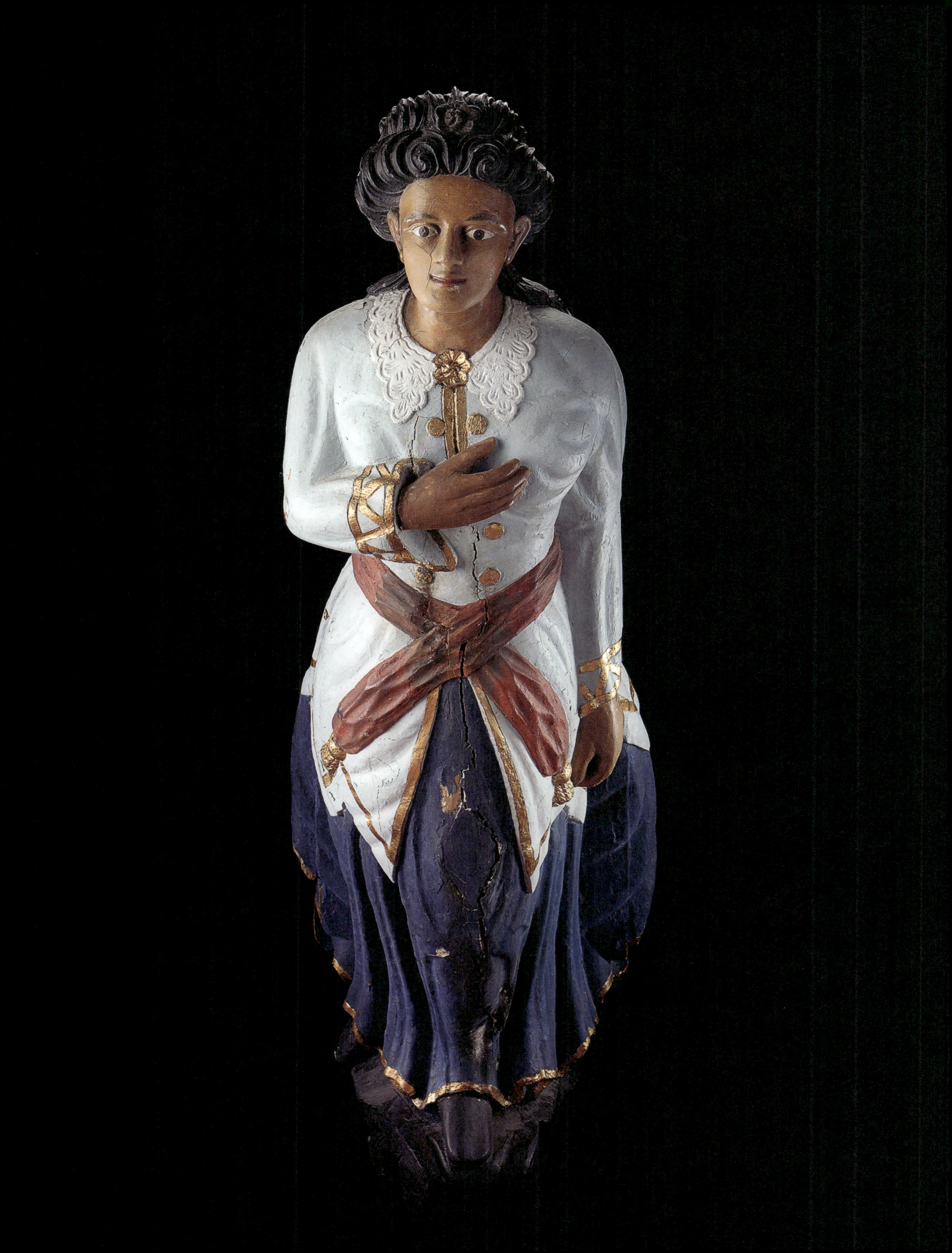

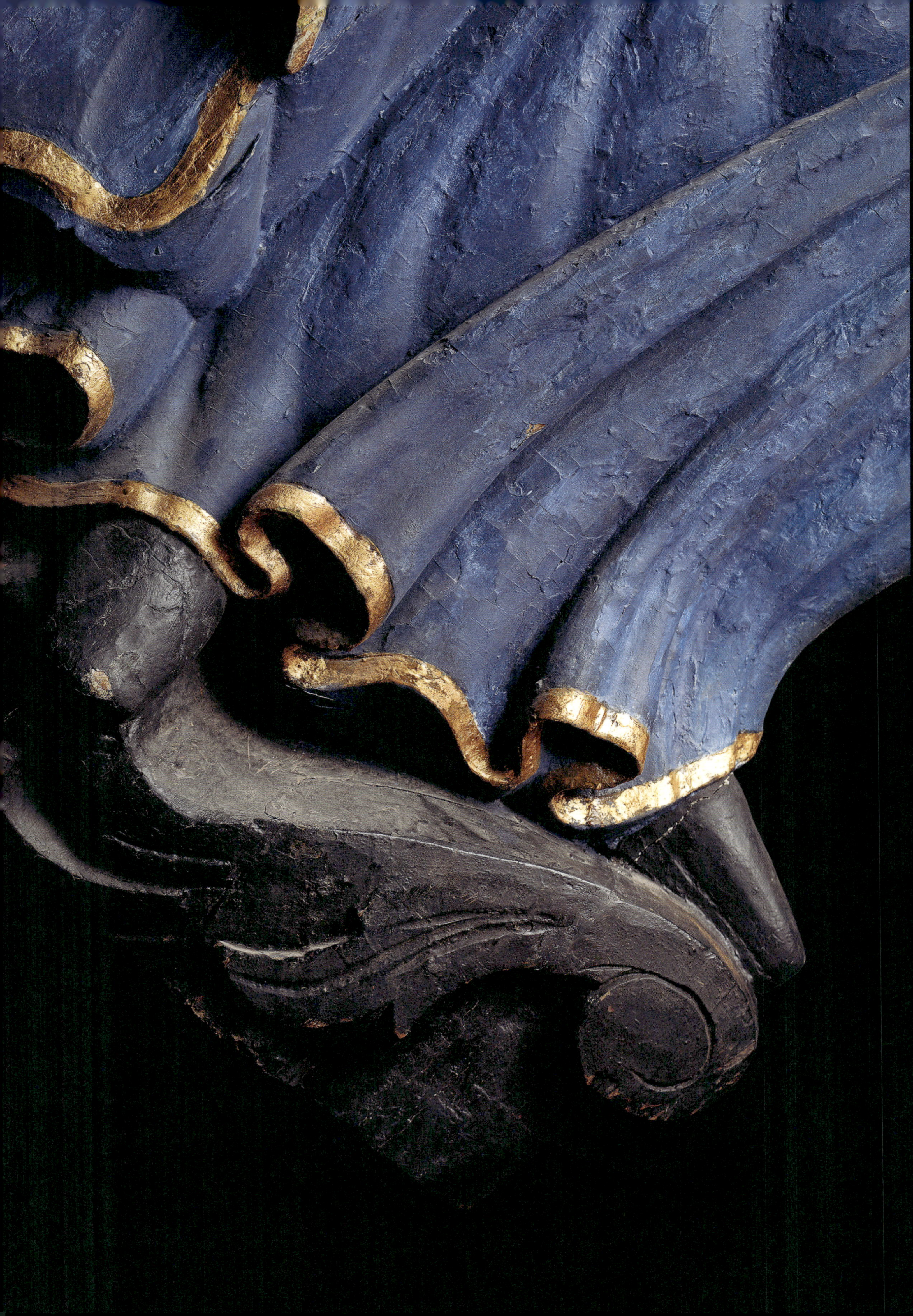

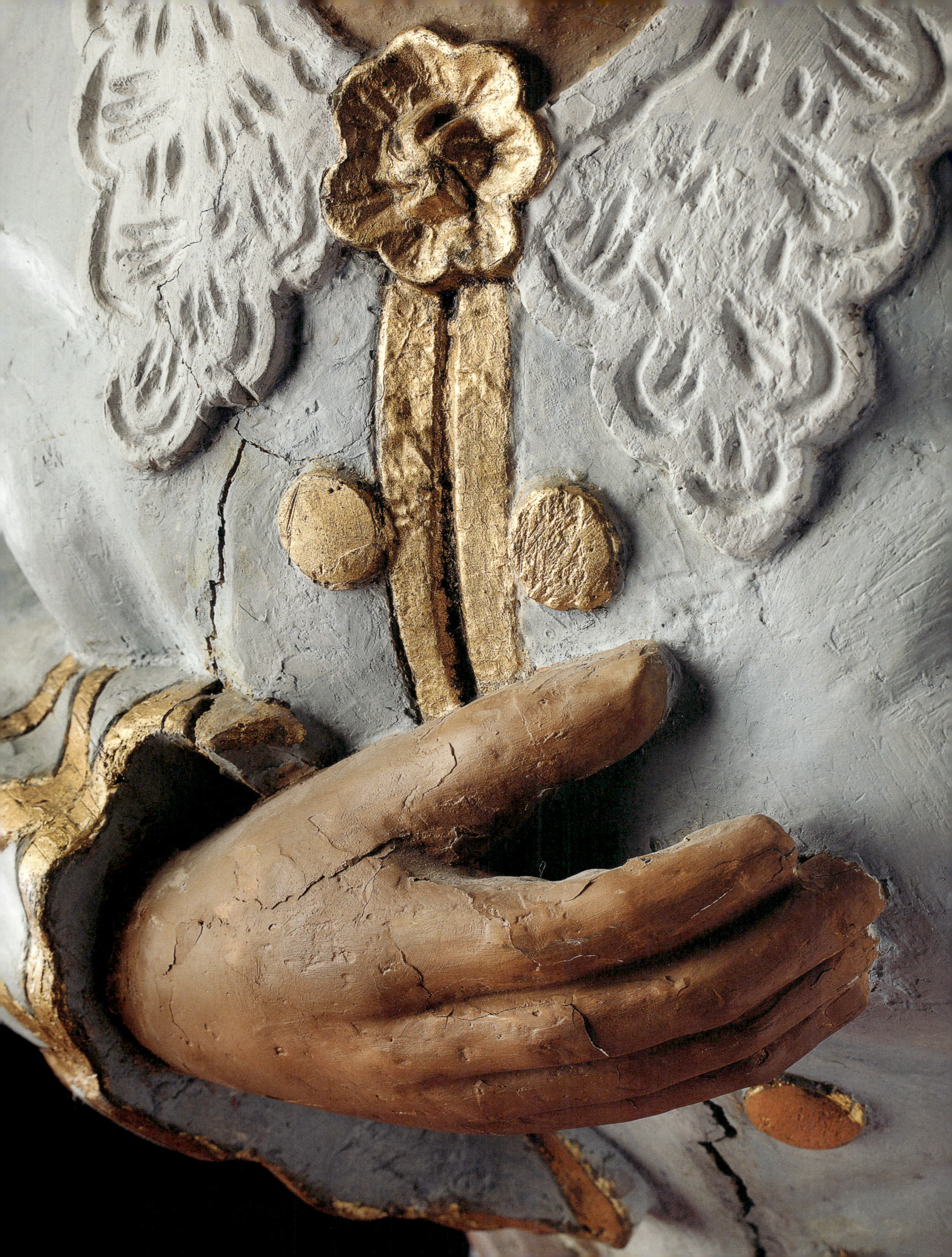

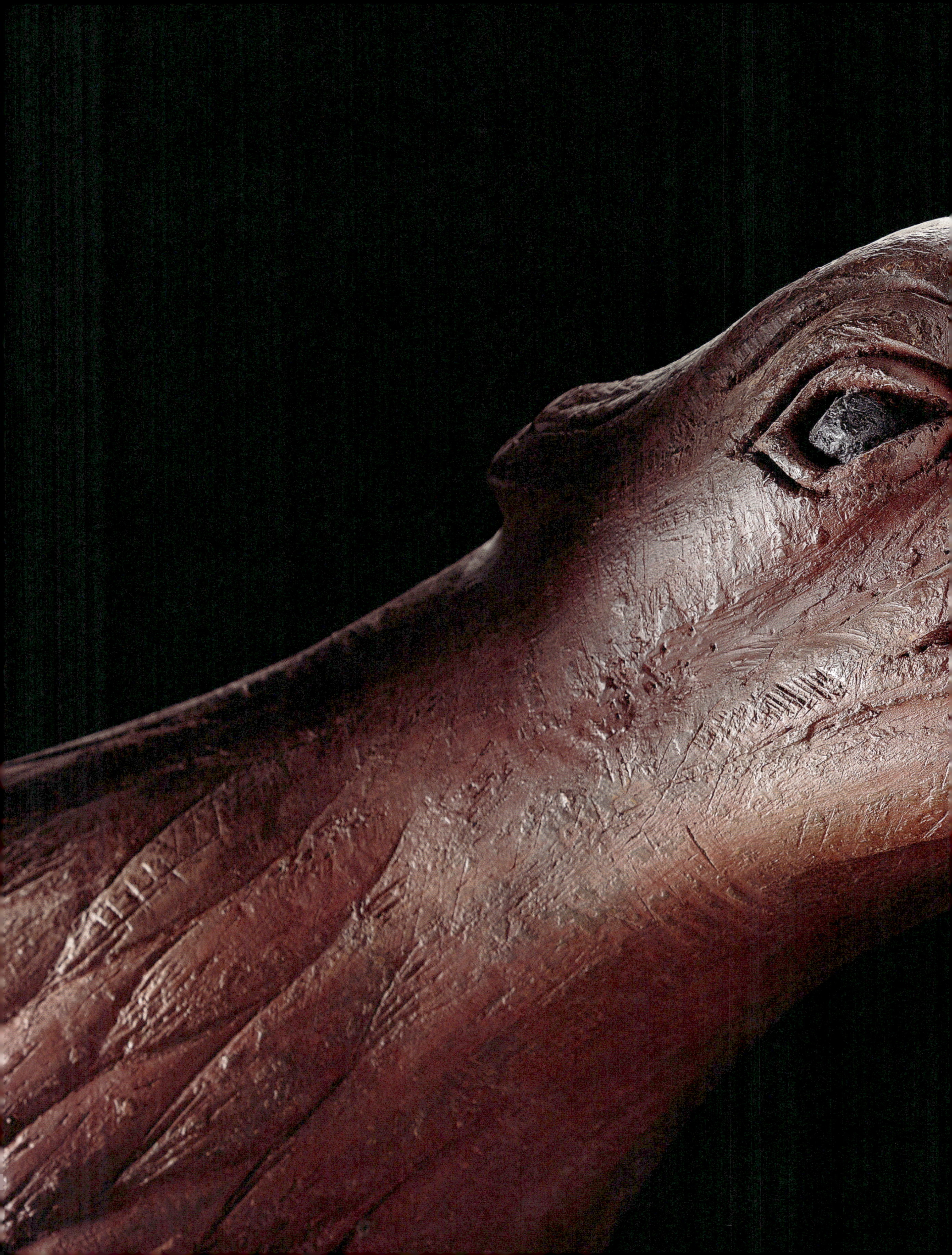

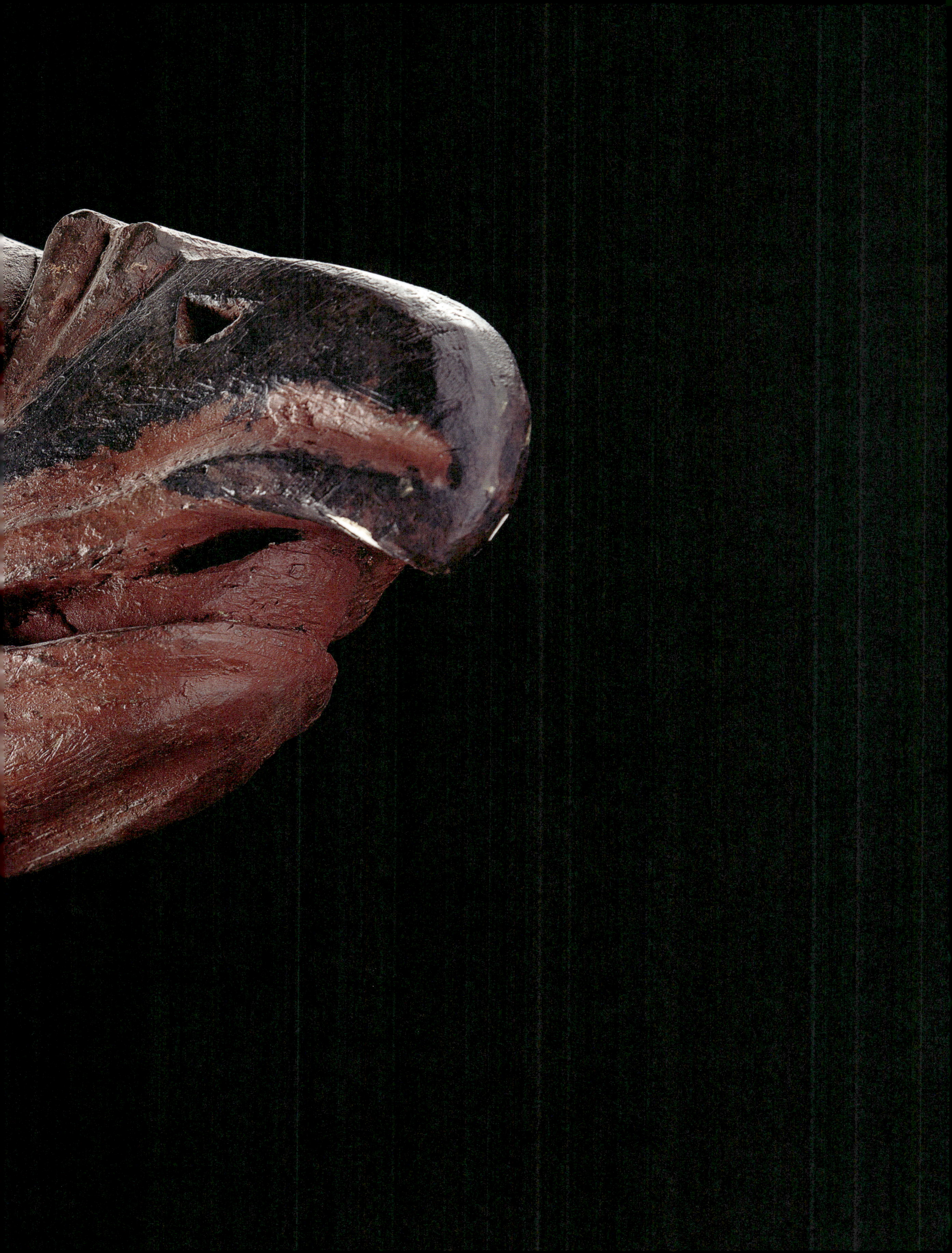

Preceding pages: Full-length figure of eagle with folded wings, painted brown with black bill. Origin unknown.

Three-quarter figure of the shipowner's wife type, dressed in reddish-brown blouse with white collar and holding bouquet of yellow roses. Origin unknown.

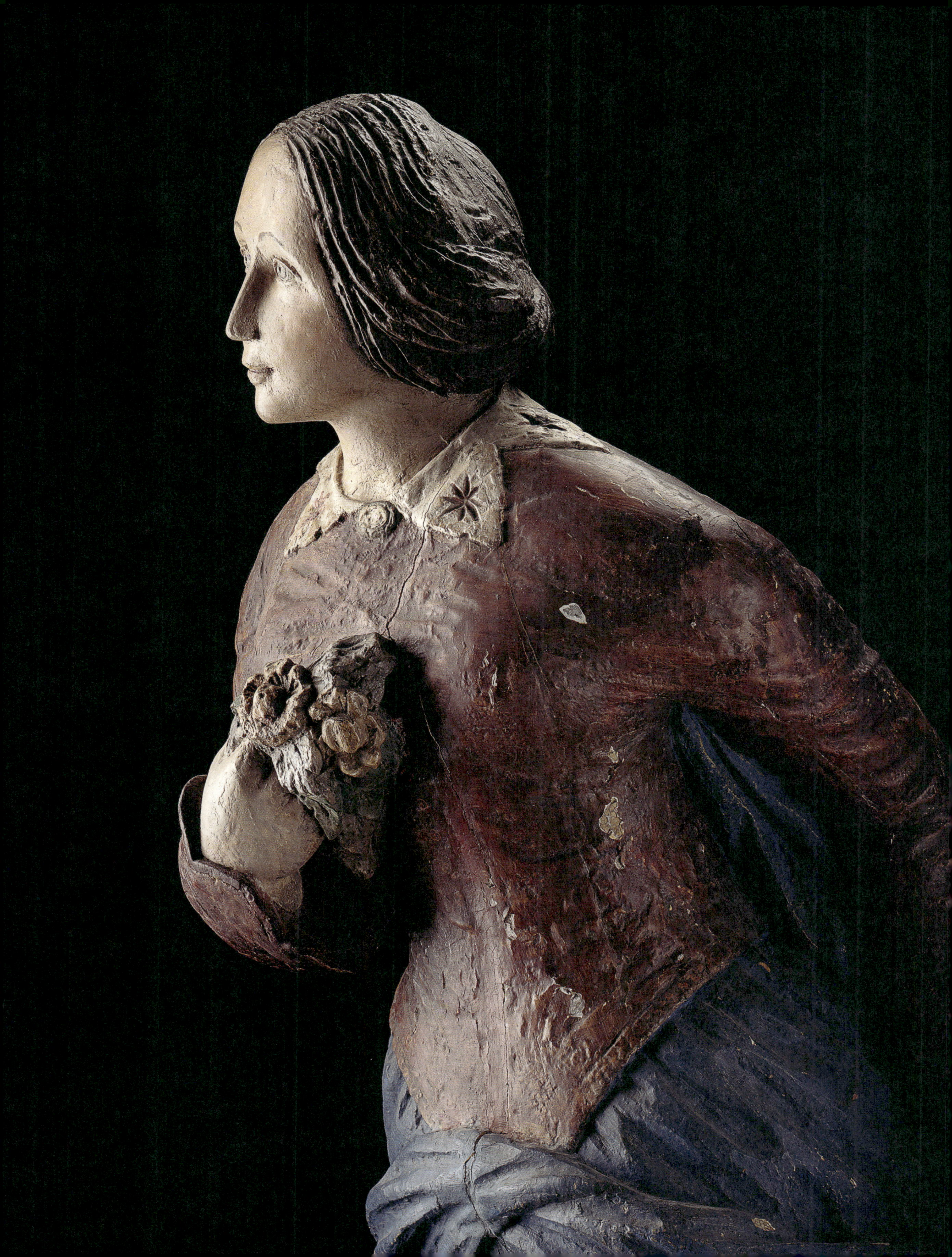

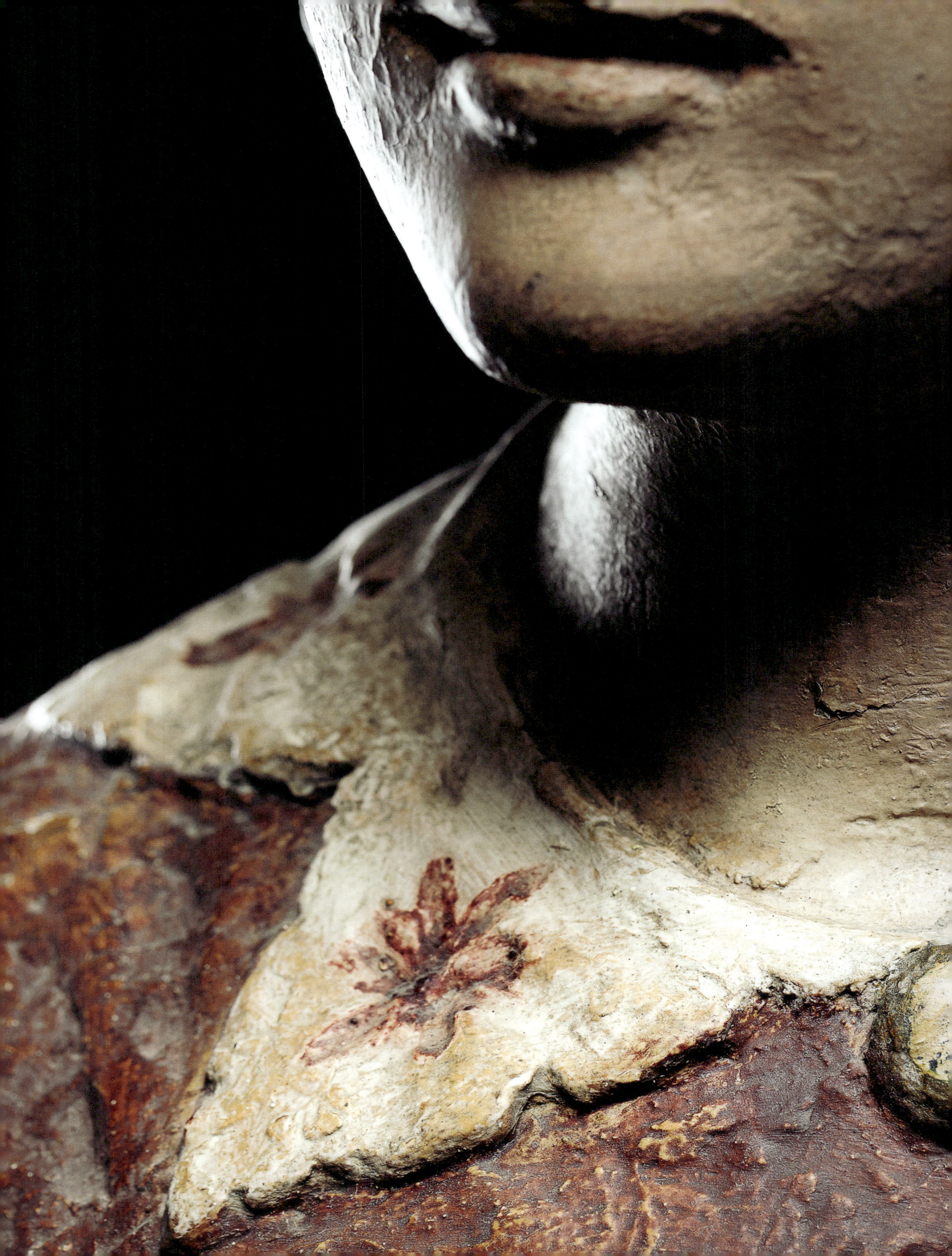

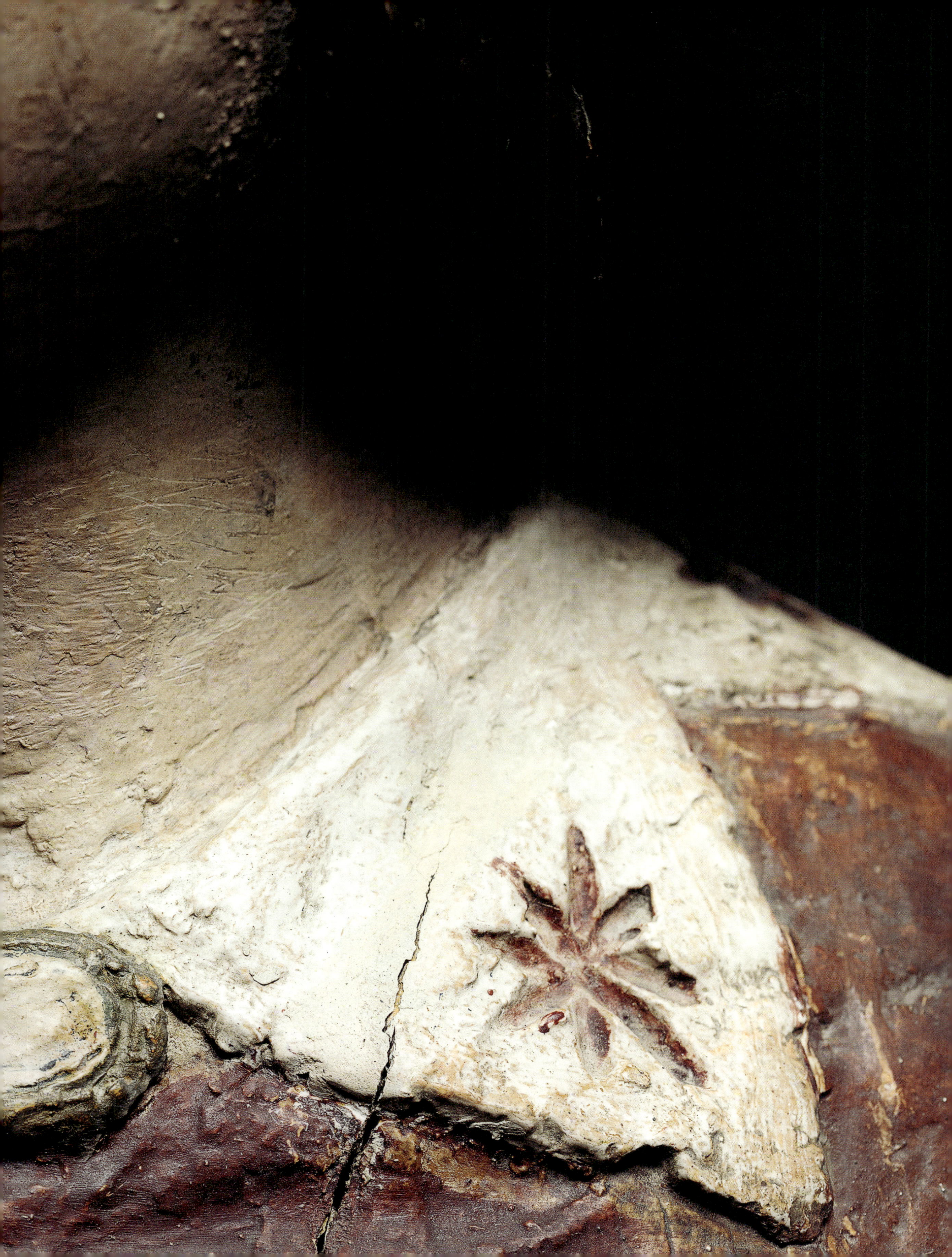

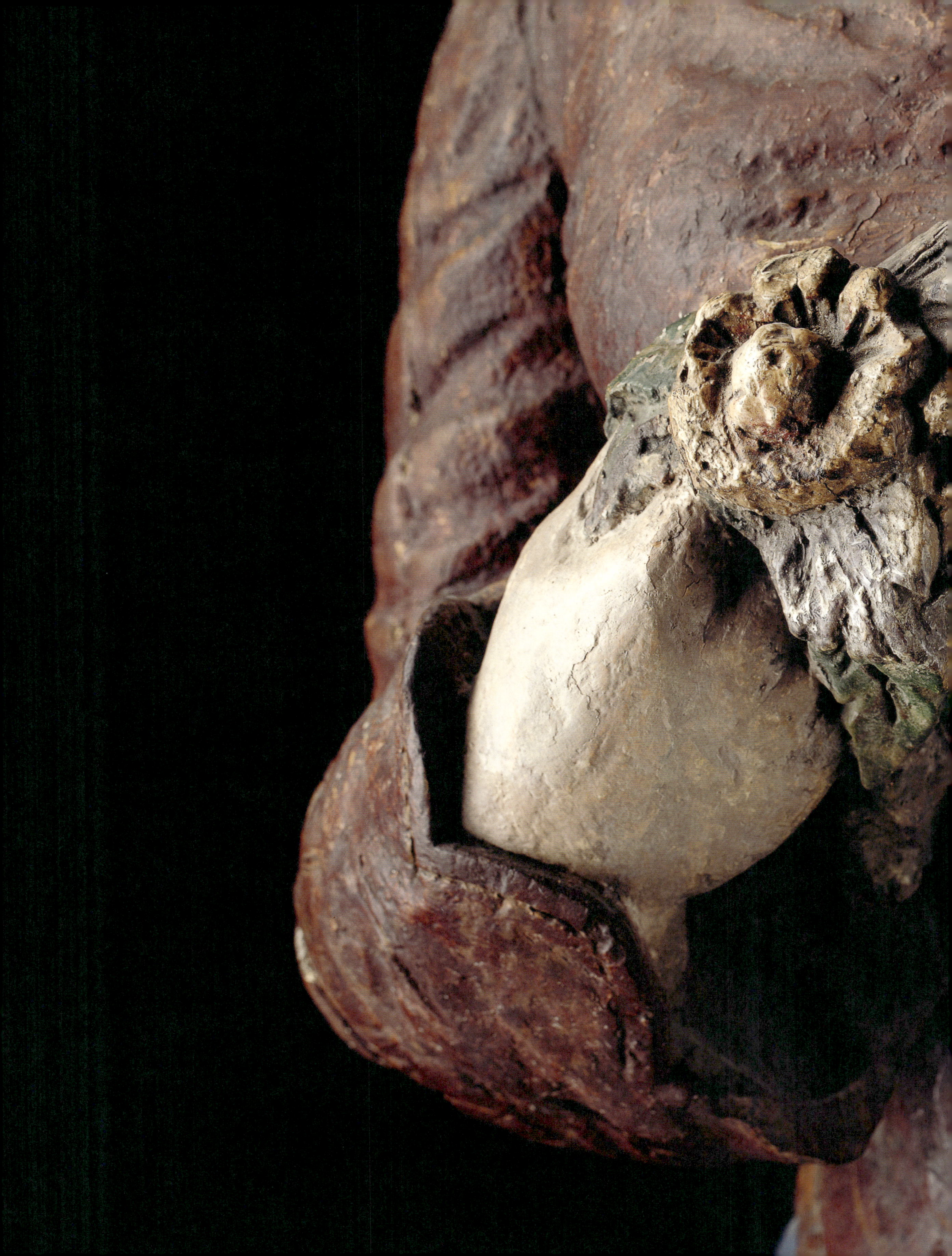

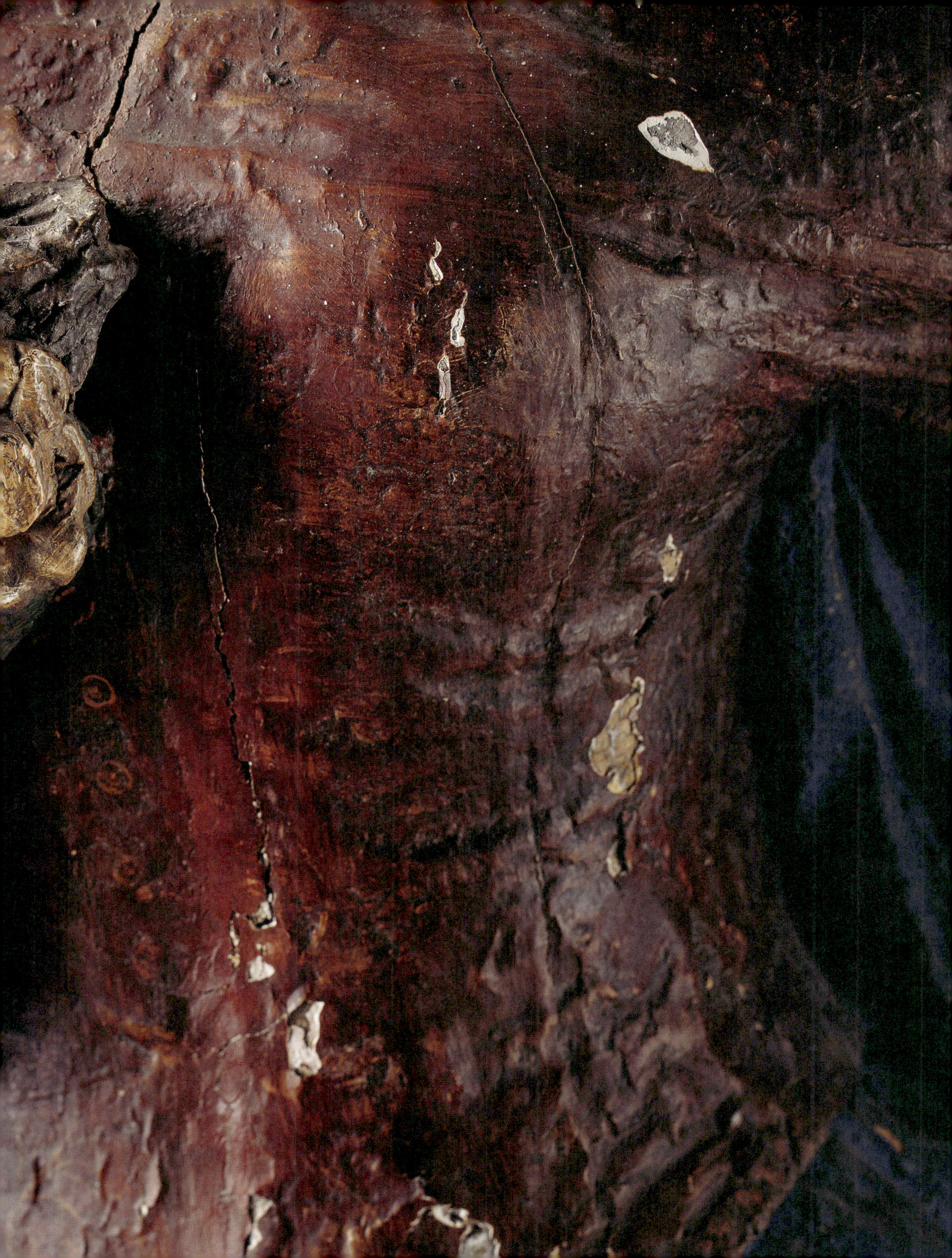

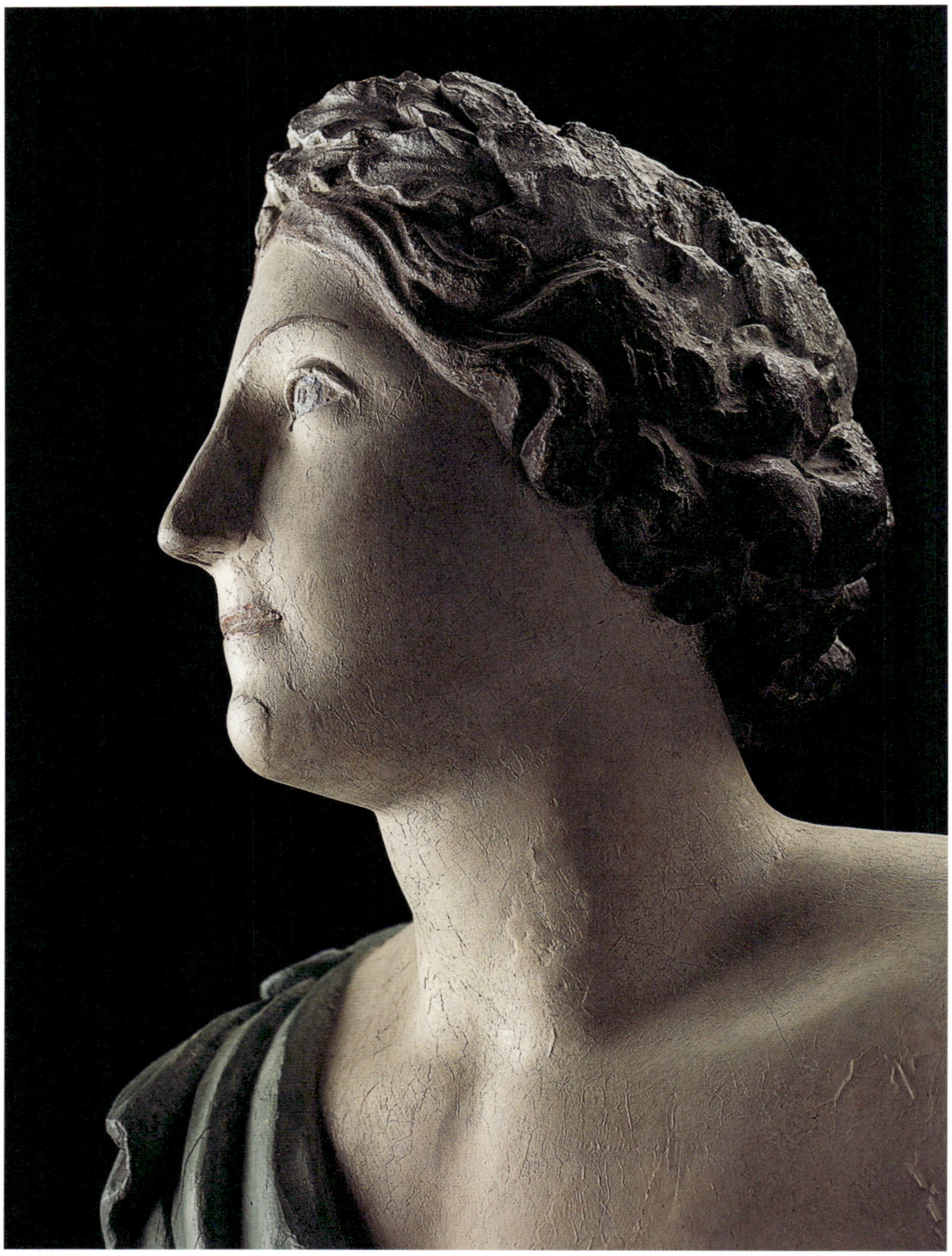

Full-length figure of Roman youth with green robe, laurel wreath and sandals. Origin unknown.

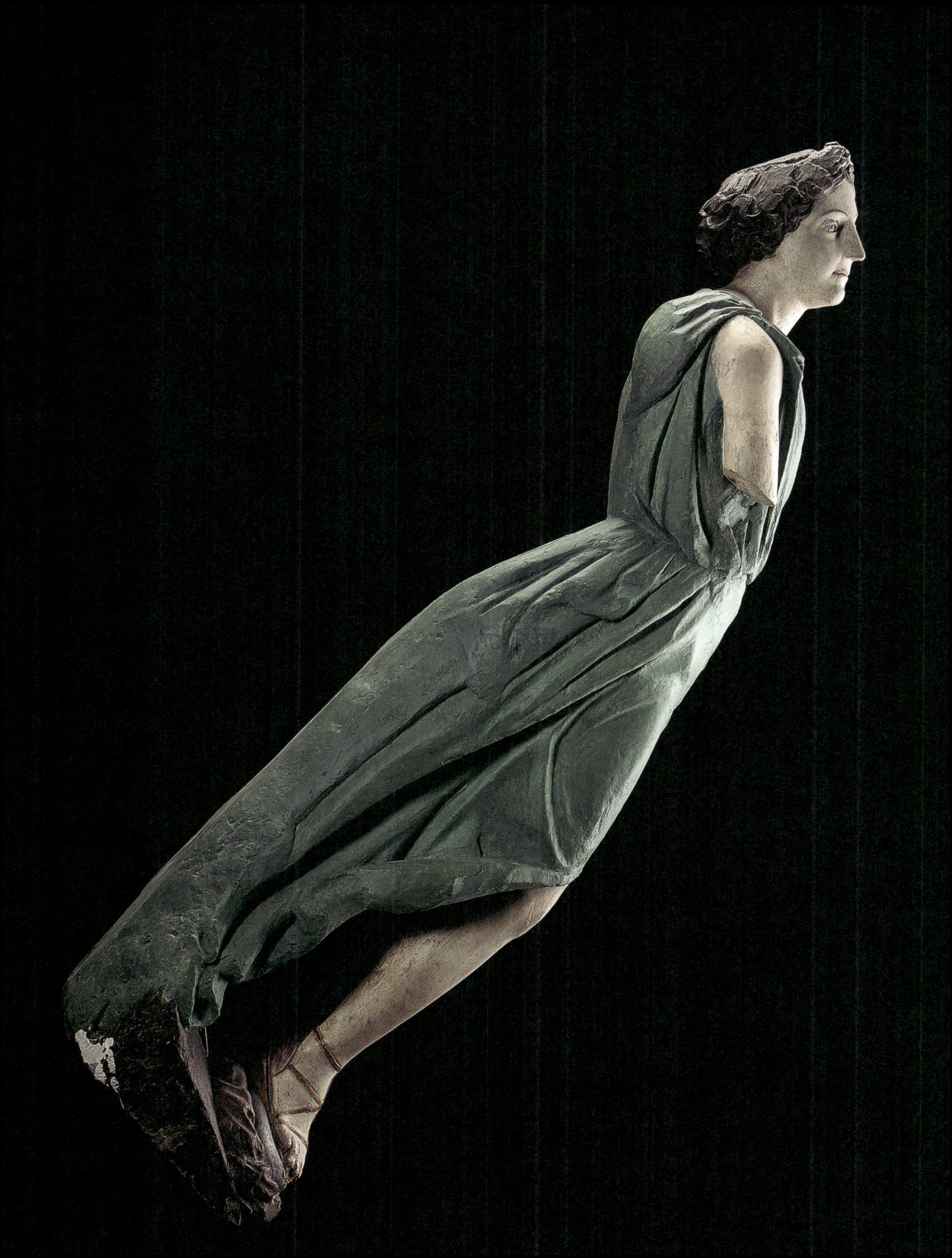

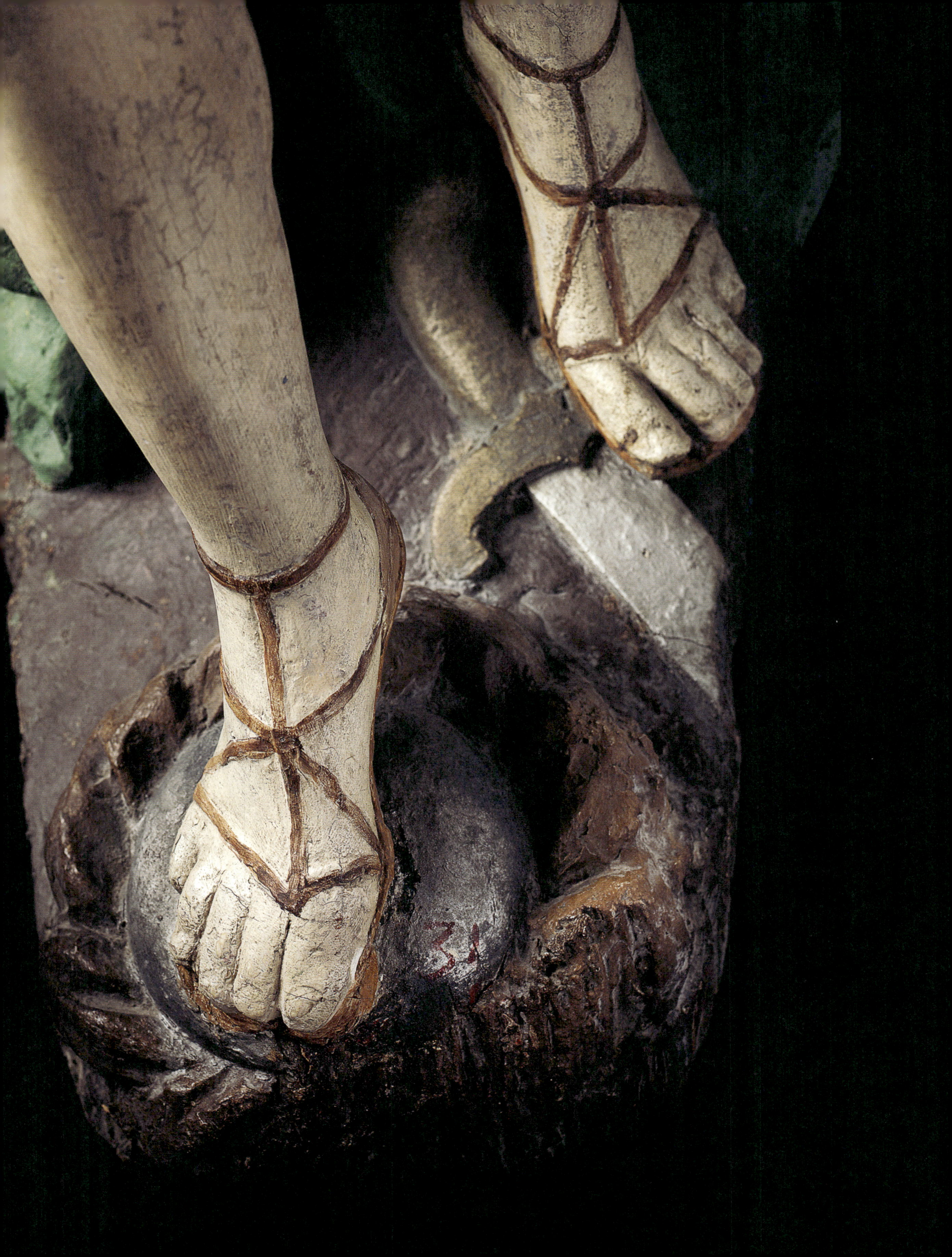

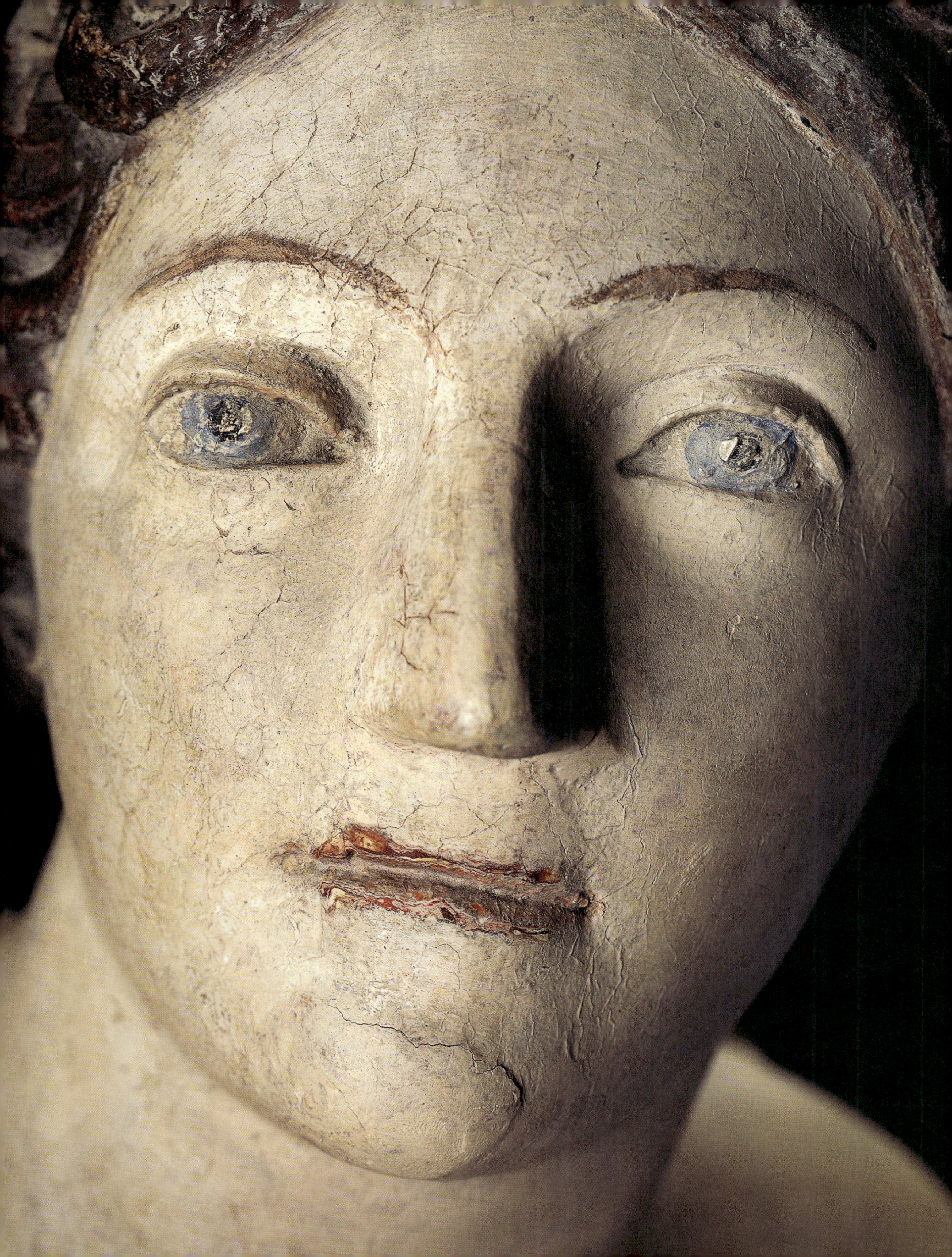

Three-quarter length figurehead of putto-like boy with dark, curly hair and draped in brown robe. From the ship *Georg* of Gdánsk, which ran ashore off Lønstrup in 1857.

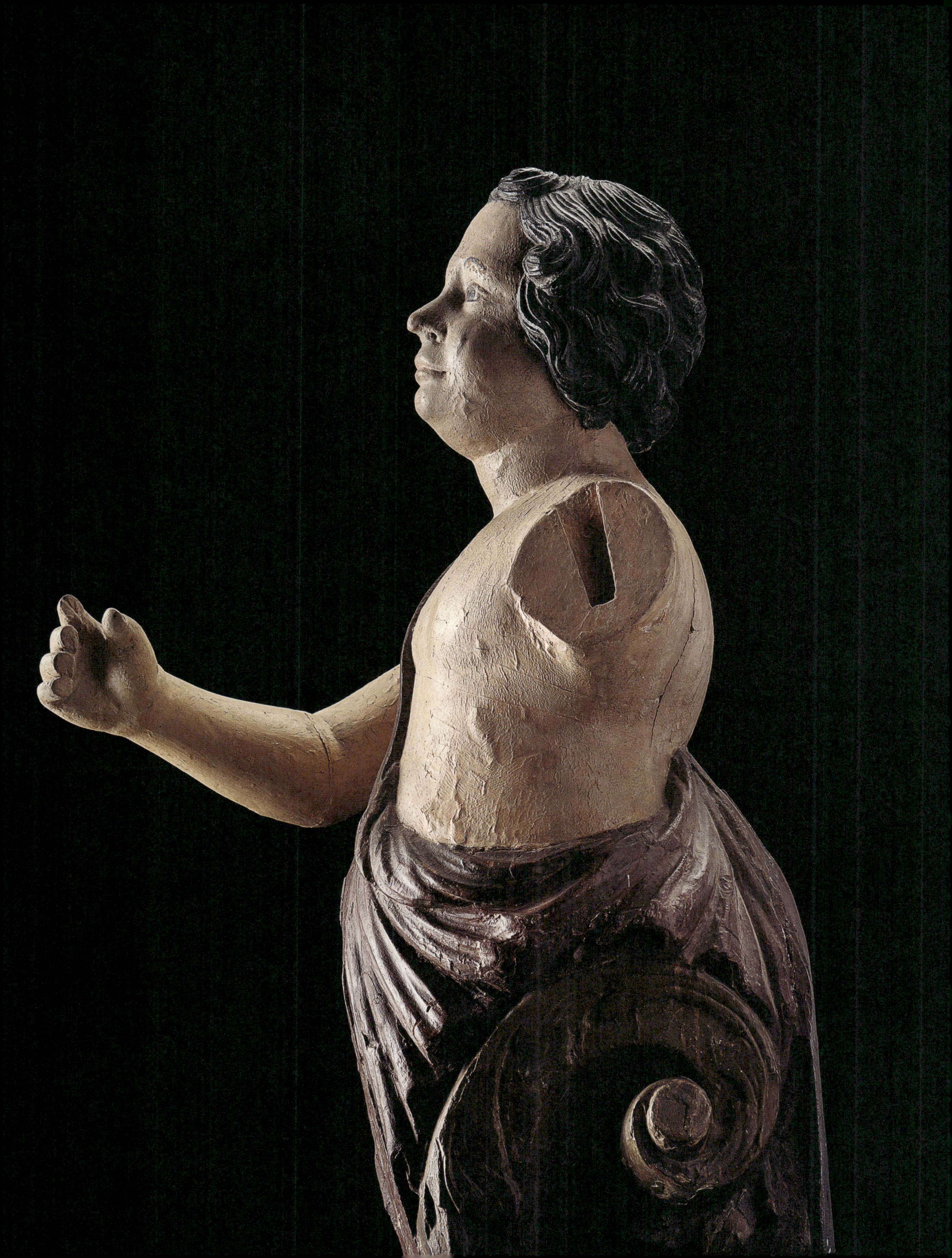

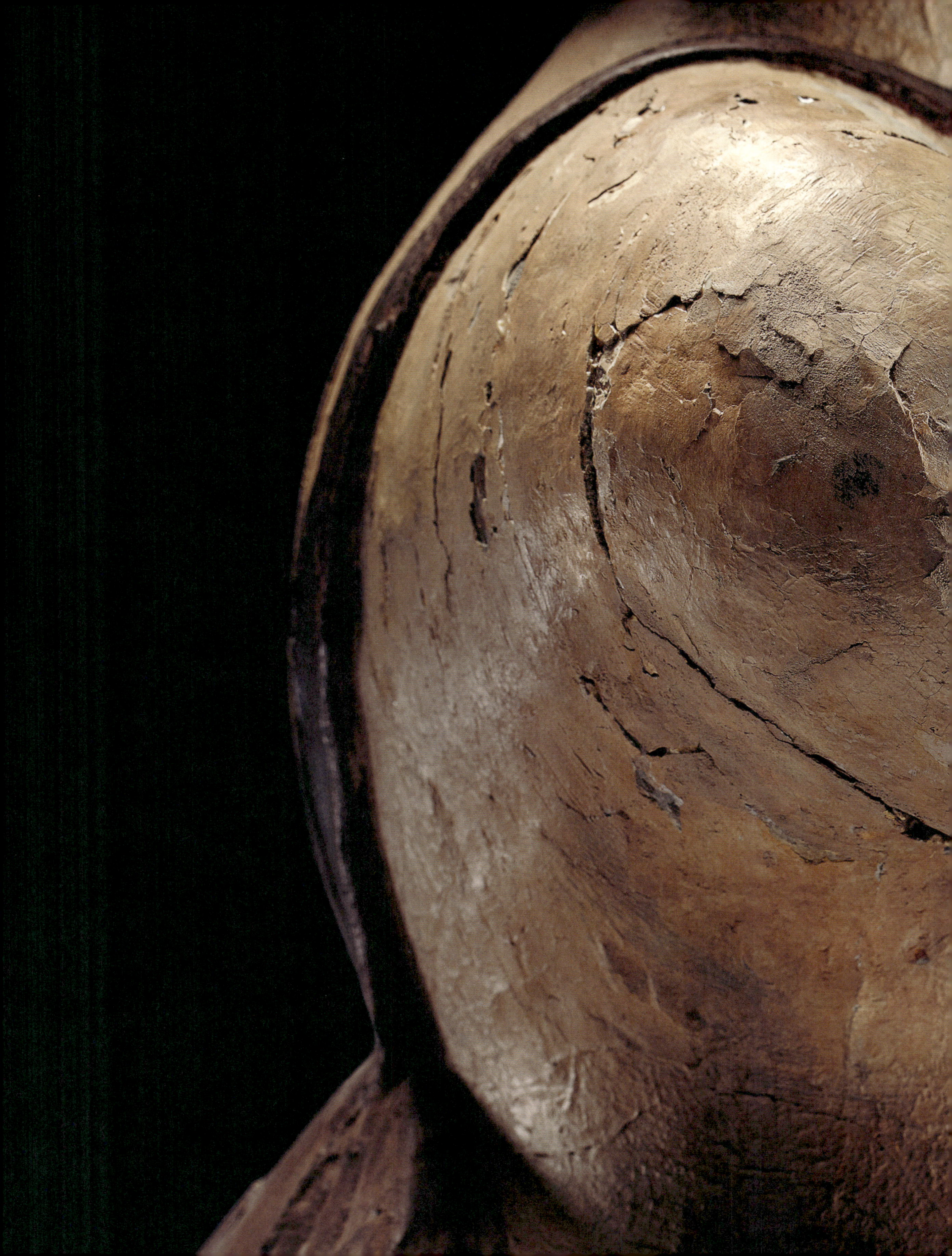

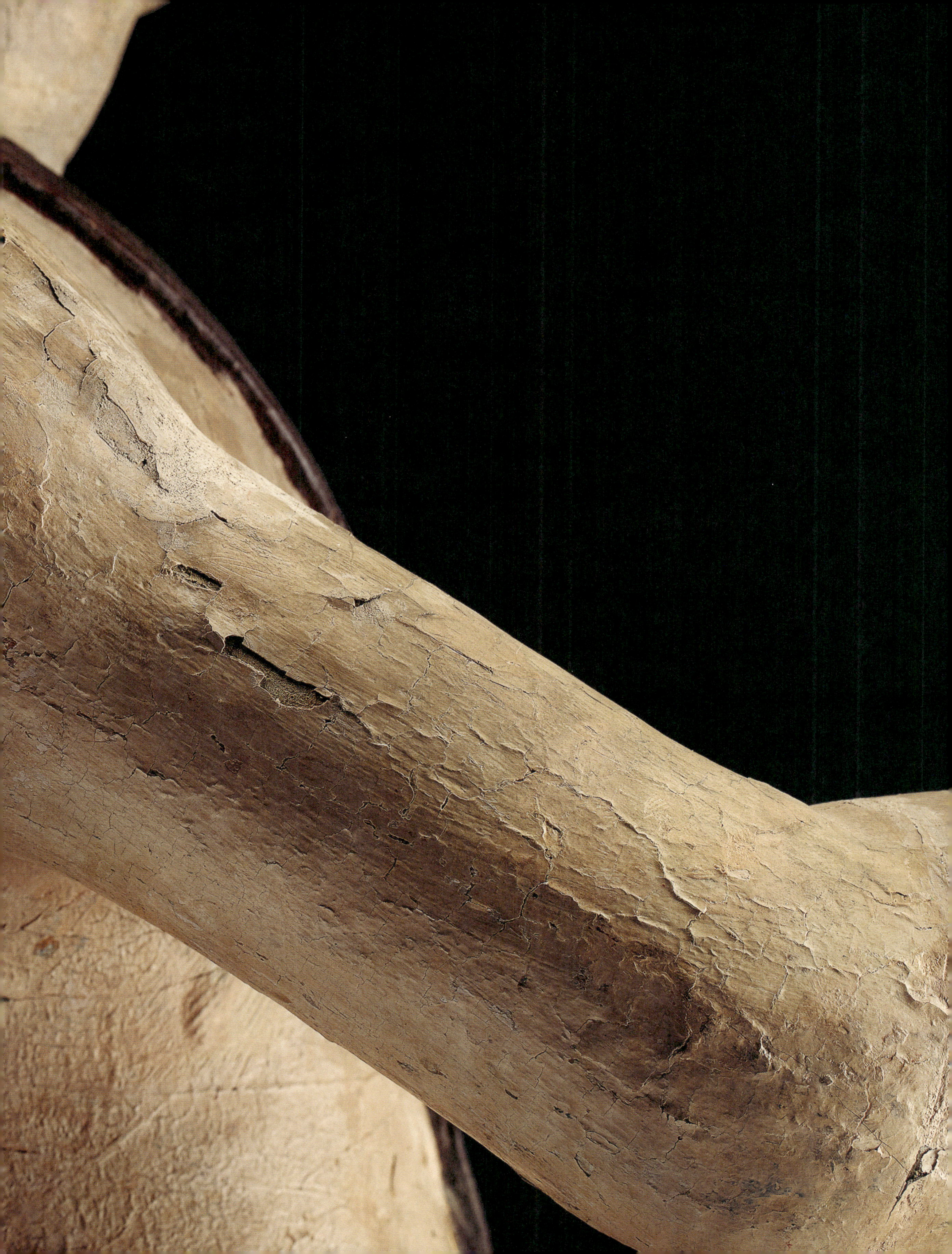

Three-quarter-length figure of man in traditional Scottish costume with tam and kilt. Probably from the schooner *Pilot* of Montrose, sunk in 1859 off Klitmøller on the west coast of Jutland.

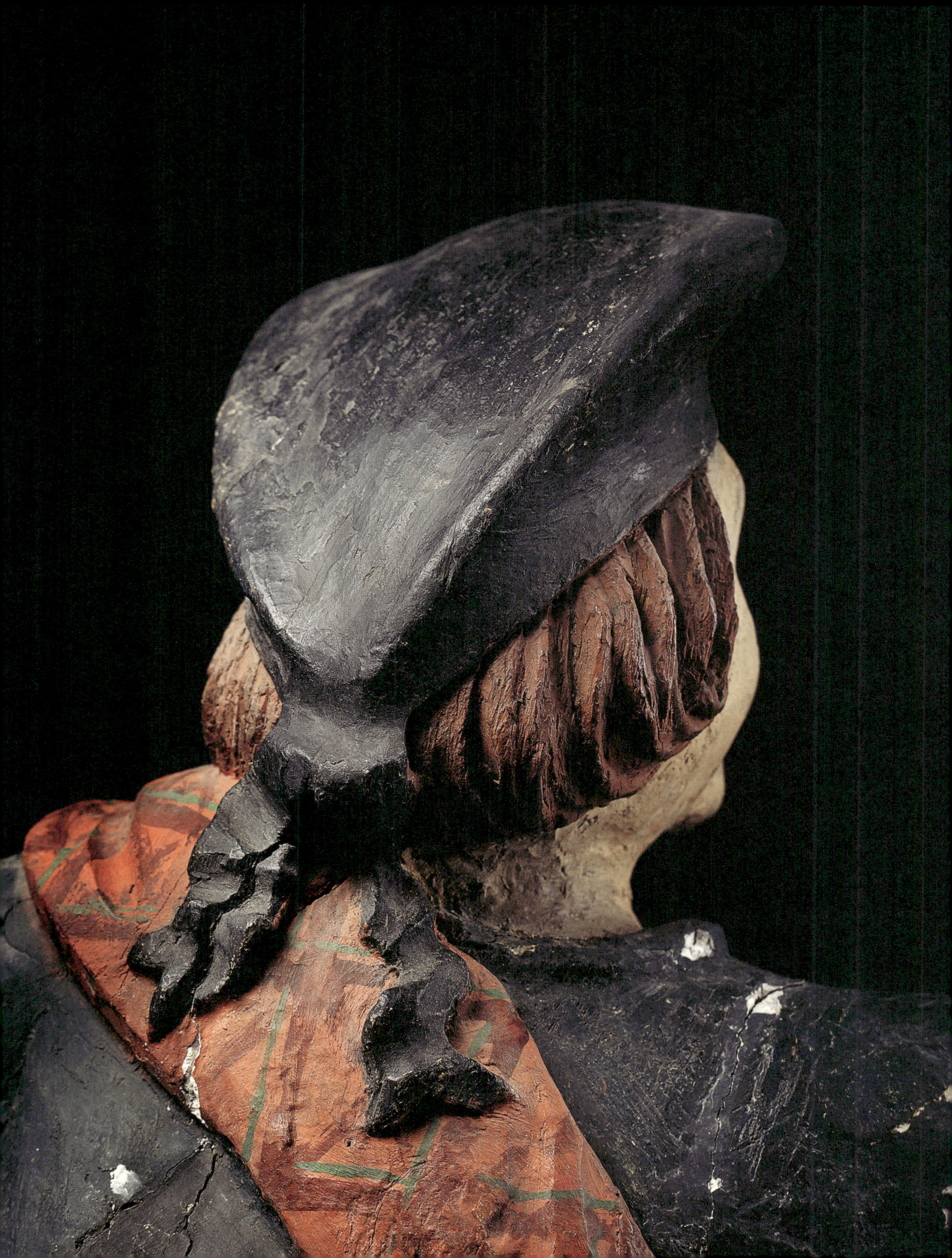

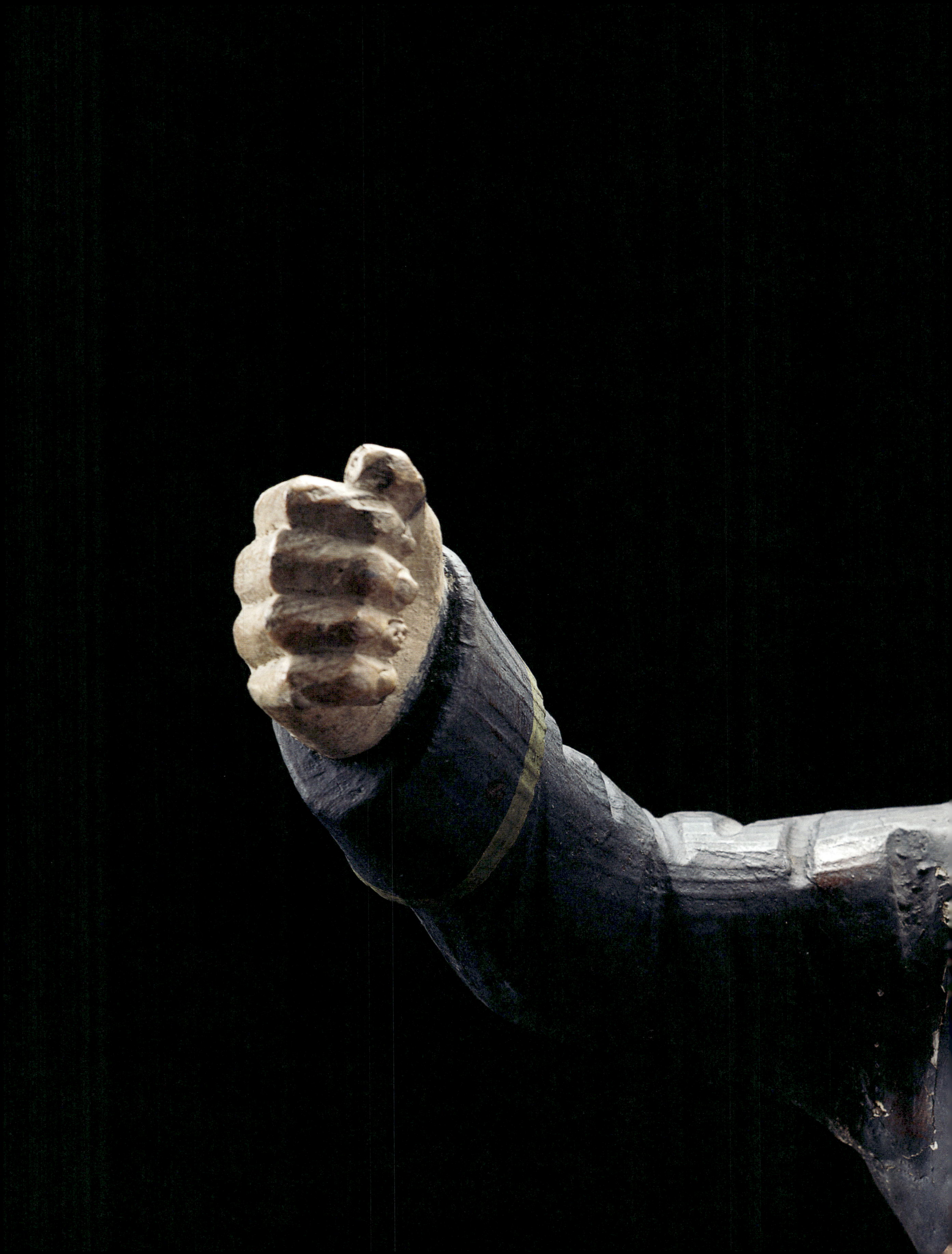

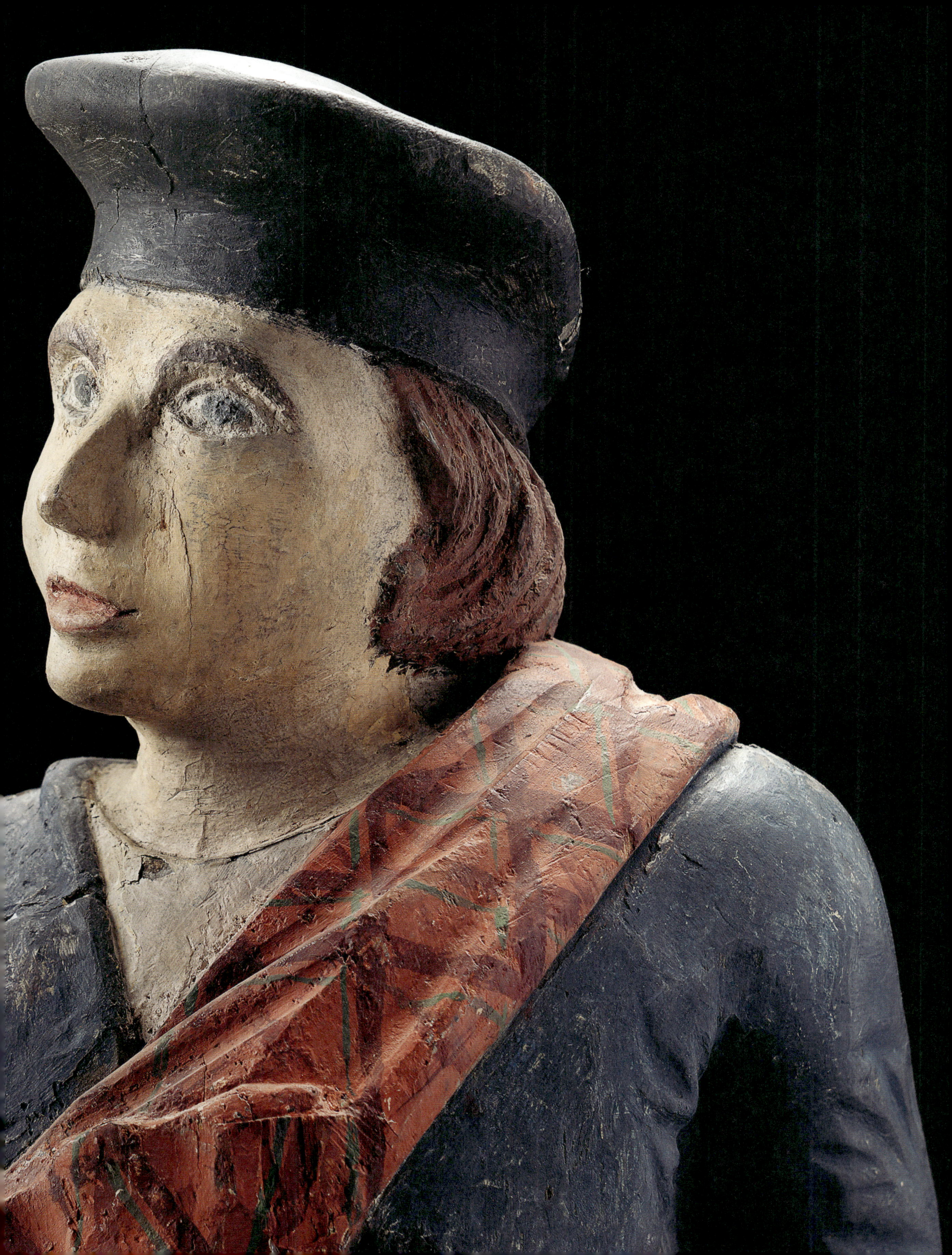

Full-length figure of dark-haired woman with flowers in her hair and dress in brown, blue and golden yellow. Origin unknown.

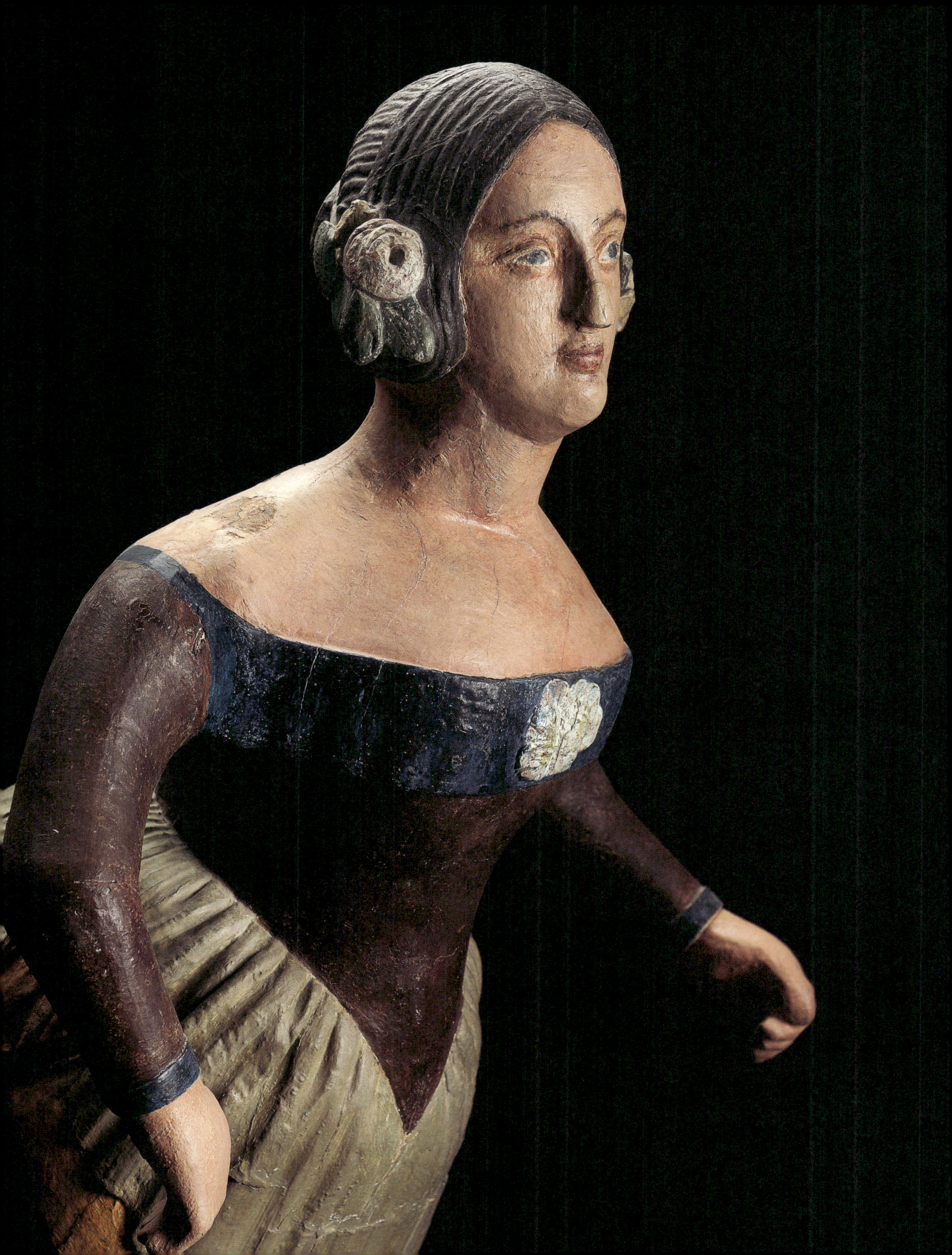

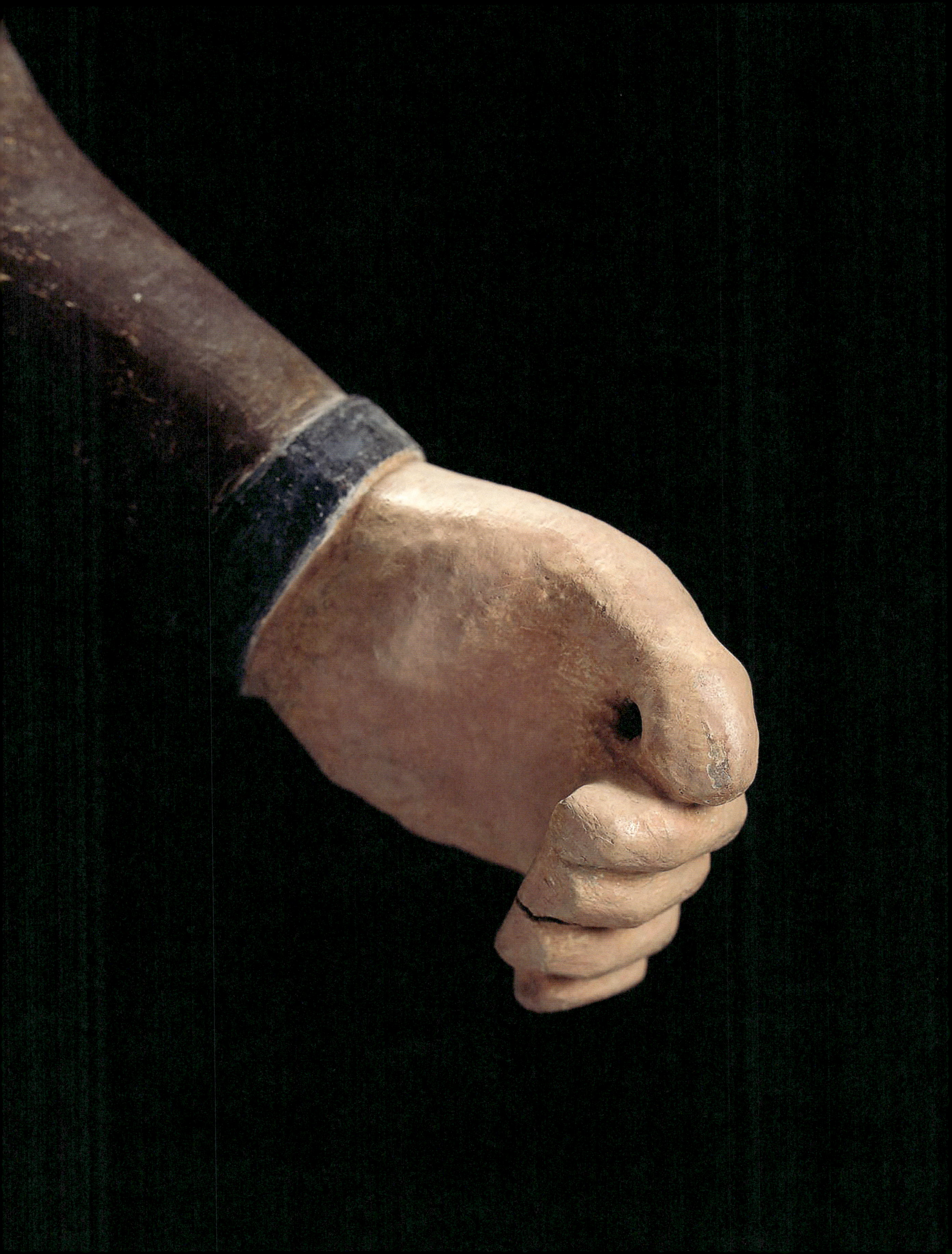

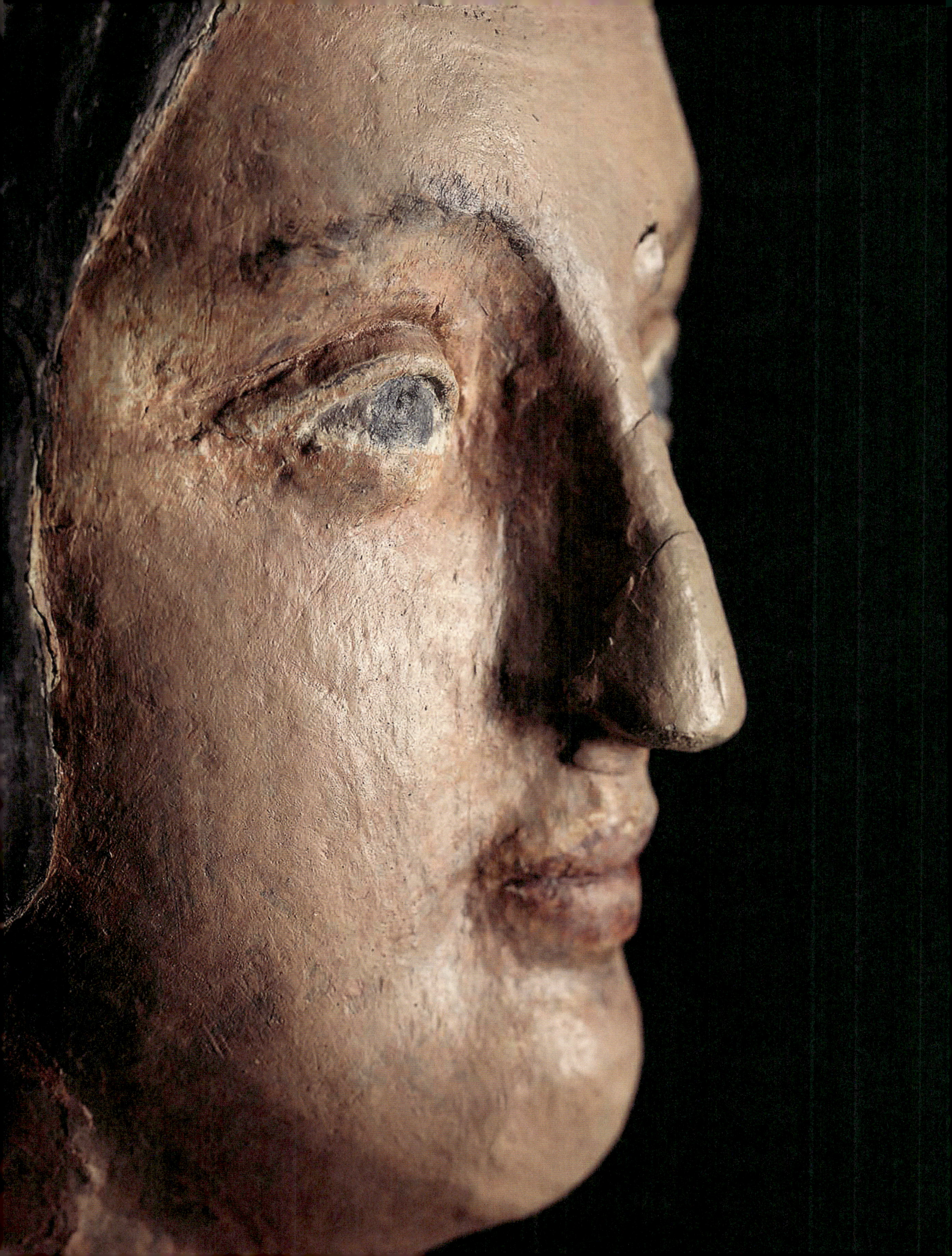

Full-length figure in black frock coat of the shipowner type. From the frigate *C. L. Bahr* of Gdánsk, which ran ashore off Skagen in 1871.

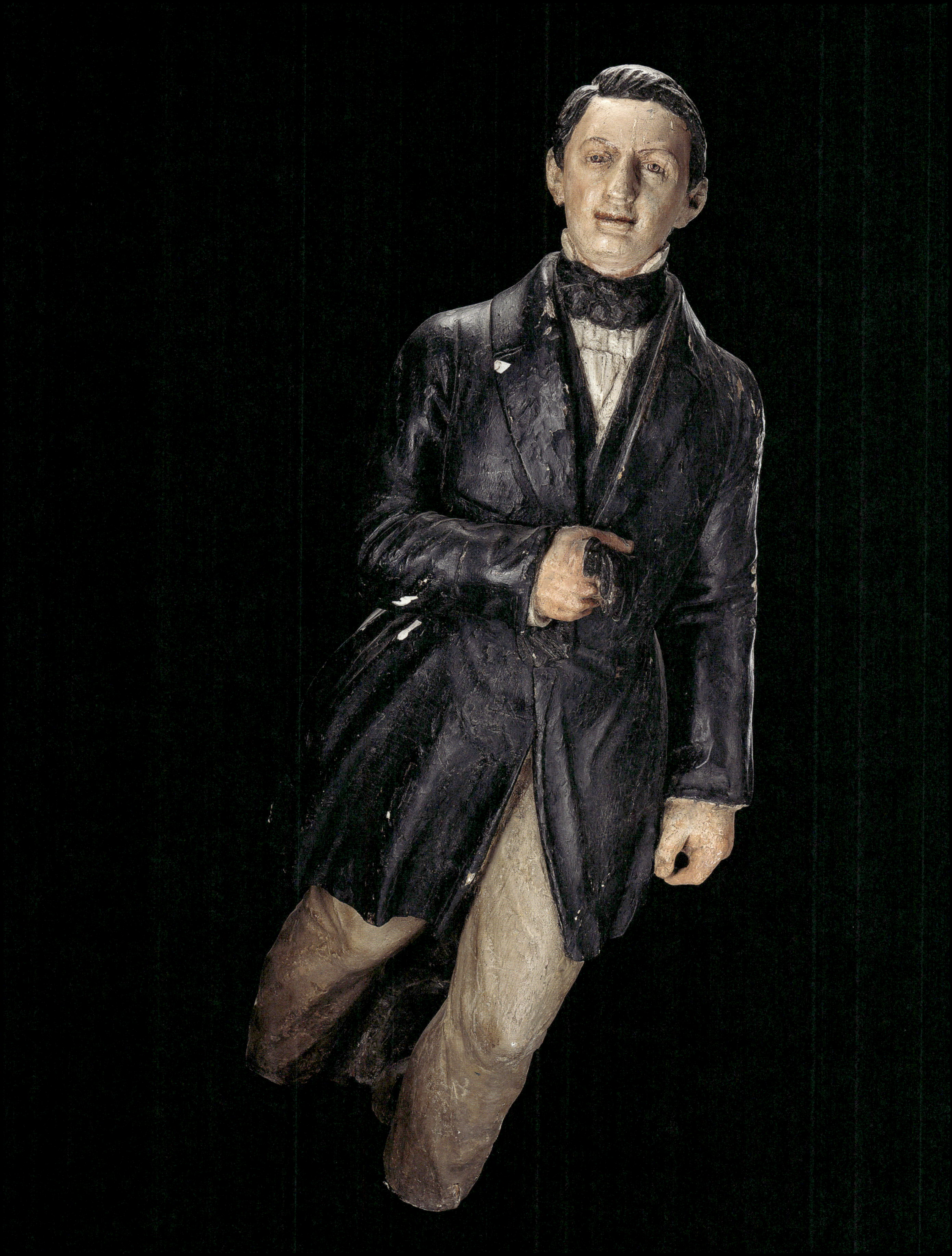

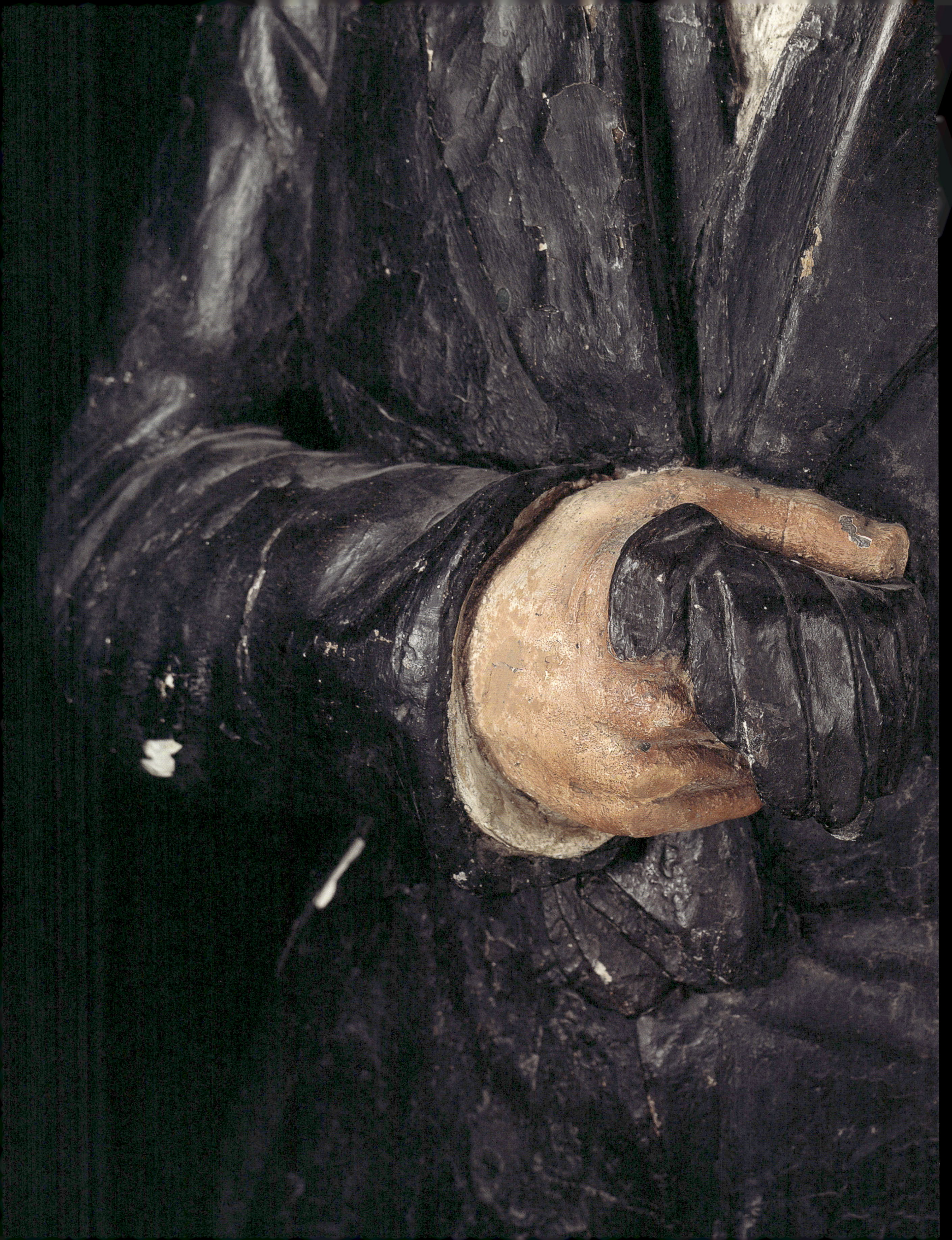

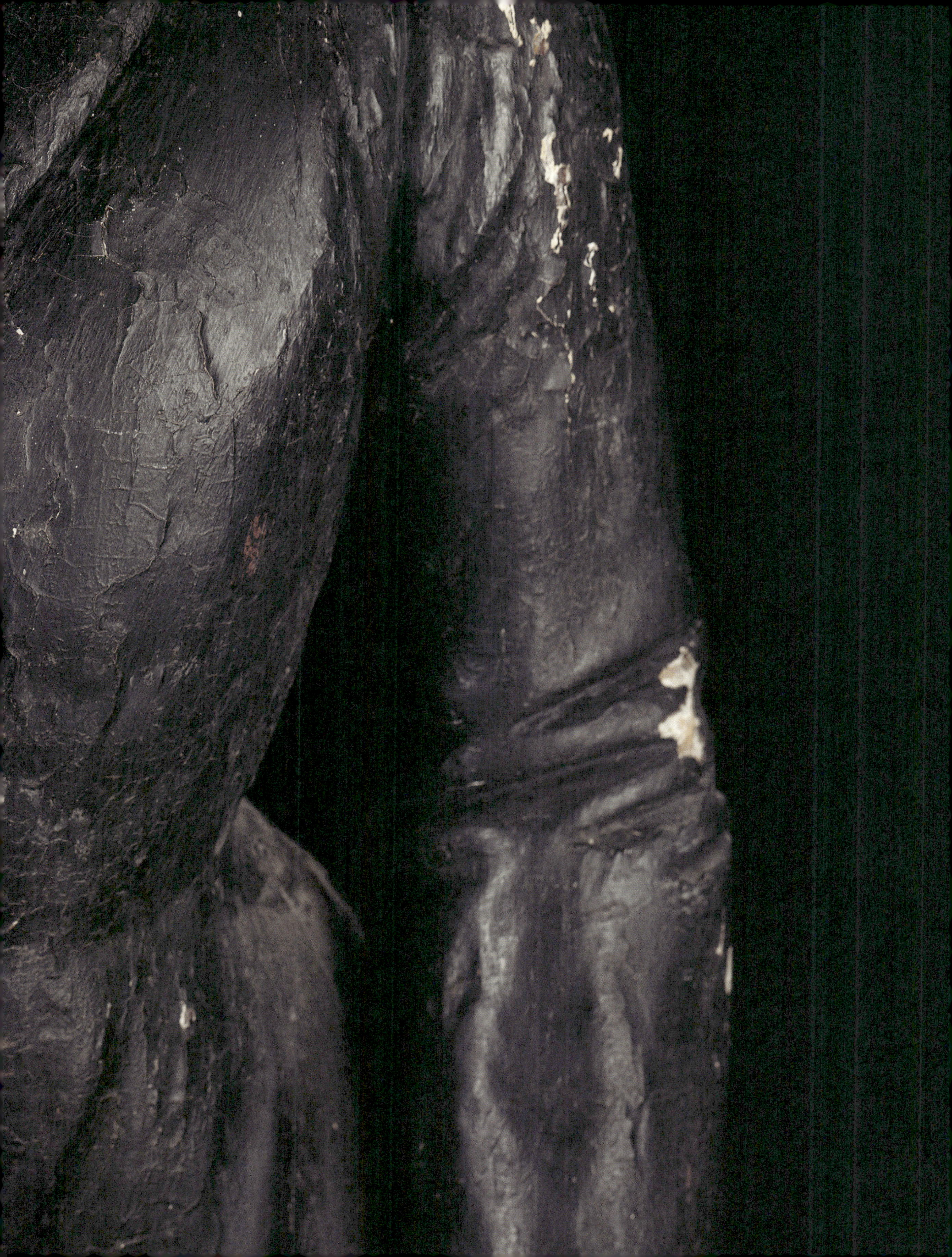

Demi figure of girl or young woman in red dress with white collar and blue bow. Origin unknown.

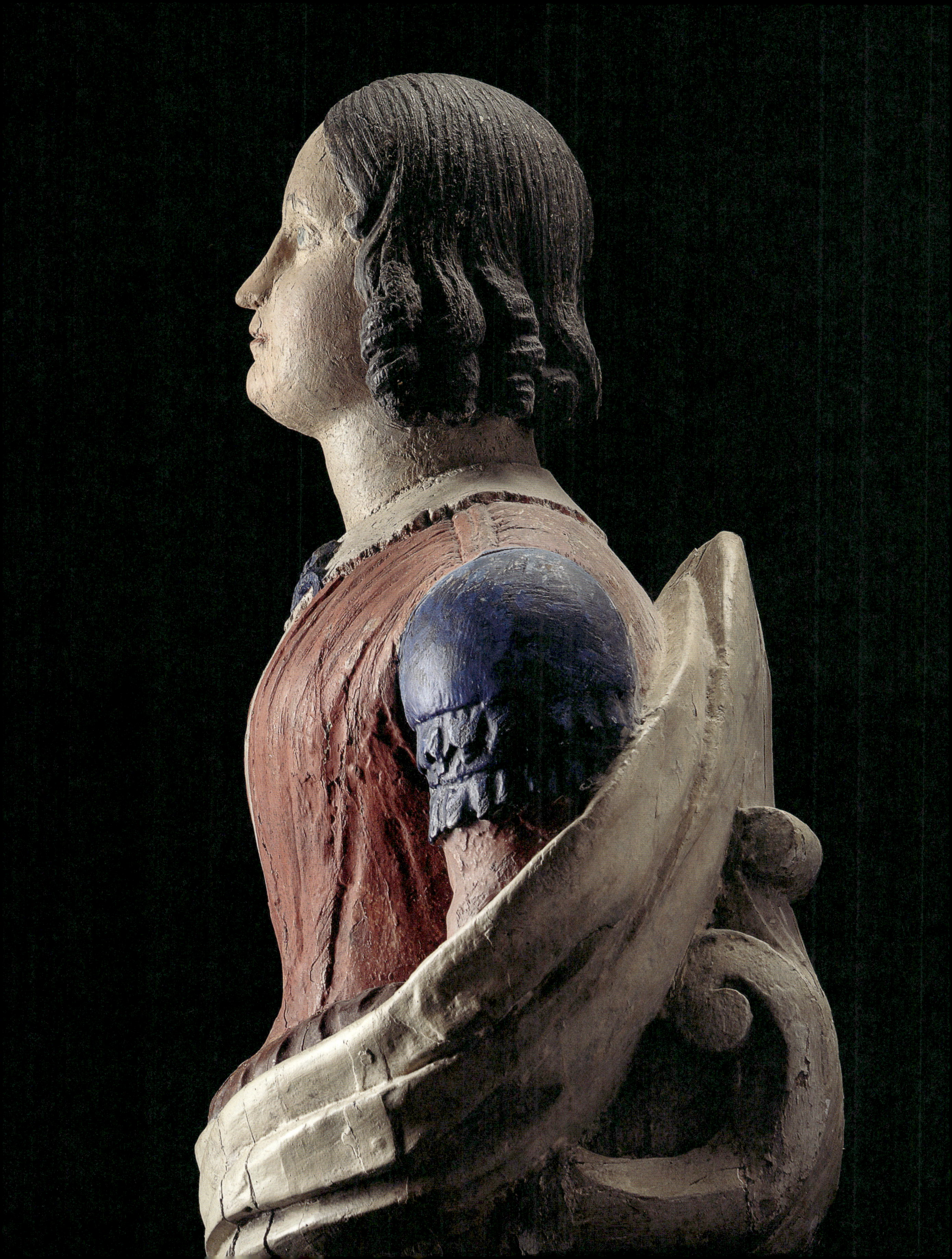

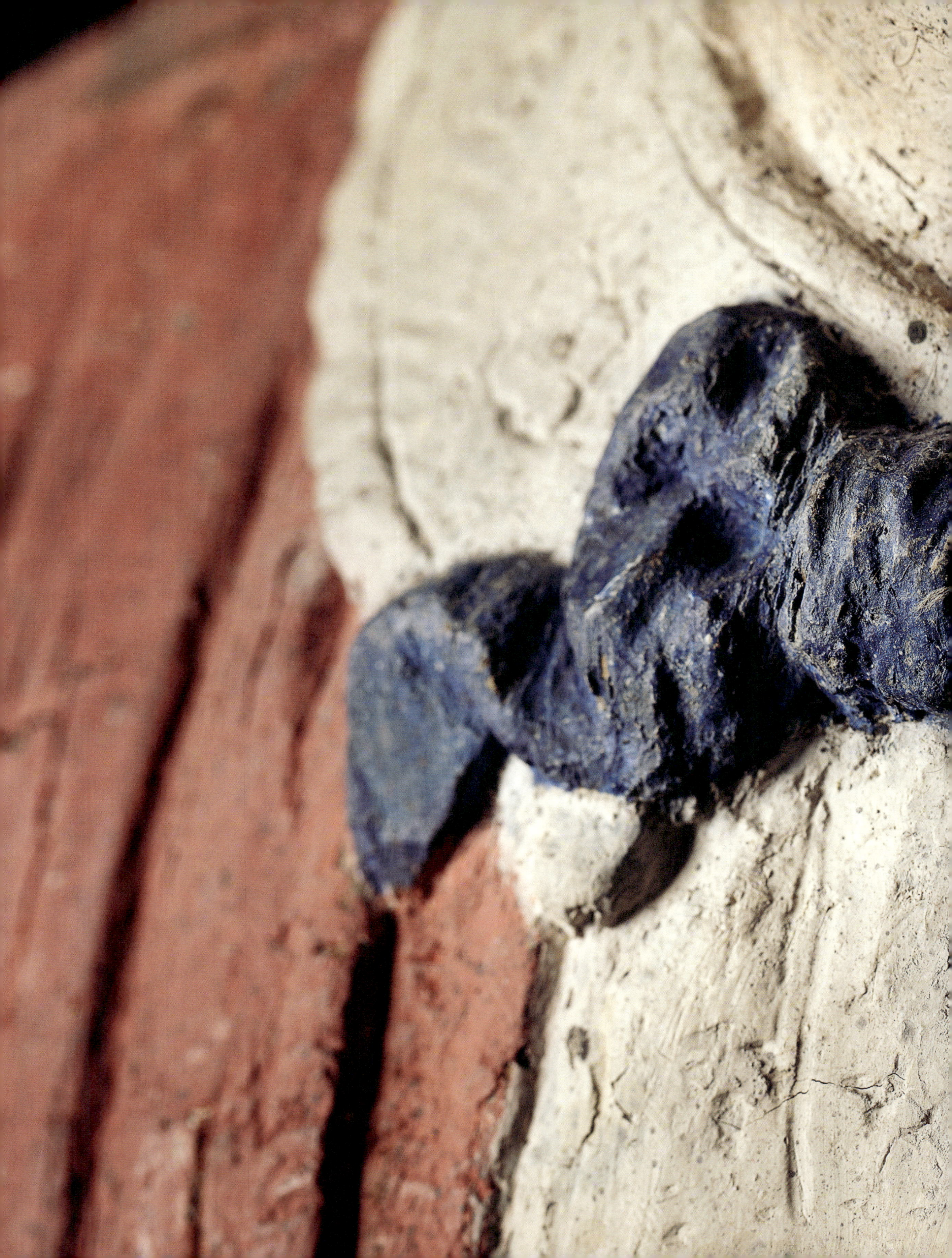

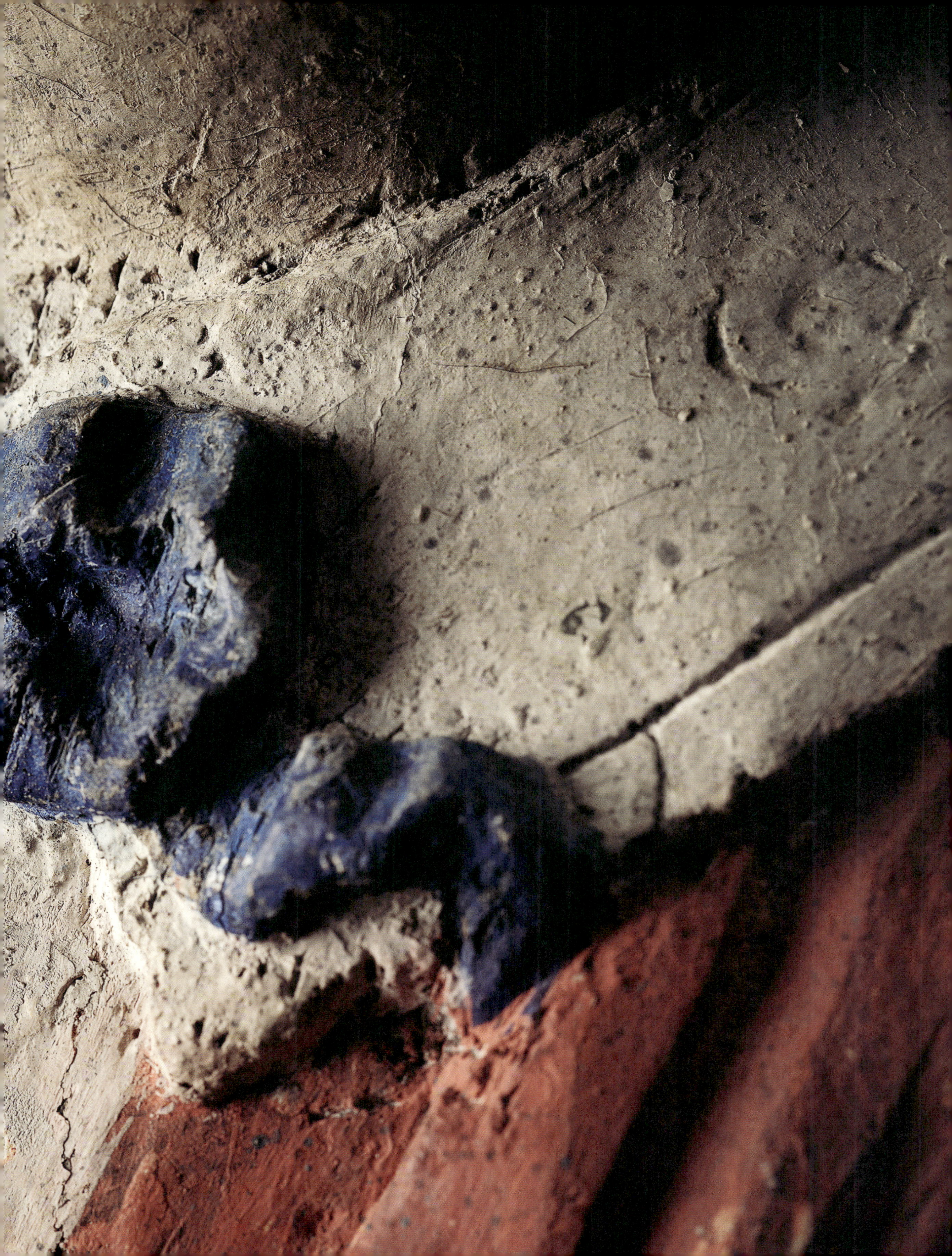

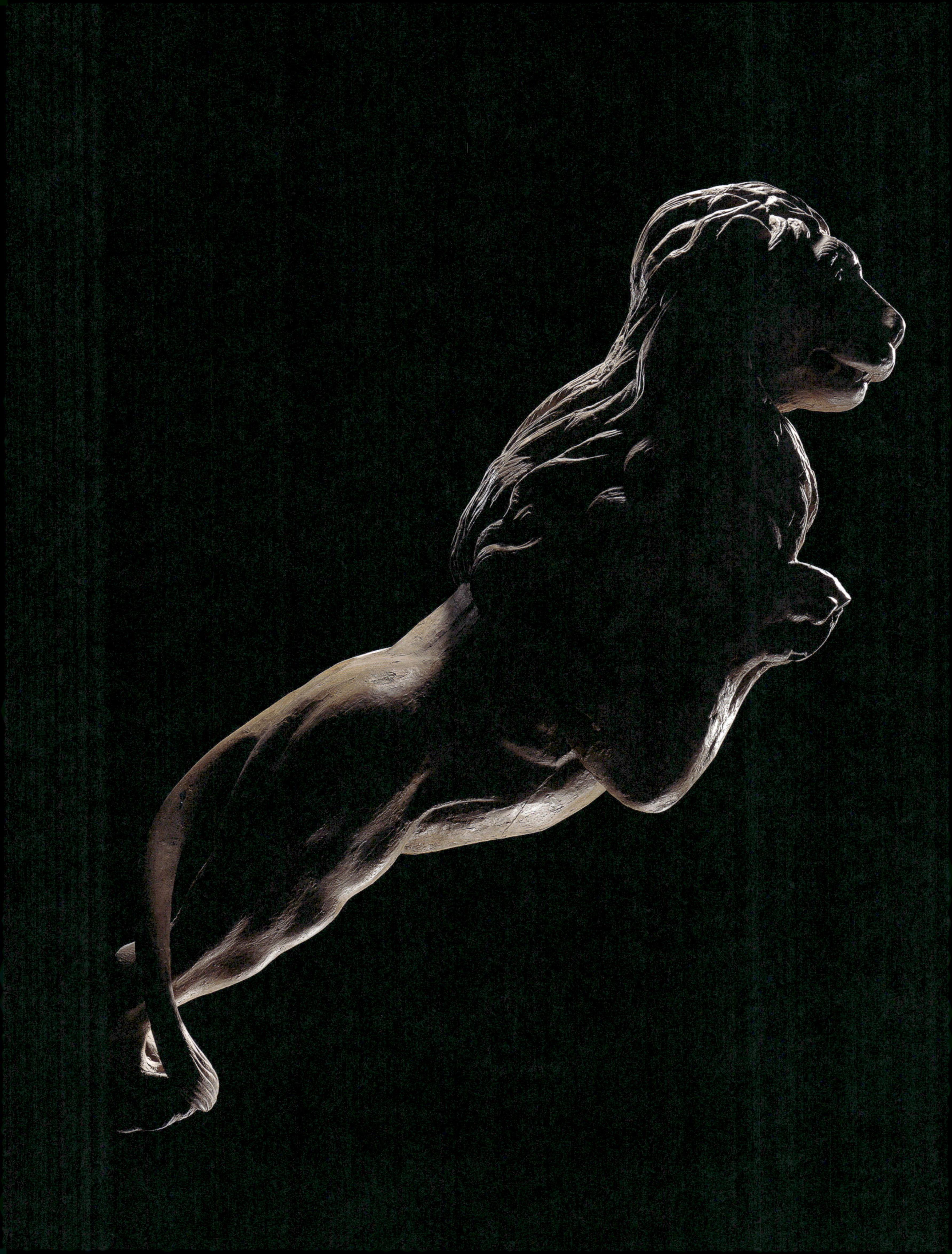

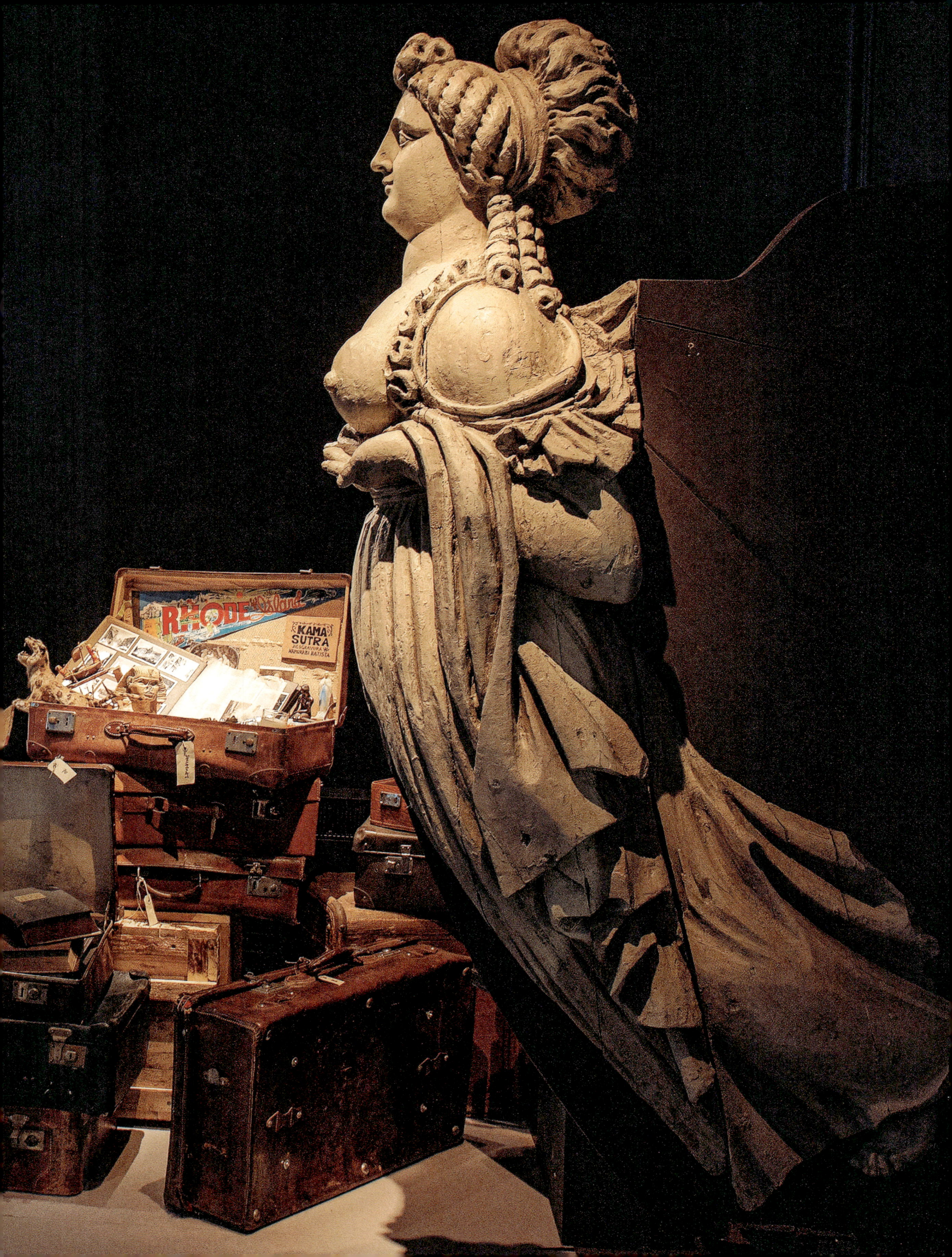
RHODE Island
KAMA SUTRA

THE FIGUREHEAD'S ADVENTURES THROUGH TIME AND SPACE

Annika Bünz

And at the bows an image stood,
By a cunning artist carved in wood,
With robes of white, that far behind
Seemed to be fluttering in the wind.
It was not shaped in a classic mould,
Not like a Nymph or Goddess of old,
Nor Naiad rising from the water,
But modelled from the Master's daughter!
On many a dreary and misty night,
'Twill be seen by the rays of the signal light
Speeding along through the rain and the dark,
Like a ghost in its snow-white sark,
The pilot of some phantom bark,
Guiding the vessel, in its flight,
By a path none other knows aright!

From: *The building of the ship*
By Henry Wadsworth Longfellow[1]

Figurehead from the travelling exhibition *Sex & the Sea*, arranged in 2015 by the Maritime Museum in Stockholm.

ONE DAY, WHILST ENJOYING an afternoon coffee in the café of the Maritime Museum in Stockholm, I happen to overhear a conversation between a child, aged three or four, and a parent. The child attracts the parent's attention and asks whether 'that' used to be a proper person. It is not clear to me what 'that' is, as my back is turned to what the child is pointing at. But the parent's answer explains things. The reply is "No, it's always been made of wood – it just looks like a real person." I turn round, and see the female figure carved in wood that they are talking about. It is a figurehead. Her dress is green, her hair is black and her skin is as white as snow. The wooden figure is hanging on the wall, above those of us sitting here eating lunch or having a coffee. As a clearly recognisable depiction of a human, the figure attracted the child's attention, just as it and the many other figureheads in museums have aroused my interest.

Figureheads are the carved wooden decorations, shaped as humans or animals, that embellished the prows of older sailing vessels. In the library of M/S Maritime Museum in Helsingör there is a book entitled *Neptunus Hölzerne Engel. Schöne alte Galionsfiguren.*[2] In another book I find the explanation of the title:

> 'Neptune's Wooden Angels' is one of the names given to the figure heads that embellished the stem of many sailing-vessels in times gone by. Our matter-of-fact world has

> not preserved a great many of these emblems, which were the object of secret, and sometimes even public, worship. [...] The Port of Hamburg wishes to show ... some of these 'wooden angels', symbols or guardians of maritime transport.[3]

In the introductory poem by Longfellow the figurehead is referred to as a pilot (the pilot of some phantom bark) – a word that recurs in the book title *Silent Pilots. Figureheads in Mystic Seaport Museum.* [4]

The type of prow decoration known as a figurehead can be traced back to the 16th century, but it is above all wooden figures created from the 17th century onwards that are preserved. The word 'figurehead' nowadays also has an extended meaning, referring to someone who appears to be in charge but holds no real power – rather having a symbolic function.

The practice of ornamenting the prow with carved decorations disappeared together with the sailing vessels, [5] but when the ships were broken up or wrecked many of the human and animal figures were found. We may now encounter them in other contexts, e.g. in museums, where this group of objects is usually allocated to the category 'Maritime collection', and when they are displayed in exhibitions this will usually be in museums described as being maritime. The carved wooden figures decorate the institutions' vestibules and cafés, and are displayed here and there in various exhibitions. You will sometimes even see a figurehead placed on a maritime museum's facade, as an icon for the building. At nearly all maritime museums worldwide there is a group of objects that are particularly cherished, namely figureheads, ships' wheels, binnacles, ships' clocks and name boards. "[Of these objects,] the figureheads attract the most attention. With their exotic designs and magnificent colours these prow adornments have always stimulated the imagination and led one to mentally traverse the oceans to foreign lands together with the sailing vessels, at the same time as representing something of the ship's soul."[6]

> Figureheads, both naval and merchant, have survived as artistic as well as historic icons in maritime museums worldwide, where they stand as testament to a tradition and craft from the earliest days of humankind's fascination with the sea.[7]

There is an abundance of vivid descriptions of figureheads in the literature, and writers emphasis the wooden figures' symbolic meanings, both then and now, for both seafarers and land dwellers. The figureheads not only bear witness to former practices and life at sea – they are very much present on land, and are current in the here and now.

Museums have collected and continue to collect objects as representations of space, time and events, and exhibition narratives are produced based on collections of objects that concern various aspects of people's lives and realities. The practices of the modern museum have developed together with the scientific disciplines, and its role has been to convey narratives of research's path from ignorance to truth. A number of types of museum have developed, e.g. geological museums, natural-history museums and art museums, and the objects collected (selected) have been arranged in accordance with the principle of making a scientific view of the world comprehensible.[8] This was, and largely still is, science's response to questions about the self and the world, and about life and death, which at the museums are arranged in collections and exhibition halls. Areas of knowledge and philosophies, parts of the world and epochs, science and cultural history, historical events and individuals' fates are compressed, categorised and re-presented.

When the artefacts end up in museums, it is the museum staff who make decisions about them, and the labels, groupings and classifications used separate the artefacts from their original (or at least previous) environment and social context, re-contextualising them in collections intended to create graspable temporal and spatial orders. [9] But objects have often been used in several different contexts. Things can be said to have their own biographies where they are used in various ways in different social contexts, thus they do not mean the same thing to everyone. When it comes to items that end up in museum collections, it is perhaps above all time and space that separate the everyday context in which they were created and used from the contexts and narratives created in museum practices. But in one and the same situation an object can also mean different things to different people.

Museums' activities tend, however, to be restricted by narrow, totalising notions of how one should deal with collections and the information about them, according to Fredrik Svanberg. Artefacts can always be observed from different perspectives and interpreted in different ways, but this is concealed in old, narrow classification systems that have bound the objects to the 'grand narratives' that the museums have participated in creating.[10] Museums are not neutral 'memory media' that convey the knowledge generated by science using arranged artefacts. They do not reflect the world through the collected objects. They instead create re-presentations and assign the artefacts values and meanings that reflect historically specific perspectives,[11] though these perspectives are hidden, for instance in the museums' classification systems,[12] and thus also in the exhibition narratives that are produced. Furthermore, museums do not only convey the knowledge created within the field of research. The collecting, the staging of the exhibitions and the public's response to the presentations contribute to the fashioning of knowledge, according to Stephanie Moser.[13]

There is thus every reason to examine what happens when the figureheads go ashore, and in the museums' practices of collecting, knowledge production and narrating they are sorted into categories and ascribed meanings. This group of objects is a good example of how artefacts can wander through time and space and between different contexts and varying meanings. And figureheads are not just artefacts carved in wood; the name also covers the use of symbolic images as icons for interests and objectives, just as figureheads themselves to a certain extent act as icons for maritime museums and maritime cultural history.

The literature that in various ways describes, explores and discusses figureheads usually starts with a more or less detailed review of their history. All the writings repeat a fairly similar narrative that states that the tradition of figureheads has its origin in various uses of prow decoration that have occurred in maritime traditions worldwide for a very long time.[14] What is asserted as being the oldest known example is an ancient Egyptian ship built about 6,000 years ago.[15] Descriptions of the origin of figureheads nearly always mention the decorations on Viking ships, and the Norwegian Oseberg Ship is usually shown as an archetype. Its prow decoration is used throughout the literature as being representative of the period's ship ornamentation and 'evidence' of the practice's 'deep roots' in history. How and whether the prow decorations lived on in Europe from the Viking period until the 17th century is little discussed in these writings. David Cordingly to some extent fills in the gap with the information that in the Middle Ages European maritime countries provided warships with heraldic embellishments and carved figures that often represented animals. During the 17th century the use of adornments on the prow and stern increased dramatically.[16]

Hallén & Rüster recount that across the ages the function of prow decorations was to protect seafarers against dangerous beings and sea monsters as well as human enemies. Some figures were intended to fortify the crew against fear and suffering, whilst the purpose of other ones was to induce the forces for good to protect the voyage. Many prow decorations have acted as auspicious magical objects that would increase the chances of a crossing being successful.[17] A figurehead could also embody a ship's soul, and sometimes enabled identification of the vessel.[18] Giancarlo Costa asserts that the reason why the practice of adorning the ship's prow with figureheads developed and persisted is to be found in sailors' inclination to give vessels faces, souls and personalities.[19]

This is a fairly unanimous explanation of the history of figureheads, which is repeated time and time again. The research presented principally takes the form of systematic descriptions and surveys of the figureheads to be found around the world, often in catalogue-like presentations.[20] Where the sculptors are known, their production is sometimes described.

Some writings differ from the rest of the literature by problematising and discussing the Swedish sculptor Johan Törnström from an art-history perspective.[21] And 2015 saw a special issue of *Sculpture Journal* (24.2), entitled *Sculpture and the Sea*, in which several writers discussed the figurehead in an art-history context.[22] The history of figureheads has also been examined in detail in a book on women who went to sea during the sailing-vessel era.[23]

On the basis of writings and books that in various ways deal with the use of figureheads, in the following section – 'A protective being and/or a political tool' – I summarise the accumulated knowledge on the subject. It has been equally important to me to draw attention to the way writers write about figureheads. There are many trains of thought and topics to monitor for their depictions and choices of words.

Under the heading 'The figurehead goes ashore' I then explore what is to be found in the literature on the figureheads' various adventures when the artefacts' life at sea comes to an end, and I also discuss practices of collecting, reusing and the creation of new meanings.

A protective being and/or a political tool

During the 17th century use of ship ornamentation increased dramatically in Europe. It was above all naval vessels and representational ships of the leading naval powers of the day – England, France, Holland and Sweden – that were filled with sculptural ship's decorations.[24] The symbolic images were used as political tools, and were intended to express the nation's power and glorify the head of state. The bigger the vessel the more important were the decorations.[25] When part of a naval fleet, the vessels could also be embellished with a depiction of the regent. 1628 saw the launch in Stockholm of the ship *Vasa*, which was decorated with 500 carved figures and had a lion as its figurehead. The central theme for the embellishment of *Vasa* is glorification of the king, whose carved portrait was an focus for the stern decorations. The majority of the figures on the ship are men – most of them armed warriors.[26] Within the merchant fleet the figureheads came into use a little later than within the naval fleet, and it was not until the second half of the 19th century that they became common.[27] The motifs varied greatly, and occasionally the merchant ships' carved wooden figures depicted a wife, a daughter ('But modelled from the Master's daughter!') or the ship owner.

Many people associate figureheads with female figures with more or less bare breasts. There is a notion, which can be traced back to antiquity, that water is a female element and that women exert a power over the oceans that men are denied. A naked female body was deemed able to allay a storm, and David Cordingly thinks this may be a reason why many figureheads represent a bare-breasted woman.[28] There is also an old tradition of considering ships to be female. Some people think this is because ships are considered to be beautiful and have female curves, and others because ships, like mothers, offer womb-like protection. It has also been suggested that seafarers see ships as being women because of the female figureheads.

If women had a special power to allay storms it would appear logical to adorn ships with female figureheads. But figureheads with female features did not become commonplace until the 19th century, and Cordingly emphasises that it is actually only because the majority of the figureheads preserved date to the 1780s onwards that is it generally assumed most figureheads feature bare-breasted women.[29] It was in the 1780s that figureheads representing women started to emerge, and during the 19th century they were as common as male ones, especially on merchant ships. Warships, which used only to be adorned with kings, warriors, marine heroes and Greek gods, could now also be accoutred with goddesses, sirens, sea nymphs and mermaids.[30]

Within the naval fleet, for reasons of prestige the numerous big wooden sculptures eventually assumed somewhat extravagant dimensions, and more emphasis started to be placed on the woodcarver's reputation and skill than on the vessel's safety. The prow decorations became incredibly expensive, but also so heavy that the vessels became difficult to handle. This led to the French and British authorities having to put a stop to the embellishments at the beginning of the 18th century.[31] Restrictions were also introduced in many other countries.[32] At the Royal Danish Naval Museum in Copenhagen it is stated that the ships' decorations were important to the kings but that the figureheads sometimes became so big and heavy that they made the ships cumbersome. It occasionally happened that in rough weather seamen had to chop the figures up, in order to get rid of the weight and regain control of the vessel.[33] In such situations, from a seaman's perspective figureheads were more of a danger than a form of protection. But when during the 18th century legislation against the figureheads was gradually introduced, and finally a total prohibition, the seafarers protested. They "could not imagine going to sea on a vessel with a naked prow. There was great bitterness, and the protests became so vocal that the law had to be repealed."[34]

K. B. Bjering points out that figureheads were extremely important to sailors – a situation reflected in the fact that the crew were particular about keeping them in a good state. Delicate parts that were removable were taken off in rough weather, and there are examples of the crew sawing the figurehead off a sinking ship under cannon fire and taking it ashore in a rowing boat.[35]

Ole Lisberg Jensen, however, states that the removable parts were taken off as soon as ships put out, and draws the conclusion that it was when a vessel was in port that the figureheads were the most important. They could there be looked at undisturbed and at close quarters. Out at sea the distances were so great that it was not possible to make out the details of the embellishment.[36] What Bjering's and Lisberg Jensen's varying perspectives above all show is that figureheads often mean different things to different groups. For the king and ship owners they mean status, prestige and political stance, and being able to impress the outside world, and in this context ship decorations would appear most important when in port, where everyone can see them. For seamen they had a completely different function.

A recurring assertion is that the figurehead was the only woman on board. Bjering submits that for obvious reasons the female figurehead had the potential to be popular amongst seamen, and they would sometimes paint a ring on her ring finger. The woman at the prow became a symbol of a normal life in a home environment, he writes, and states that all seamen probably considered themselves engaged to her.[37] Gerd Wietek mentions in passing that "the artistic quality of the figure head ... never had the slightest bearing on the direct, personal relationship between the sailor and his 'doll', as it was sometimes called regardless of its appearance".[38]

In both these examples the figurehead assumes the role of a woman with whom the seaman can have a relationship. In other depictions the figurehead is a protective being. Giancarlo Costa talks of a notion widespread in the 19th century that a figurehead was the only way of protecting yourself against death if you encountered the notorious Flying Dutchman – a harbinger of death and destruction: "In such an event if a seaman were to survive he had to hold onto the figurehead with all his might, the figurehead being the very soul of his own ship."[39] And in a short text I find an account of what the sailors perhaps did when the ship was in port: "Sailors often chipped a little wood from the figurehead for a good luck piece when going ashore – that is, providing the mate didn't happen to catch them in the act."[40]

These quotes are largely what is to be found in the literature on seamen's relationship to figureheads. What they have in common is that the authors do not state any sources for their information, and it is thus hard to relate to them. This is maybe because the accounts are hearsay, and because it is hard to produce sources to verify this type of statement.

Information on the creation and purposes of naval vessels' decorations is all the easier to produce, thanks to the fact that the regents' actions are well documented in the state authorities' archives. [41] And preserved figureheads from both naval and merchant fleets allow study of the figures' styles, artistic trends and the woodcarvers' skills. The seafarers' beliefs and relationships to the figureheads are harder to get a handle on – a little more detective work and other approaches are probably needed in order to trace and substantiate them. Through targeted studies of preserved figureheads one could maybe investigate whether there actually was a tradition of chipping off a small piece of the figurehead. By way of an exciting but small contribution, I can report that I once discovered an interesting detail on a figurehead at M/S Maritime Museum in Helsingör. On its left ring finger someone had chipped off a small piece precisely where a ring would normally be found. Whoever created this little notch maybe saw it as a good way of killing two birds with one stone …

Someone has carved out a small section of the finger where an engagement ring is usually worn.

A recurrent link in the literature is that between the figurehead and a protecting eye. On ancient Egyptian ships an eye was painted on either side of the ship's prow, so it would be able to find its way through the water.[42] Painted eyes are also to be found on Greek vessels dating from the 8th century.[43] There was a belief that an eye would drive away evil forces. The tradition persists to this day, above all on Chinese junks,[44] but also in some areas around the Mediterranean, where eyes are painted on fishing boats.[45] According to Hanne Poulsen there was a perception that the figurehead acts as the vessel's eye, seeing through storms, high seas and the darkness of the night, and bringing the ship and its crew safely into port.[46] With its gaze directed straight ahead in the direction of travel, the figurehead was maybe the one who was always awake and was able to see everything the ship encountered. Description of the figurehead as a pilot, as in the introductory poem by Longfellow ('the pilot of some phantom bark') and the title of *Silent Pilots,*[47] a book about figureheads, relates to the carved figure's perpetual gaze, which sees far into the distance and can detect the possible dangers ahead.

In one account the figurehead's eyes are covered over so it avoids seeing what is happening:

> In 1778 when the Channel Fleet was in full retreat from the French Navy – a disgraceful situation – a boatswain's mate in HMS Royal George slipped over her bows with his hammock, which he lashed over the eyes of the figurehead. This was a carving of King George II, who was a courageous man and the last English monarch to personally lead his troops into battle. When challenged about his action, he is said to have replied, 'We ain't ordered to break the old boy's heart, are we? I am sure if he was to turn and see this day's work, not all the patience in heaven would hold him a minute.'[48]

In this quote the figurehead is not just a portrait of the king, but has also been given the king's gaze and feelings – and maybe the entire ship has been ascribed his personality. Gordon Marshall points out that figureheads "… embodied the 'soul' as well as being the 'eyes' of the vessels they headed".[49] And goes on to state that they in a sense represented 'the forward looking age of European discovery and expansion'.[50] Marshall thereby includes both the protective being and the political tool in the wooden figure's forward gaze, whilst at the same time adding his own interpretation of the figurehead's significance in terms of the emergence and expansion of Eurocentric modernity.

When regents and ship owners invested in costly figureheads it was because the latter had aesthetic, artistic and artisanal values that demonstrated wealth, as well as symbolising the ship, the maritime environment and the owners' status.[51] To make sure of getting high-quality ship's decorations, trained artists were engaged to design and make the

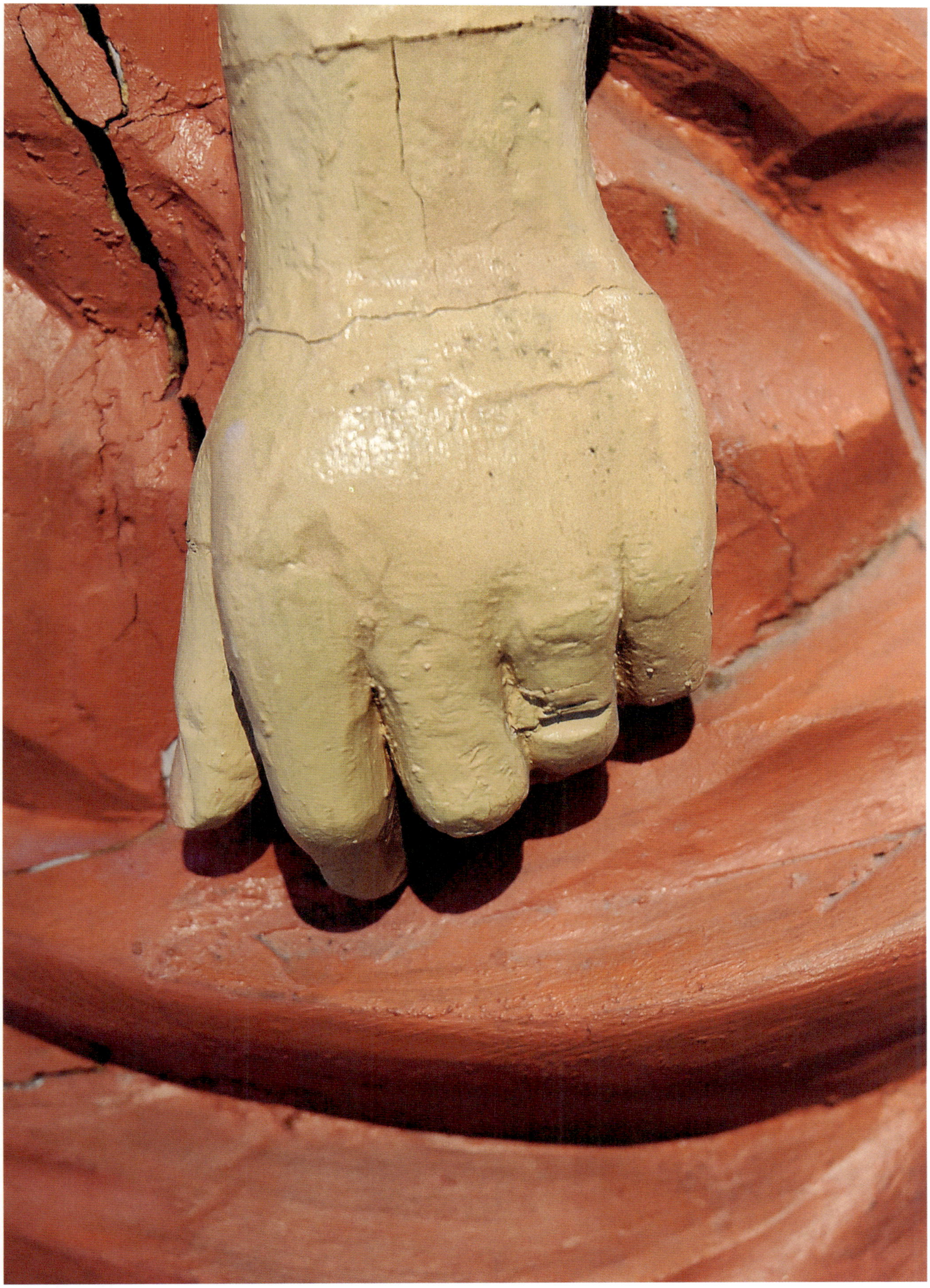

figureheads. The carved wooden figures thus also arouse art historians' interest, and many figureheads are talked about and discussed as sculptures with artistic qualities.

The period 1750–1815 is given as the time when the figureheads created by artists attained their highest artistic level.[52] Cordingly stresses that French female figureheads during this period were especially impressive and reflected the quality of French monumental sculpture.[53] Sculpted allegorical female figures in the neoclassical style symbolised Fame, Honour and Fortune.[54]

The Naval Museum in Karlskrona displays enormous wooden figures made to adorn Swedish naval vessels. Many of the figures were created by 'the artist behind the figureheads in Gustav III's and F. H. af Chapman's proud naval fleet' – the royal sculptor Johan Törnström.[55] The shipbuilder af Chapman is also included in the museum's narratives, both in the form of text and as a white figurehead. Johan Törnström is considered to have been one of the most gifted figurehead sculptors,[56] and was very familiar with the neoclassical designs that were all the rage in the late 1700s.[57] Together with a collection in Paris, the collection in Karlskrona is according to Bjering 'the world's most distinguished collection of its kind from that century'.[58]

Much has been written about Törnström, as regards both his life and his work, and his skill is described in detail.[59] Bjering writes:

> His figureheads are imbued with classical dignity, confidence and peaceful monumentality, which in times of misfortune are well suited to symbolising strength, decisiveness and the will to defend. The secure lines and the powerful play of muscles often create a stimulating impression of restrained colossal strength.[60]

Danish figureheads also maintained a high artistic quality,[61] and during this period 'the extraordinarily skilled' F. C. Willerup carved figureheads for the Danish fleet in the neoclassical style.[62]

Kings and regents in other words chose ship decorations with care, whilst many of the merchant fleet's ship owners maybe to a greater extent took what was on offer, not setting as much store by motifs and execution. But despite the small shipyards' somewhat more artisanal figureheads not constituting what is known as high art, they displayed 'a decoration technique that is in many cases admirable'.[63] Hallén & Rüster stress that many of the merchant ships' figureheads reflect a genuine provincial culture and 'tell about a folk art that has now completely died out'.[64] In the larger port towns skilled craftsmen made ships' decorations, but there were also professionally trained figurehead sculptors.[65] Bjering is of the opinion that from a cultural-history standpoint these figureheads display an immediate originality that is more interesting than the recognised artists' 'somewhat inhibited specimens'.[66] According to Peter Norton, during the 19th century many of England's best figureheads were made by local carpenters – a far cry from the trends prevailing within the field of art.[67] However, the figureheads adorning the larger and more expensive vessels that went on long trips to other continents were influenced by international trends,[68] though it was rare for ship's embellishments to be signed. It is Lisberg Jensen's contention that one reason may have been that they were regarded as handicraft rather than art (Lisberg Jensen 1984:10–11).[69]

What is clear is that the preserved figureheads across the globe have inspired research and lively discussion. The interesting thing about these conversations is that writers contrast what is known as high art with folk art, and decorative techniques against artistic skill, categorising some of the figureheads as art and others as artefacts. Julia Kelly, however, observes that as regards carved embellishments (sculptures, as she calls them) in maritime contexts, there is an overlap between handicraft, design and so called popular art forms. The objects are often both functional and decorative, which according to Kelly means they challenge the prevailing subdivision between artwork and artefact by unifying complex networks of patrons, designers, artists, craftsmen and woodcarvers.[70]

A glance at the dates of the publications to be found in libraries shows that the work of describing and cataloguing figureheads has largely come to a halt and academic discussion has ceased. One exception is the woodcarver Andrew Peters' review of ship's decorations during the period 1630–1780.[71] Commissioned to carry out a preliminary investigation ahead of the building of the East Indiaman *Götheborg*, Peters studied the artistic movements that influenced European ship's decorations during this era, so as to then be able to create the decorations and the figurehead that now adorn the ship. A further exception is the art historians who in 2015 were brought together in the above-mentioned special issue of *Sculpture Journal*, which deals with maritime sculptures from the perspective of seafarers, travellers, observers and land dwellers.

Discussion of figureheads as a practice and historical artefacts is in other words no longer as lively, but the wooden figures have not for their part fallen silent. They have gone ashore and are very much present amongst us as eloquent forms of expression in the here and now. And the word figurehead has gained the extended meaning of a symbolic leader who lacks real power. Regardless of whether figureheads are ship decorations made of wood or human beings made of flesh and blood, they are used as icons to communicate power, status, knowledge etc. and thereby achieve a specific objective, e.g. conveying trust or winning an election or war.

The figurehead goes ashore

> The bronze man and the wooden man wandered around the shipyard and completely forgot to look for the boy. Like two old salts, they relished everything they saw. They finally reached the yard where the figureheads from old naval ships were set out. The boy had never beheld a sight more peculiar – they were mighty, terrifying faces, full of proud daring. The bronze man then said:
> 'Doff your hat, Rosenbom! They have all been in combat for the fatherland!'
> (from *Nils Holgerssons underbara resa genom Sverige*).[72]

Back during the sailing-vessel era many figureheads were found and reused in various ways. Gerd Wietek talks about how people looking for usable wreckage on the beach could hit upon what was deemed to be an even more valuable remnant of the vessel:

> On rare occasions, the eyes scanning the water for timber would detect a figure head, the symbolized embodiment of a ship's personality which represented, not only in the eyes of the crew but of those living ashore as well, a much higher value than just a structural part of the ship or a piece of useful material. Such a figure head was frequently saved from further destruction and in some cases it was even put into a casing especially made for it and set up like the mediaeval figure of a saint.[73]

Treatment as a medieval depiction of a saint is exactly what happened to a female half-figure washed ashore several centuries ago, according to Wietek. When it was rescued from the sea it was provided with a specially constructed cabinet and was then kept in a barn. When it was subsequently taken care of by a museum, it was displayed there in the same way as it had stood in the barn.[74]

The connection between figureheads and depictions of saints is also obvious from the arrangement whereby figureheads were displayed at Karlskrona Naval Base Lindholmen, before the wooden figures were moved to Karlskrona Naval Museum's newly built premises. In groups or individually, the figureheads stood in small, semi-open structures that framed them in the same way as depictions of saints in a cabinet.

Many of the recovered figureheads were, however, used slightly less ceremoniously as garden decorations, or were set out in the landscape as wooden statues.[75]

> Not that many years ago there were many figureheads up on the outcrops on the islands and along the banks of the fjords in Bohuslän, and to this day you still come across the odd 'Neptune' or 'Hermes' in the gardens of fishermen's cottages. Other images have ended up in museums or fallen into the hands of city dwellers, who have made a more or less disrespectful or crazy profit from them.[76]

One figurehead was discovered in an attic in 1994, obscured and almost forgotten. The person who found it first took fright at the sight of a white hand peeking out from amongst the clutter. The owner of the house "explained that the hand belonged to a scarecrow that had been used on the farm. In the fields, in the moonlight, it had frightened not only birds but also people, and had thus been hidden away in the barn. There, in the loft, the scarecrow had been standing for perhaps a hundred years."[77] The old wooden figure eventually ended up at an antiques dealer's, who spent ten years in search of its origin. The search led to the clipper *Nightingale*, named after Jenny Lind, and the figurehead proved to be a portrait of the world-famous singer.[78] The wooden figure, which when it went ashore was first put to use as a scarecrow, and was then hidden away in a barn, once again entered the limelight. And its value now shot up dramatically: "On Saturday 'the Swedish nightingale' is to be sold at Sotheby's in New York. The reserve price is $100,000-150,000. Not bad for a scarecrow."[79]

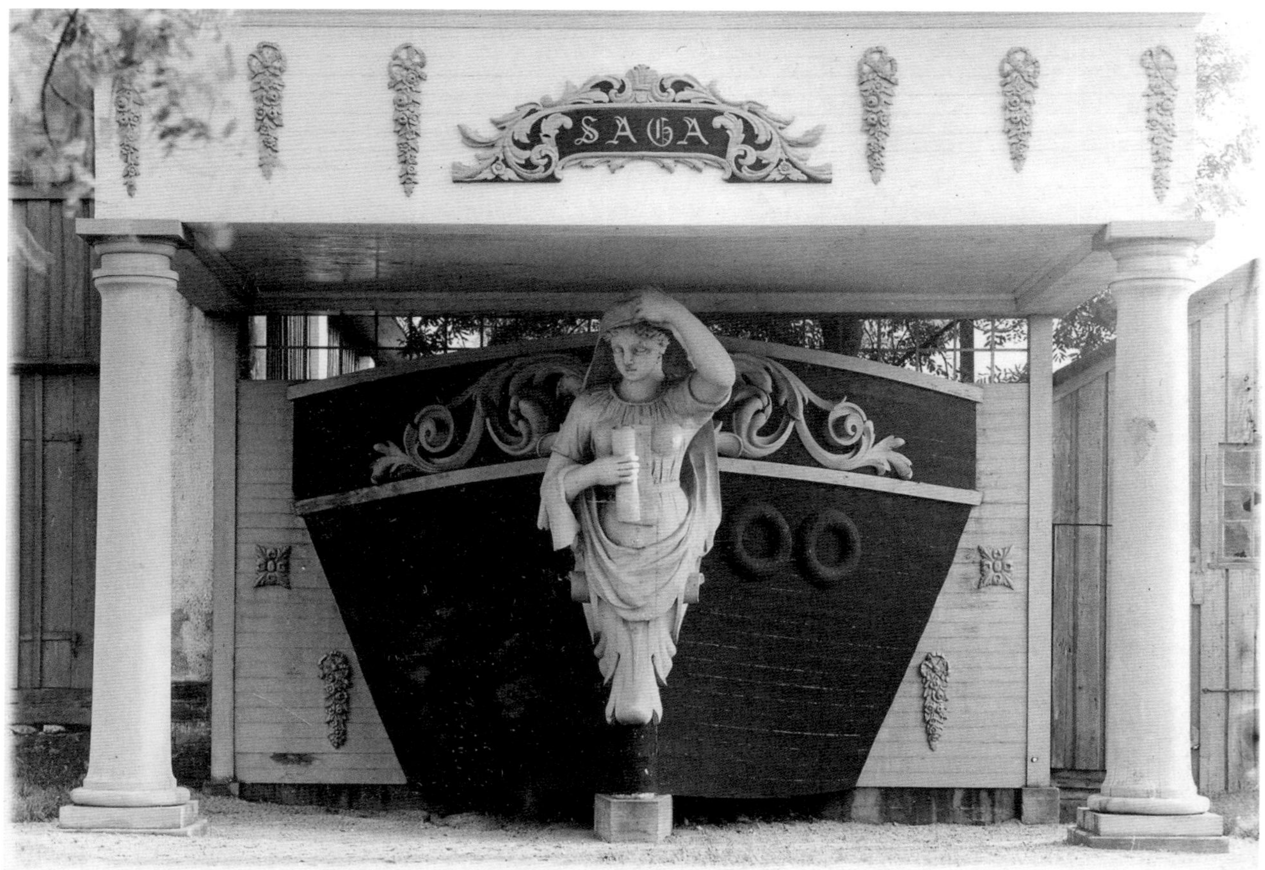

The steam corvette *Saga's* figurehead from 1935.

But it is by no means always possible to trace the origin of figureheads that have been washed up ashore. They can instead be assigned a new identity when they come ashore, alone and anonymous. At the Maritime Museum in Stockholm there is a figurehead that goes under the name of Ulla Winblad.[80] The background is that at the end of the 19th century it had an annual role as Ulla Winblad in the Bellman Procession on the island of Furusund – a tradition that continued until 1907. There is no information on the vessel for which the figurehead had been made. And when after its time as Ulla Winblad it was taken into the care of the museum and was incorporated into the collections, it was instead the story of the Bellman Processions that was recorded by way of provenance. At the museum the figure was then displayed in various exhibitions over the years, for a while it hung in the café, and at the time of writing it is being used in an art installation about seamen and sex.[81] In other words the various adventures of 'Ulla Winblad' when she went ashore do not have that much to do with the artefact's 'real' background, and the exhibitions in which she has been displayed have not much concerned figureheads per se.

But in some old exhibition catalogues I find exhibitions that have had the figurehead as their theme. In the early 1960s the Altona Museum in Hamburg sought information on all the figureheads to be found preserved in West Germany. The figureheads that were located were brought together to form an exhibition that lasted three months.[82] Wietek describes his experience of the way the wooden figures were arranged in the hall:

> Going by the direction into which their widely opened eyes looked, they were placed … side by side or opposite each other, a strange gathering of beings who, in spite of their enchanting substantiality, seemed to belong to a remote magical world.[83]

But at the same time as Wietek sees in the figureheads a mixture of magic and depictions of saints (see above on figureheads in cabinets), he emphasises that they primarily constitute a type of what he calls popular art, which has been widespread amongst the coastal population and whose styles unify seafaring traditions across national borders.[84] Gordon Marshall points out that a collection of figureheads hanging in a museum hall is a display of both folk art and history. The art form is, or at any rate has been, living, and has been 'used' far more than contemporary statues.[85] Giancarlo Costa in turn notes that they are 'the only real "living sculpture" in the world'.[86] The keyword in Marshall's and Costa's statements is 'use(d)'; it is the practices – the use of the figures – that is the interesting thing, and the practices persist and assume new forms.

Many of the figureheads are thus referred to as artworks – as wooden sculptures. According to Martin Heidegger an artwork is based on a unification of directions and relationships that has its roots in the world of historic persons, and is both spiritually and physically linked to its surroundings. An artwork is thus lost if it is moved from its original location.[87] This is very obvious as regards the figureheads. When the prow decorations were on a ship they were part of its architecture and appearance, and perhaps illustrated its name. Their degree of inclination is aligned with the shape of the vessel they were made for, and considering the exposed location right at the front of the bow, compositions with parts that projected too much were avoided. Many figures have a bound composition with their arms kept very close to their body. Another solution was to make the projecting parts removable.[88]

The sailing vessels were of a variety of types and architectural shapes during the different epochs, and the figures that adorned their prows reflect these varying lines. Components belonging to the figure sometimes also spread over the ship's stem and bow. The boundary between vessel and wooden figure became fluid, and it was not really possible to differentiate one from the other. When a figurehead was removed from a vessel, then a piece of the ship's hull and its form, as well as the notion of the vessel, went with it.

As visual expressions and historic artefacts that bear witness to a tradition and a craft, the collected wooden figures subsequently often hang with their backs to a wall, and the impression of the ship is then sometimes hidden to the beholder. In other instances a figure's adaptation to the shape of the prow can also be seen from the front. Visible material components from the vessel's prow and bow often remain, but the ship and its maritime environment are lost, and remain so. The figurehead as artwork has been separated from its roots and directions – and from spiritual and physical relationships in the world for which it was designed. The original artworks have been lost, and are now instead artefacts in private barns or museum repositories.

I said earlier that a shipbuilder who had ordered and chosen the design of a figurehead had a relationship with it and its design, and the seamen who then sailed the ship in question ascribed to the figures completely different meanings and symbolisms. Those who subsequently salvaged and collected the figures furthermore have other relationships with them, and probably create new meanings and narratives. When a figurehead goes ashore and lives on amongst land folk in gardens and museum exhibitions, it can thus be seen both as an artwork that has been separated from the situation in which it was created and as an artefact that has changed context and meaning. The individual figureheads have their own life stories, in which they roam between different historical and social situations and are placed together with other objects, out at sea, by the waterside and in the garden, as well as in museum repositories and exhibition premises, where they are included in the institutions' surveys and stagings of the world and reality.

Regardless of whether a figurehead was created by a trained artist or a shipyard carpenter, and whether it ends up in a museum collection or a private garden, it is on the basis of its power as a visual expression and human communication that the object and its use must be understood. The aesthetic and visual qualities that were important to kings, shipowners and seafarers were equally important when the wooden figures were taken care of and used on land in various contexts, and they are equally important today in museum collections, at auction and in the antiques trade. And their use has been, and still is, in living practices that involve both seafarers and land dwellers, as well as regents and ordinary people.

This is a translation of a text that was first published by Nordic Academic Press in *Galjonsfigurens många öden och äventyr* (2018).

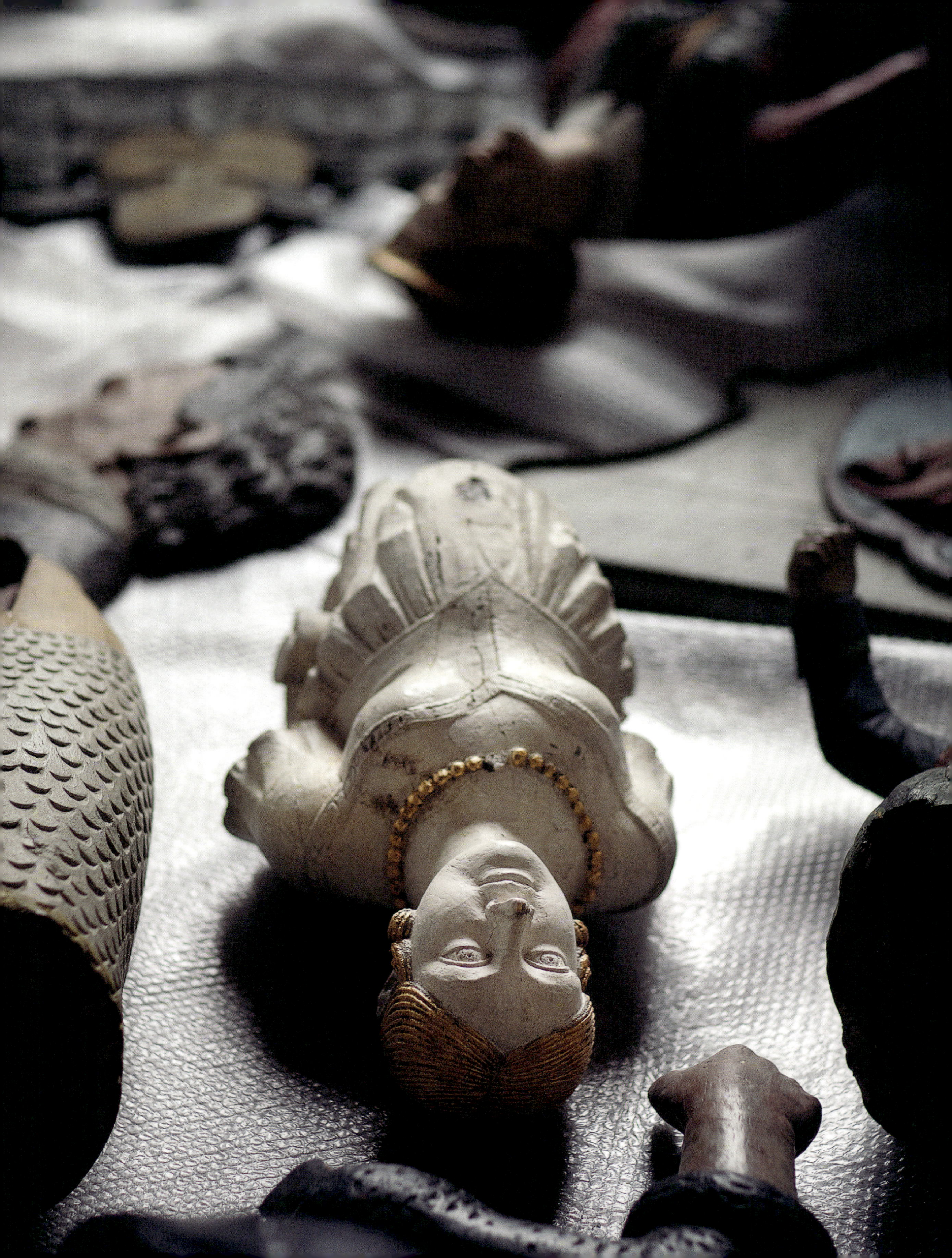

THE FIGUREHEADS AT ENGELSBERG

Svante Helmbaek Tirén

SINCE TIME IMMEMORIAL the coastal waters off North Jutland in Denmark have been known for being dangerous – not least around Skagen, at the transition of Skagerrak into Kattegatt, where many vessels have met their fate thanks to treacherous reefs and violent storms. In the archives one finds traces of hundreds of shipwrecks together with descriptions of rescue missions, heroic deaths and tragedies.

Trade in flotsam was at times an important source of extra income for the local population. Following accidents, wreck auctions were held at which objects that had been washed up or rescued went under the hammer to the highest bidder. On other occasions flotsam was collected by passers-by, and to this day there are a host of artefacts and stories preserved from the era of sailing vessels, all providing a picture of the tough life at sea and the significance of the many contacts navigation brought with it.

The figurehead collection at Engelsberg Ironworks is a remarkable and thought-provoking monument to our maritime history. It comprises thirty-two figureheads plus a number of ornaments and nameplates. It used to be known as the Skagen Collection, since the objects originate from vessels that sank off the North Jutland coast. Most of them date to the early

Left: Parts of the collection being photographed in conjunction with its move to Engelsberg Ironworks, where work on conservation and documentation began.

Below: Collecting and sorting wreckage on one of Skagen's shores. This was a common sight, and many people stood at the ready to salvage items for sale. Woodcut by Carl Neumann published in the Danish magazine *Illustreret Tidende*, 1861.

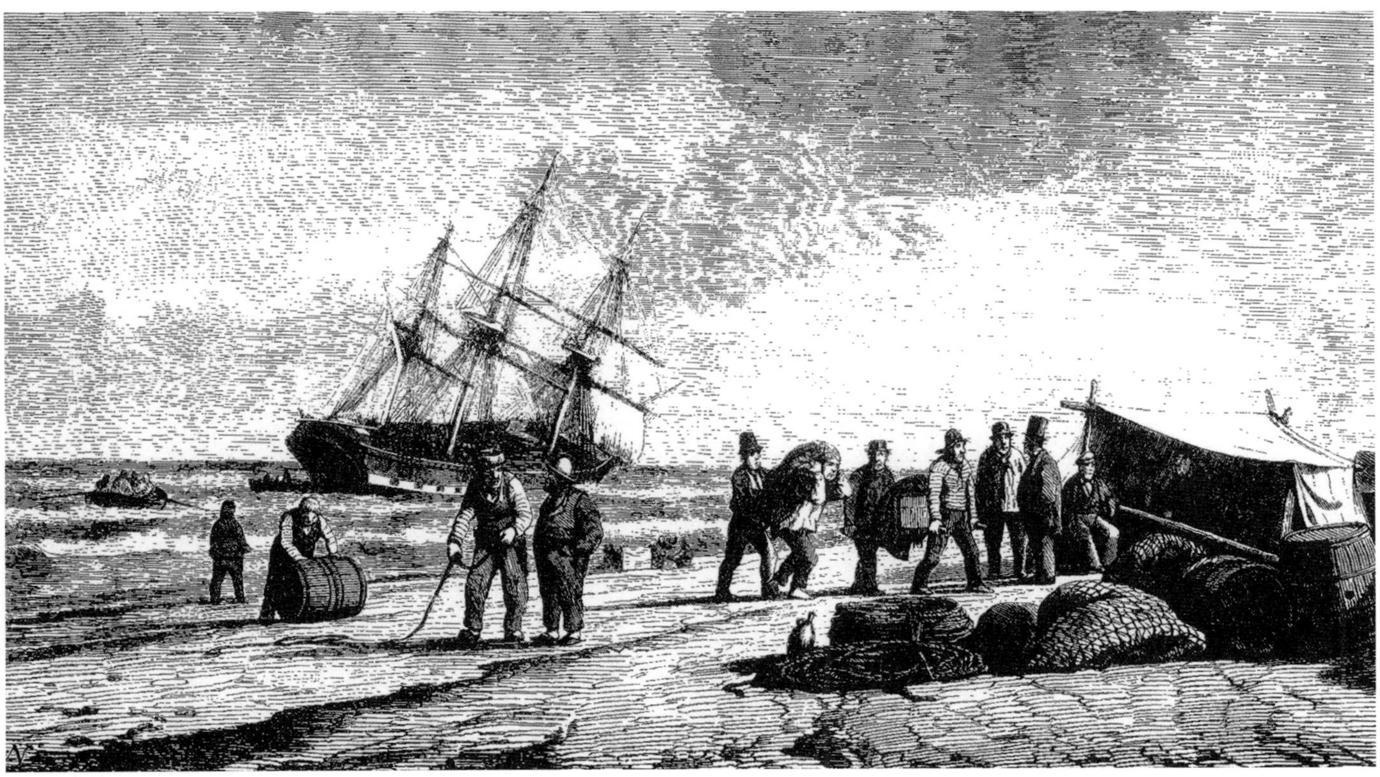

and mid-19th century – the period that became the last phase of the heyday of sculpture within the art of shipbuilding. The final generations of seafaring commercial ships are represented here in the form of sea gods and goddesses, virtuous shipowners' wives, roaring lions and a medieval knight in shining armour. The collection captures the wide range of motifs and the artistic breadth that prevailed during this transitional period, when traditional shipbuilding in wood was gradually starting to languish. On the more modern steamers and the new steel vessels there was no longer a natural place for figureheads, and this pictorial tradition thus died out.

The Figurehead Dining Room at Skagen Hotel, taken before the figureheads were sold in 1938.

Some of the collection's figureheads were created applying great artistic precision and realism, whilst others are more coarsely sculpted and sometimes slightly comical or grotesque. A number of them are figures from the worlds of the sagas and ancient mythology, contrasting sharply with those representing men and women of the period in fashionable clothes. But they all express the symbolic importance of figureheads. Their striking design as it were brought a vessel's identity to life, and was key to the emotional experience. Only a dozen of them can reliably be linked to a specific vessel, but future research will doubtless be better able to shed light on their origin.

The history of the collection's origins is an important example of the way interest in maritime cultural objects has developed. The collection was originally created in Skagen by the Danish hotel director Ludvig Emil Andersen (1859–1910). He appears at an early stage to have been fascinated by figureheads and flotsam, and he realised their historical and nostalgic value. For a number of years he bought figureheads and nameplates, probably at wreck auctions and in similar contexts. When in 1893 Andersen became the owner of the Hotel Skagen the figureheads were placed in a dining room, which was thereafter called the Galleon Hall or Stranding Hall. The room was widely spoken of as one of Skagen's major tourist attractions. In a German magazine the hotel was in 1905 described as a House of Death, as several of the figureheads were witness to catastrophes with a fatal outcome. A later hotel guest was the author Karen Blixen, who spent a winter at the hotel. Her 1942 short story "Peter and Rosa" describes a drama about the jealousy between a captain's wife and a figurehead – which represented Blixen. Figureheads' infallible ability to spark off the imagination has thus indelibly influenced literary history.

In the 1930s the Hotel Skagen was beset by financial difficulties. It was at the time run by Andersen's daughter Ingeborg Østergaard (1886-1969), also known as 'Skagen's rose', and a much talked-about figure in Skagen's social circles. She finally decided to sell the collection, and that led to heated discussion of the famous figureheads' future. It was hoped that a Danish collector or patron could be got hold of to buy them, but no-one suitable was found. In autumn 1938 the collection was instead sold to the consul general Axel Ax:son Johnson (1876-1958), who was himself a shipowner and had a great interest in collecting. The basic idea was to stop the collection being split up and then exhibit it at a planned museum.

It was, however, 1982 by the time this aim was realised. The collection was then deposited at the Maritime Museum in Gothenburg, where parts of it were exhibited up until the museum's closure in 2018 for renovation. It was subsequently incorporated into the maritime collections at Engelsberg Ironworks, which are kept together with extensive archives – and collections of objects linked to the history of navigation. In this context light can be shed on the figureheads, thanks to detailed knowledge of the field and a fresh perspective on the many layers of history concealed behind their exciting imagery.

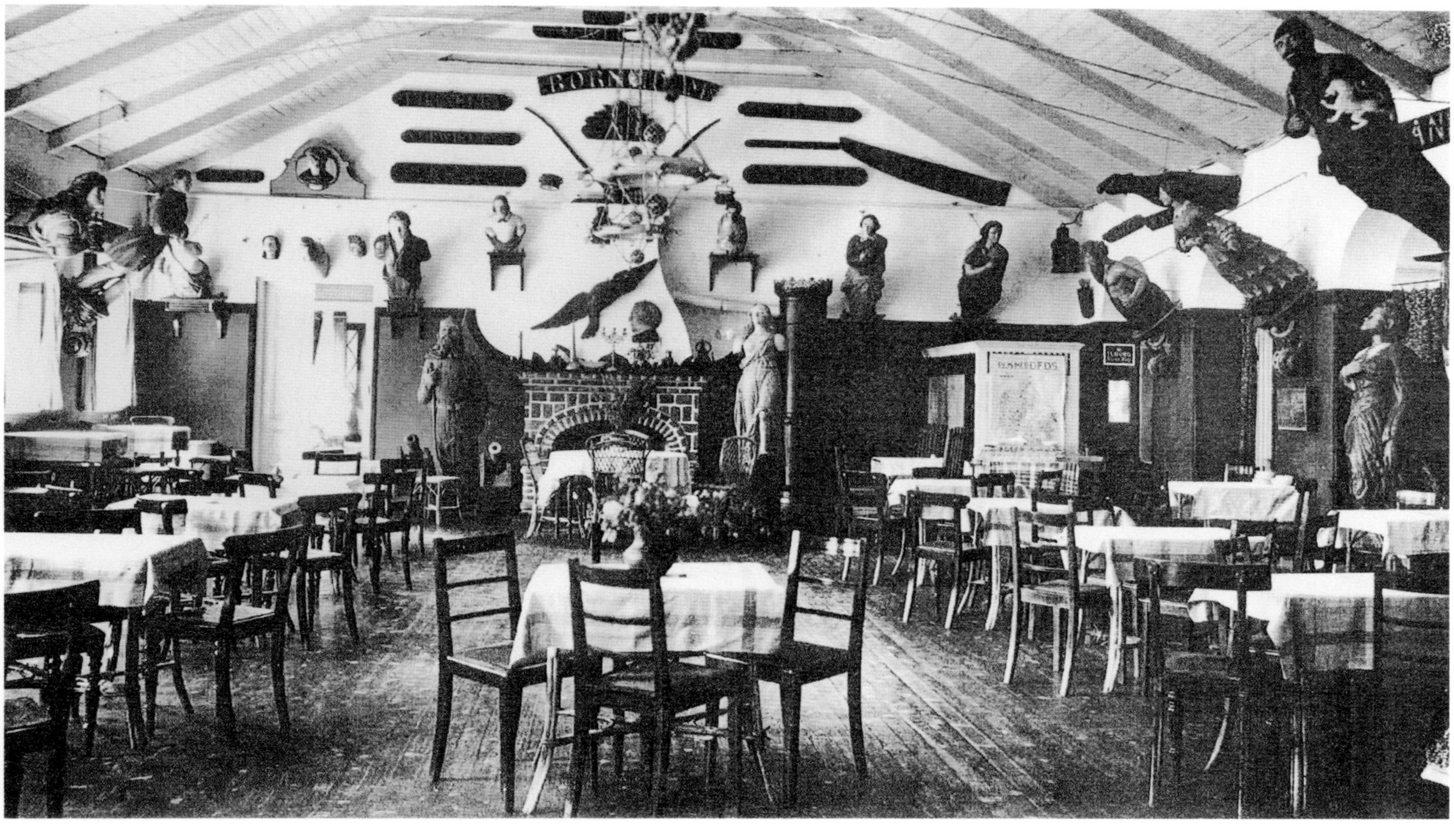

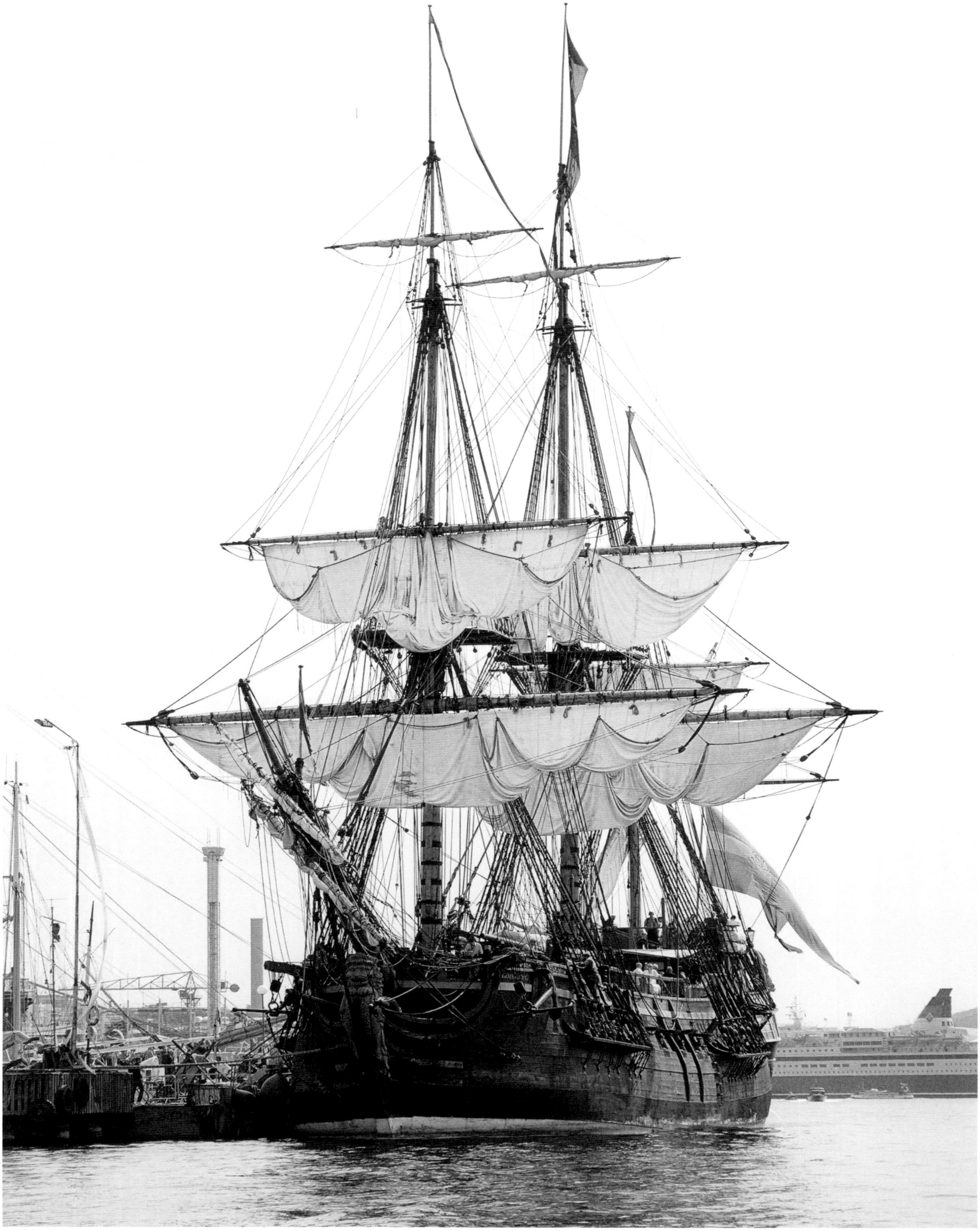

SWEDEN AND THE SEA – FROM THE 16TH CENTURY TO THE PRESENT DAY

Leos Müller

The East Indiaman *Götheborg*. Replica of the 18th century ship (built in 1995–2003 in Gothenburg).

THE RELATIONSHIP OF SWEDEN and Swedes to the sea is in many ways contradictory. On the one hand the sea is something that goes without saying, and a strong component of the Swedish identity – from summers in the west-coast province of Bohuslän, the Stockholm archipelago or Österlen in the south to shanties, films, TV series and, not least, the dream of a house with a sea view. "... Nära havet vill jag bo" (I want to live by the sea), in the words of singer Ulf Lundell – and most Swedes do actually live close to the sea. Stockholm, Gothenburg and Malmö are furthermore old ports. One explanation of this situation is of course Sweden's geography: it is a peninsula, and has one of Europe's longest coasts. Most Swedes in fact live just a few kilometres from the coast.

On the other hand the sea is in many ways invisible. Its significance is scarcely mentioned in discussions of the country's economic development. As a historian I am also surprised how absent the sea is from stories about Swedish history. Most people admittedly know of a few naval battles and major catastrophes at sea such as *Vasa*, *Kronan*, *Mars* and *Estonia*. But who is aware that Sweden had one of the world's biggest merchant fleets in the 18th century, or that Swedish shipping companies and shipyards were world leaders up until the 1980s?

The historian Jan Glete spoke about the sea's role for Sweden in two senses: the sea as a bulwark and the sea as a bridge. The first meaning refers to the nature of the sea as protection against the enemy – provided you have a navy to protect you, of course. The second one emphasises the sea's importance as a transport route, i.e. that it facilitates contact between people, migration and trade in goods. The sea route has always been the cheapest way of transporting goods, and remains so to this day. It is also often more reliable and quicker than transportation over land.

Focusing on the first meaning – the sea as a bulwark – Sweden was only able to break away from Danish dominance when it was capable of confronting the Danes at sea. In other words, the birth of the Swedish navy in 1522 also marks the beginning of independent Sweden. Gustav Vasa and his sons were well aware of the navy's role in Sweden's struggle for independence and the enormous resources invested in its development. During the Northern Seven Years' War from 1563 to 1570 the Swedish navy was probably the biggest in Europe. Gustav II Adolf and his heirs adhered to the same marine strategy and invested in the navy. The royal ship *Vasa* is an excellent illustration of this. The Baltic Sea was to serve as a transport bridge for Swedish armies but also as a bulwark for Sweden's enemies. Russia, Poland-Lithuania and the German emperor were to have an encounter on the other side of the sea. The only state in the Baltic area that used the same marine strategy was Denmark, and as is widely known the wars between the two Nordic neighbours often played out at sea. This is something that clearly differentiates Sweden and Denmark from their land-based neighbours to the east.

Sweden during its time as a great power was a markedly maritime state – with its capital in the middle of the Baltic Sea and waterways going in all directions. Lake Mälaren linked Stockholm to the province of Uppland and the Bergslagen district, whilst the transport routes

Drawing depicting the Battle of Öresund on October 29, 1658.

to the Gulf of Bothnia, Finland, the Baltic Sea and Pomerania interconnected the kingdom to the north, east and south. It was not until 1809, or actually 1814, after the loss of Finland and Swedish Pomerania, that the modern territorial Sweden developed. All at once the capital moved from the middle of the kingdom and became its eastern outpost. The proximity to Åland, where the Russians were starting to build a naval base (Bomarsund), shows how exposed Stockholm suddenly became. This exposure decreased in the middle of the century after Russia's loss in the Crimean War in 1856, and above all during the 20th century, when Finland and the Baltic states became independent. Fear of the enemy coming across the water remains to this day, however. The dread of U-boat violations well illustrates this.

But the sea not only acts as a transport route within Sweden. It has also linked the kingdom to other countries. Sweden has always been dependent on extensive trading with the outside world – more so than countries with better agriculture. This dependence has furthermore increased over recent centuries. Today's Sweden is an export-oriented global economy that is greatly dependent on sea transportation. Over 90 per cent of Swedish foreign trade still uses waterborne transportation, though this development has not been constant.

It is well known that trade and shipping over long distances played an important role during the Viking period. During the Middle Ages trading across the Baltic Sea was handled by the German Hanseatic League. Stockholm and Visby then grew into important ports. Salt, textiles and other valuable goods were imported, while iron from Bergslagen and agricultural products such as butter, cheese and fish were exported. But freight costs were high, and trading was limited. During the 16th and 17th centuries the Dutch took over the role of the Hanseatic League, and the most important markets for Swedish export goods consequently switched from the Baltic Sea (Lübeck and Gdánsk) to the North Sea (Amsterdam, London and Hamburg).

The shift took a long time, but in the mid-17th century Holland dominated both Swedish foreign trade and its transportation. This dominance can be said to be not only an expression of Holland's political and economic power within Europe but also a consequence of the common interests of Sweden and Holland. As Protestant powers, the countries were on the same side in the Thirty Years' War against the emperor. Another common enemy was Denmark. During the first half of the 17th century Holland strove to weaken the Danish dominance in the Baltic Sea. A balance of power between Sweden and Denmark was the best solution for the Dutch. Amsterdam's loans helped Sweden rearm and wage war. Payment was in the form of export products such as copper and iron bars. Between 1610 and 1650 Sweden's foreign trade increased tenfold, and the credit for this commercial revolution largely goes to Holland. The Dutch invested in Swedish ironworks and gun factories, reorganised the copper industry and copper exports, and shipped the Swedish export goods using vessels built, insured and manned in Holland. Many of the imports into the fast-growing Swedish capital, however, came from Amsterdam. We find Dutch entrepreneurs as shipbuilders at the navy's shipyards and shipowners in the field of Swedish shipping, and they were also involved in Sweden's colonial exploits in Africa (The Swedish Africa Company) and America (the colony New Sweden).

But in the middle of the century the situation began to change. Sweden's financial dependence on Holland was criticised by those in government, not least by Axel Oxenstierna, and after 1650 domestic economic policy switched to a mercantile approach. The Swedish National Board of Trade, which was to promote Swedish foreign-trade interests, was founded in 1651, and a number of measures were taken to support the development of a Swedish merchant fleet.

The second half of the 17th century was marked by several naval wars between England and Holland, and between Louis XIV's France and England. What characterises Swedish developments during the period is a marked expansion of tonnage under the Swedish flag. This growth can in part be explained by the mercantile policy, but an even more important

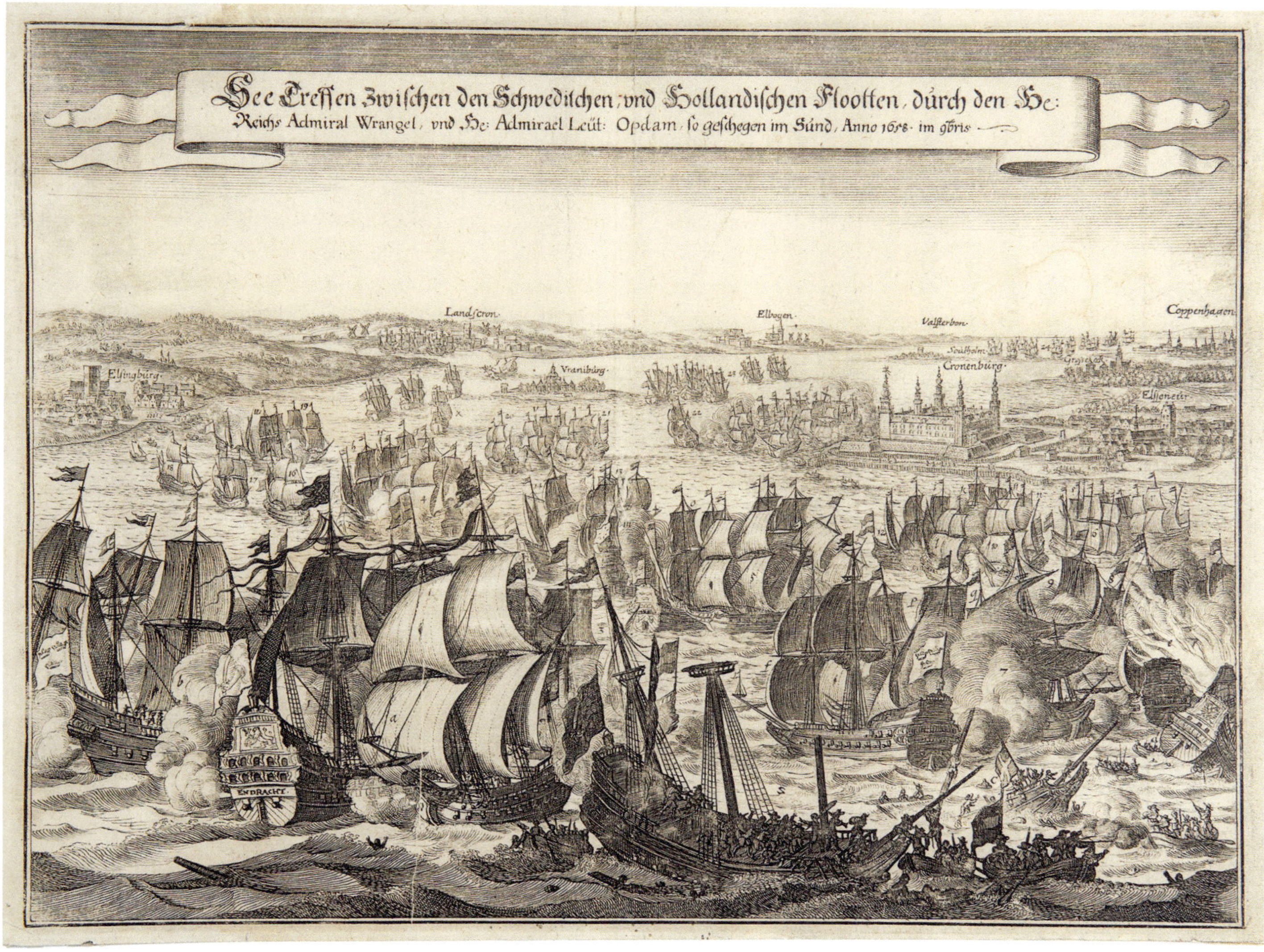
See Treffen zwischen den Schweditchen, vnd Hollandischen Flootten, dúrch den He:
Reichs Admiral Wrangel, vnd He: Admirael Leût: Opdam, so geschegen im Súnd, Anno 1658. im 9bris
Landscron
Elbogen
Valsterbon
Coppenhagen
Elsingbúrg
Vranibúrg
Soulholm
Cronenbúrg
Elsseneür
ENDRACHT

explanation is probably the fact that Sweden made itself scarce during the naval powers' conflicts. The number of Swedish vessels registered for foreign navigation rose from just under a hundred in 1660 to 750 in 1693. In the 1690s nearly all Swedish exports were transported using Swedish ships. Many vessels also started sailing outside the Baltic Sea and the North Sea. The so-called *spanienfarare*, for example, were Swedish fluyts that carried salt from the Iberian peninsula and the Mediterranean. The final decade of the 17th century can be described as the unknown golden age of Swedish merchant shipping.

The 18th century started with a long and abortive war. Not only the Baltic provinces and Sweden's position as a major power in northern Europe but also much of the merchant fleet and foreign trade were lost. After the Great Northern War the Swedish parliament started changing its shipping policy. First came the introduction of the Commodity Ordinance, which was modelled on the English Navigation Acts. The Commodity Ordinance decreed that import goods could only be carried on vessels from the product's producing country or Sweden. The ordinance was principally aimed at the Dutch, who during the 1700–1721 war took over freight travel to Sweden. The Convoy Office was also set up, and was responsible for convoying Swedish vessels outside the Baltic Sea. In practice the Convoy Office became an authority that was above all to protect Swedish vessels against highjackers from North Africa.

The new policy was successful, and the number of ships under the Swedish flag increased. Swedish ships carried Swedish export products such as iron bars, copper, tar, pitch and sawn

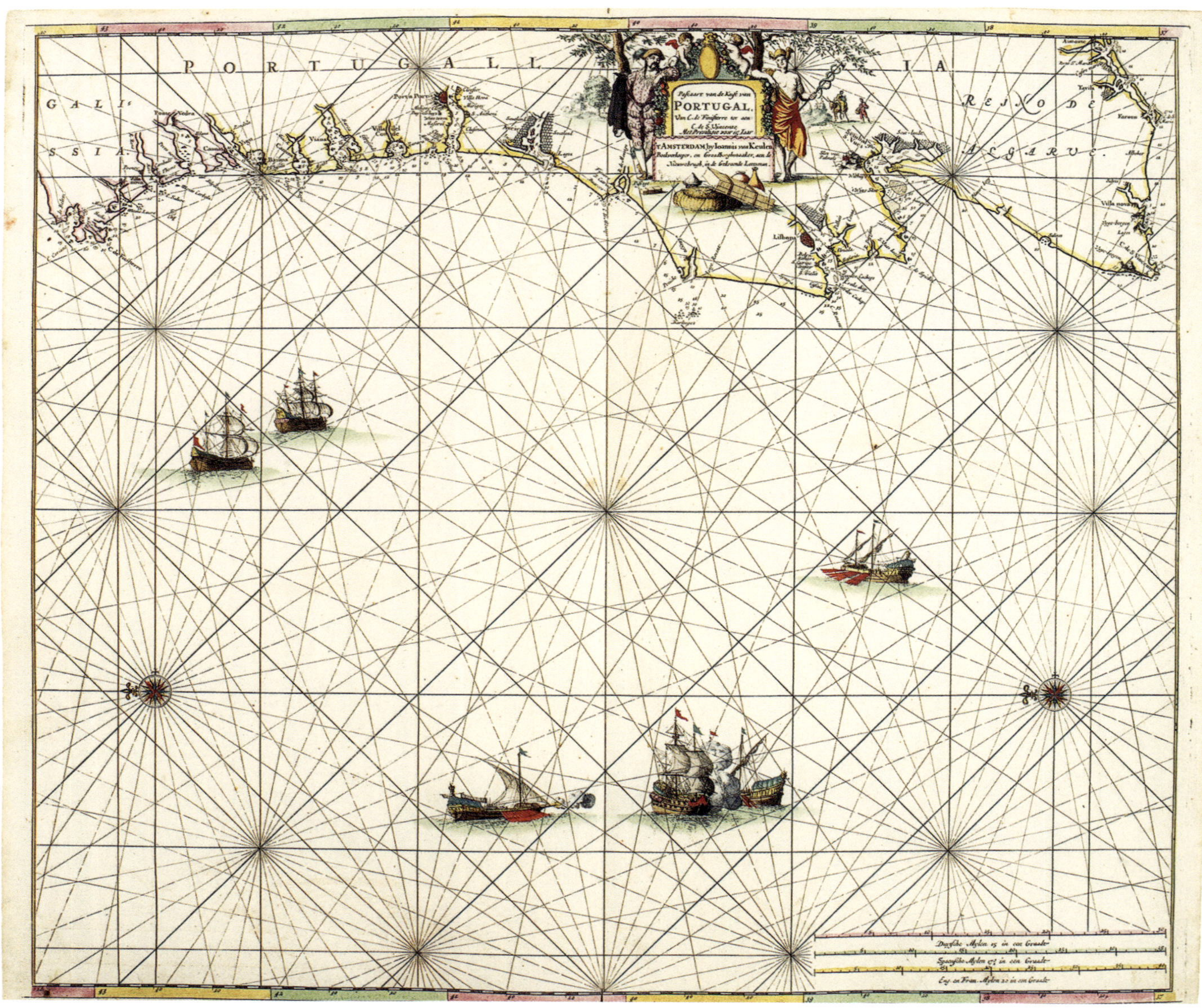

Nautical chart of Spain and Portugal from around 1700. During the 18th century, Portugal and southern Europe were important destinations for Swedish seafaring.

goods. Holland's importance as a destination market for Swedish goods was already on the decrease before 1700, and the iron bars were now exported to England. France and southern Europe, however, also became important destinations. Southern Europe's growing importance is clearly reflected in the Commodity Ordinance, the convoy protection and investments in Swedish consular presence in the Mediterranean.

But southern Europe was not just an important destination market. Nearly all the salt consumed in Sweden was sea salt from Portugal and the Mediterranean countries. It was the Dutch who back in the 16th century introduced sea salt in the Baltic region. It was ideal as a return load on Dutch fluyts, and was perfect for salting of fish and other foodstuffs. Swedish subjects became so accustomed to sea salt that nothing could replace it, and in the 1740s the west-coast county of Bohuslän saw the commencement of herring fishing, which required huge quantities of salt. Salt imports from Portugal and the Mediterranean took up nearly all the incoming tonnage from southern Europe. After grain, salt was the import requiring the greatest tonnage. It should, however, be mentioned that grain came from the Baltic countries on the other side of the Baltic Sea, whilst the salt was shipped all the way from Portugal and the Mediterranean.

As from the mid-18th century the Swedish merchant fleet began to reach out internationally in search of tramp trade. The tramp trade involved Swedish vessels seeking cargoes abroad and earning money going from port to port. The link with the transport requirements of the export trade decreased accordingly.

There are several explanations of this development, one being the low wages paid to Swedish seamen. Another – institutional – explanation is Sweden's peace treaty with North Africa, which made Swedish ships safer and meant low insurance premiums. A third explanation is that the winter months in the Baltic Sea were the low season for sailing, whilst the Mediterranean offered the opportunity to also utilise transport capacity during the winter. A further factor I wish to emphasise is Sweden's neutrality during the wars between the major powers. During the seventy years from 1750 to 1820 a number of wars played out: the Seven Years' War (1756-63), the American War of Independence (1778-83) and the French Revolutionary and Napoleonic Wars (1793-1815). Wars were waged for a total of 34 years. Up to 1806 Sweden more or less successfully steered clear of the wars. Under a neutral flag Swedish vessels were able to transport their own export goods to the warring parties, and act as tramp vessels for foreigners. It was risky but also highly profitable work. Maritime statistics clearly show how Swedish activity increased during the wars, whilst it stagnated or decreased after they had ended. It is also evident that Swedish ships sailed far beyond the Baltic Sea and the North Sea, and even beyond the Mediterranean. During the 1780s and 1790s we see many vessels under the Swedish flag carrying cargo between the West Indies, the USA and Europe. There were also Swedish ships carrying cargo on the Indian Ocean. In around 1785, following a boom during the American War of Independence, Sweden's international fleet was estimated at over 1,200 vessels, making it the fifth-biggest in Europe, after England, France, Holland and Denmark.

And where were the vessels registered? Stockholm was the foremost maritime city, accounting for about half the registered tonnage in the middle of the century. But the capital's role diminished over time. Finnish shipping saw major expansion after 1765, and Swedish Pomerania assumed a prominent role as a maritime province right at the end of the century. Gothenburg, with its links to the North Sea and the Atlantic, burgeoned as Sweden's second maritime city and a port for herring export. It was of course also important that the city was home to the headquarters of the Swedish East India Company.

The long French Revolutionary and Napoleonic Wars mark the peak of Sweden's early-modern period of shipping. This successful period came to an end with the catastrophic Finnish War, the loss of Finland and the change of regime in 1809. During the 19th century shipping played a role different from that during the 18th century. Tonnage continued to increase, but the operating radius of vessels under the Swedish flag shrank. The Mediterranean as a destination now largely disappeared. Once again Swedish ships above all transported Swedish export goods and imported whatever was needed at home. This meant that vessels employed for international shipping chiefly sailed to Great Britain and Germany, and transported iron, steel, sawn goods and oats. The latter became Sweden's major export product. For several decades after the Napoleonic Wars the USA replaced Great Britain as the most important market for Swedish iron bars, with exports going in Swedish boats. Swedish vessels also plied their trade in Brazil and Argentina, where sugar, coffee and hides were bought.

The brig *Felix* of Norrköping, piloted by Captain O.M. Royter in 1861. The ship is depicted with all sails set. At the fore is the large union flag ("the herring salad"). From the stern flies a Swedish flag with the union mark.

In Sweden the transition from sail to steam that constituted the major shift in shipping during the 19th century was slow. Whilst Sweden's foreign trade – above all export of sawn goods, pulp and paper – saw considerable expansion, export products were transported on foreign ships. Swedish shipping increased far more slowly than that of Norway, for example, and as late as in 1890 it was still dominated by sailing vessels. But in around 1900 a new, golden age of Swedish navigation started.

There are certain similarities with the situation at the beginning of the 18th century. The state's politics, which were now protectionist, once again played an important role. At around the turn of the 20th century Sweden's economy expanded and the country's protectionist policies required export goods – everything from ore and paper to new engineering products from LM Ericsson, SKF and Separator – to be shipped in Swedish boats. The vessels under the Swedish flag were needed to transport export products to foreign markets, not to carry cargo for foreigners. But starting up new shipping companies called for major resources. The small, family-owned shipping companies in the smaller cities that had dominated shipping in the 19th century did not survive the transition. Foreign navigation now disappeared from previously large ports such as Gävle. The new large-scale shipping business was now concentrated on Stockholm and Gothenburg, and developed from two family companies: Axel Johnson's shipping companies, based in Stockholm, and Dan Broström's shipping companies in Gothenburg. Johnson's business operations are today primarily associated with the shipping company Nordstjernan, which established a number of transatlantic lines. Broström's business is chiefly associated with the shipping company Ostasiat (Swedish East Asia Company), which plied routes to Southeast Asia, China and Japan. Typical of the Swedish liner shipping companies were their modern, motorised vessels built in Sweden.

The expansion of navigation around 1900 is also reminiscent of the 18th century in another respect. As a neutral and alliance-free country, Sweden avoided the consequences of the outbreak of World War I, and was able to take market shares both during and after the war. By 1925 the Swedish merchant fleet boasted a gross tonnage of 1.2 million tonnes, and was the world's seventh-biggest fleet. This positive development continued after World War II, when the European economies started to grow and there was a big demand for oil. Swedish shipping companies to a large extent focused on transportation of oil, and the shipyards in Gothenburg, Malmö, Landskrona and Uddevalla supplied the tankers. But now it was once again all about tramp trade, and the interests of the export industry and shipping went their different ways. At the time of the oil crises in the 1970s Swedish politicians were most worried about the shipyards and shipyard employees – not about the shipping companies. The exclusive focus on oil transportation also proved to be a short-sighted and unsustainable strategy.

More than in the 1970s or 80s, Sweden is now a global economy, and more dependent than ever before on the export industry. Apart from cars, telecoms and engineering products, there are still large quantities of paper, pulp, steel products and ore to be transported by boat to foreign consumers. But nearly all transportation uses vessels owned, chartered and manned abroad. Little remains of the Swedish shipping industry, though the sea is still there as a bridge to the rest of the world, and Sweden is as ever just as dependent on it.

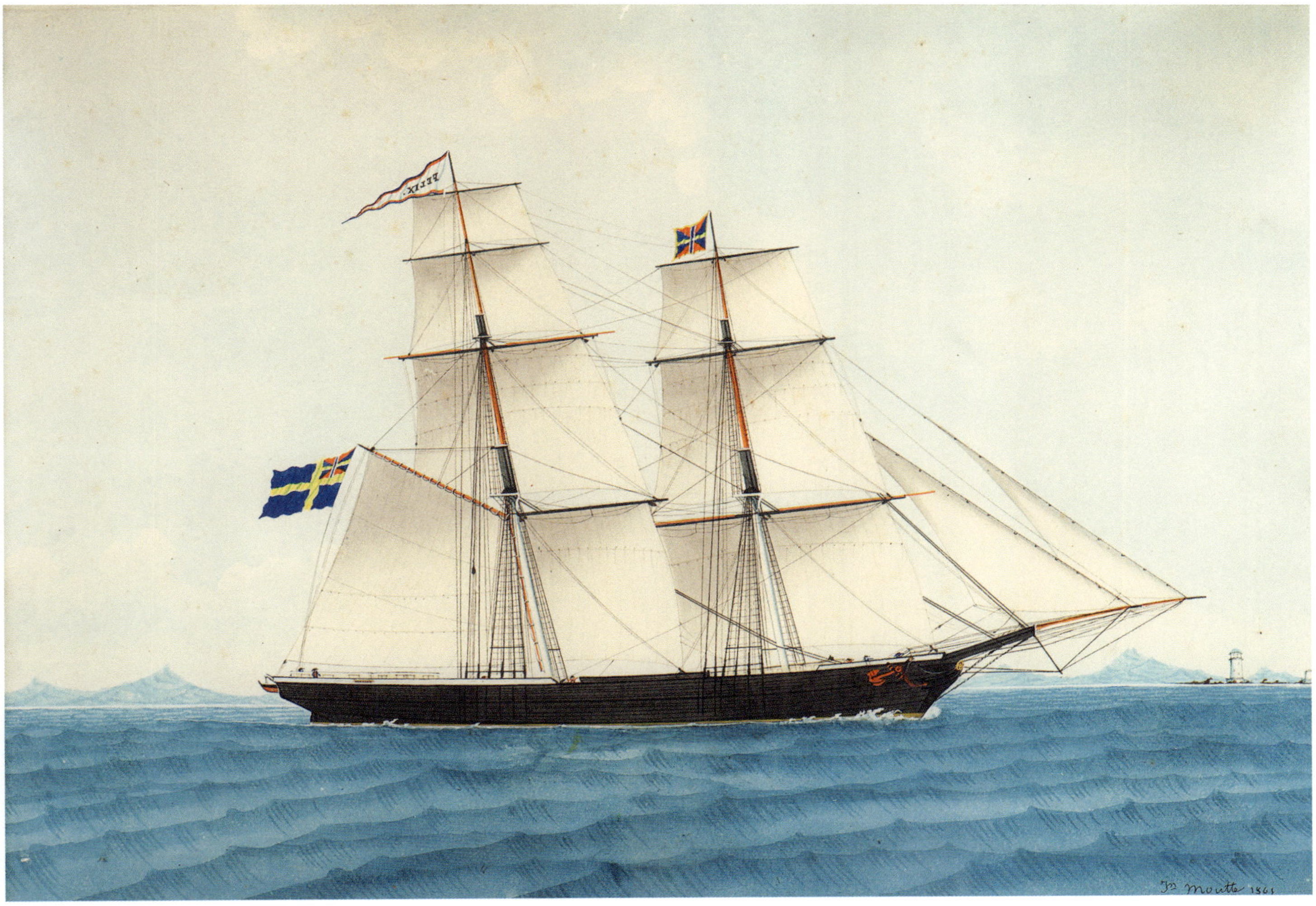

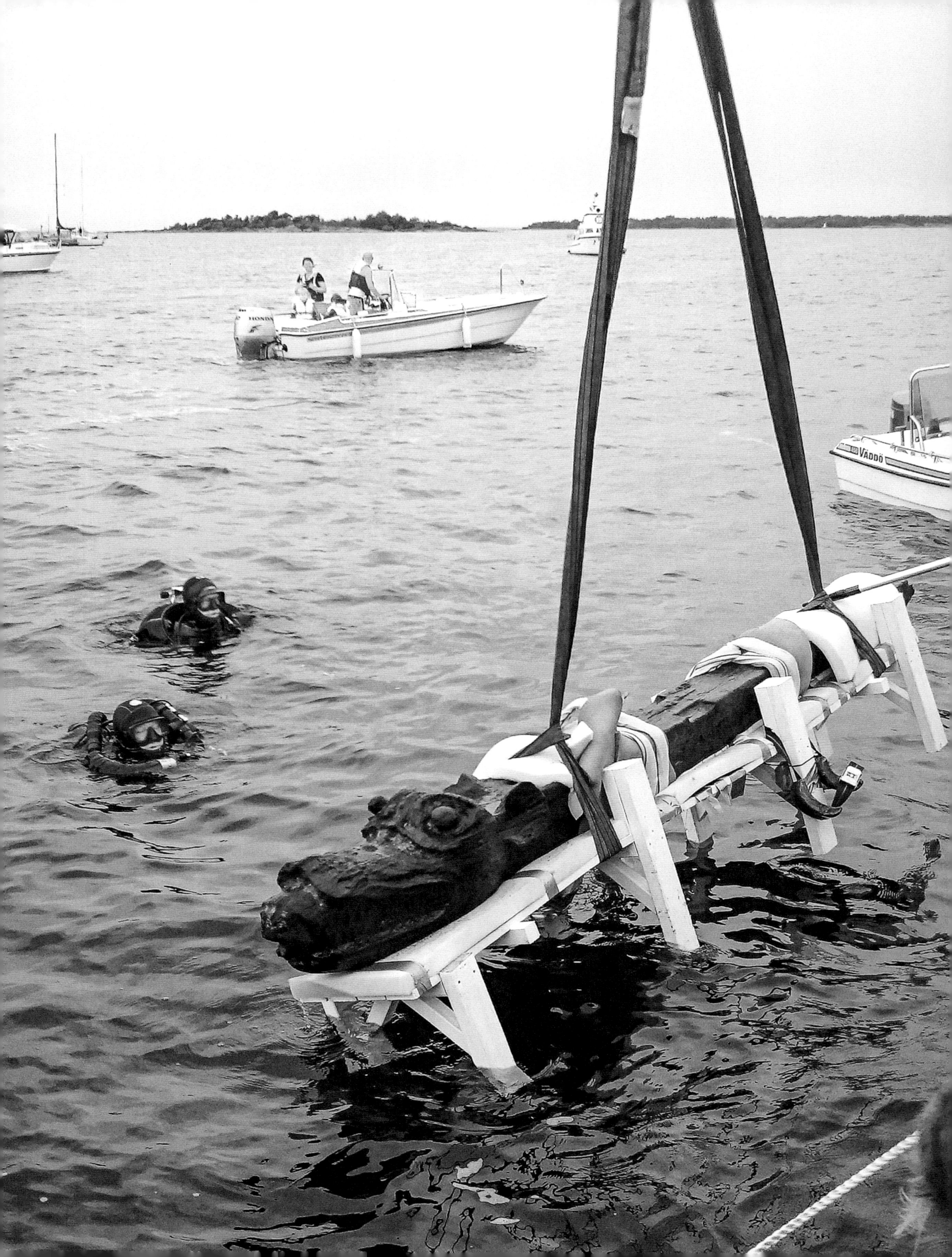
VÄDDÖ

THE DANISH MONSTER

Johan Rönnby

THE RECOVERY OF THE LARGE wooden figure down in the murky green of the sea bed has been in progress for several days. Marine archaeologist divers have cautiously turned the beam in front of the stem of which the sculpture forms part. A framework has been created and the heavy timber has been carefully attached to it with rope and straps. The actual lifting is being performed using powerful winches on board the catamarans *Svanen* (The Swan) and *Gnarly*. The working boats are moored side by side above the 15th-century wreck. Around them thirty or so leisure boats and hired tour boats holding press and excited observers have gathered. Under the watchful eyes of the accompanying divers, the large beam is slowly approaching the surface. The spectacle when it finally comes up is extraordinary. A long, gaping orifice with pointed teeth, an almost crocodile-shaped head with ears and two funny little wooden horns. A human being is sticking out of the beast's mouth, semi-devoured by the monster and seemingly shrieking with fear. It is over five hundred years since the wooden figurehead last saw the light of day. When the ship she belonged to, the *Griffen*, descended to the sea bed Leonardo da Vinci was in full swing inventing new weapons of war and painting his masterpieces, and Machiavelli was busy advising the princes of his time on the tricky art of exercising power.

Left: The salvaging of *Griffen* off the coast of Ronneby in 2015 attracted great interest.

Below: The so-called *Brandskogsskeppet* ('Brandskog ship') is a Bronze Age rock carving more than four metres in length outside Enköping in the Swedish province of Uppland.

Following pages: After more than 500 years at the bottom of the sea, the figurehead from the wreck of the *Griffen* above resurfaced.

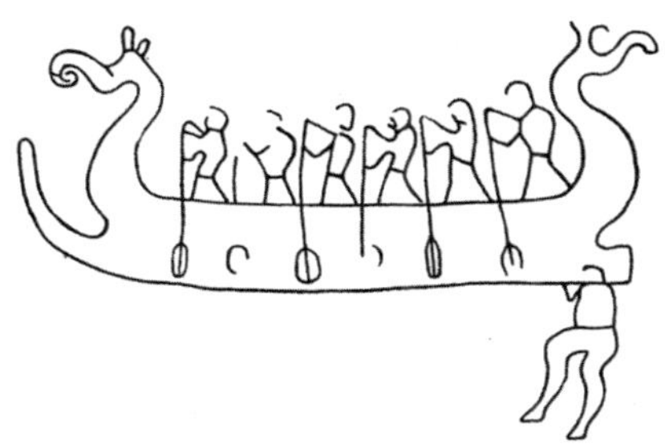

There is a very long tradition of placing some kind of figure on the bows of ships and boats. Examples from as early as the 4th millennium BC are to be found in the rock art of the Nile Valley, in what is now the Nubian Desert. They are drawings that probably represent large reed boats with animals' heads on their bows.[1] In the Nordic countries there is a wealth of examples of Bronze Age carvings depicting various types of stem figure. An unusually expressive example is the so-called Brandskogsskeppet (Brandskogen Ship) from the province of Uppland. Stem figures are to be seen here that could perhaps be interpreted as being unusually long-nosed elks. Ships on which some form of animal or dragon figure was placed on the bow are also frequent during the Iron Age and what we usually call the Viking Age. In 17th- and 18th-century Europe there was a predilection for placing gilt lions on the bows of big warships. The figureheads of recent centuries have often been mythical creatures or gods of antiquity, and during the latter half of the 19th century there was a preference for more or less bare-breasted female figures. Art-history analyses can highlight the artists' traditions and stylistic contexts of which these sculptures and figures form part. They can also be seen as an expression of mentality, ideology and communication, thus constituting a kind of material language – something important during periods of low literacy levels.

The figures on vessels are of course also connected with the ship in question. On a very general level one can maybe see the use of such figures as a kind of will to animate and anthropomorphise the things we use and have around us. This perhaps in particular applies to objects we deem to be important to us – things with which we are closely involved. Ships are often described as the most complex artefacts that people created prior to industrialisation,

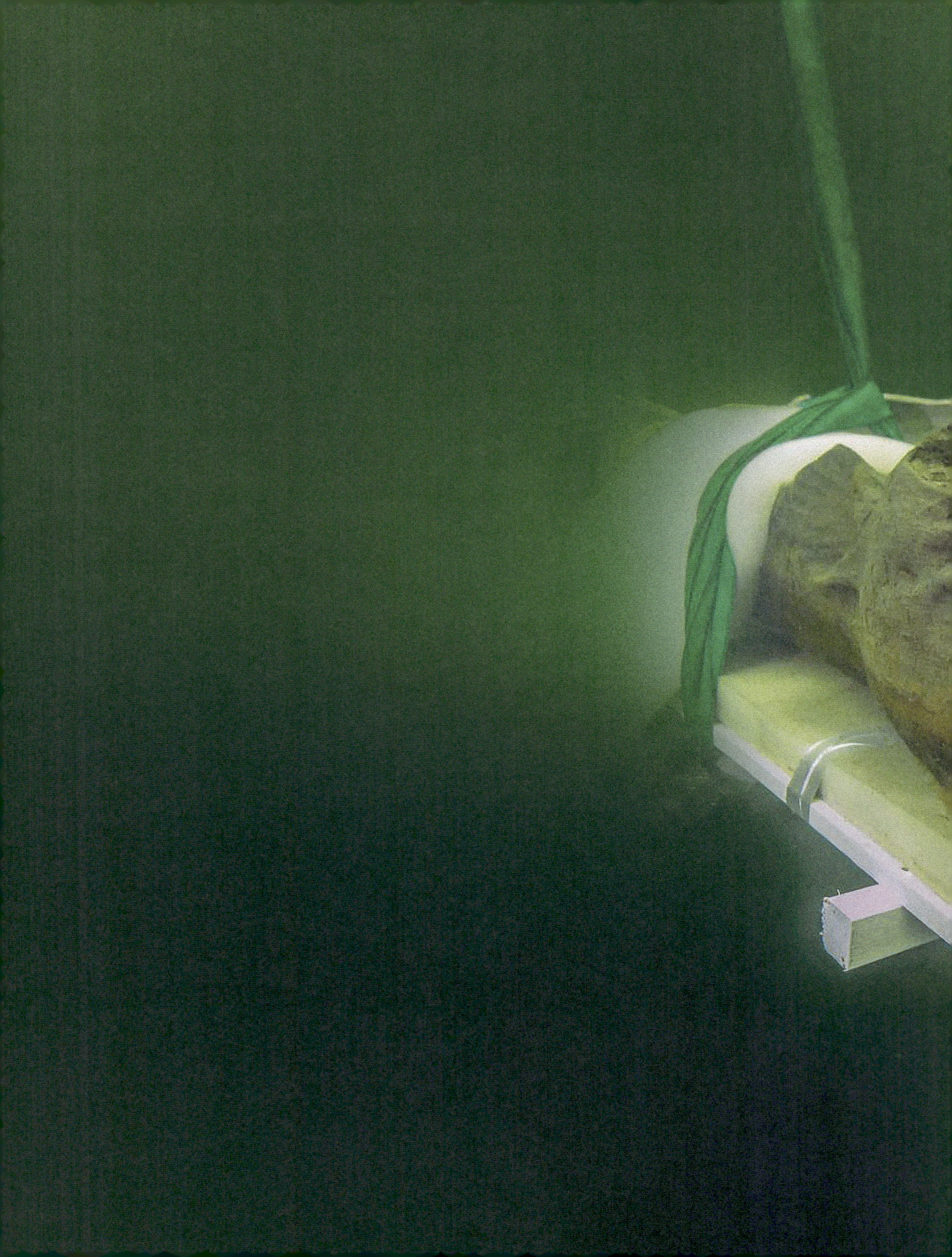

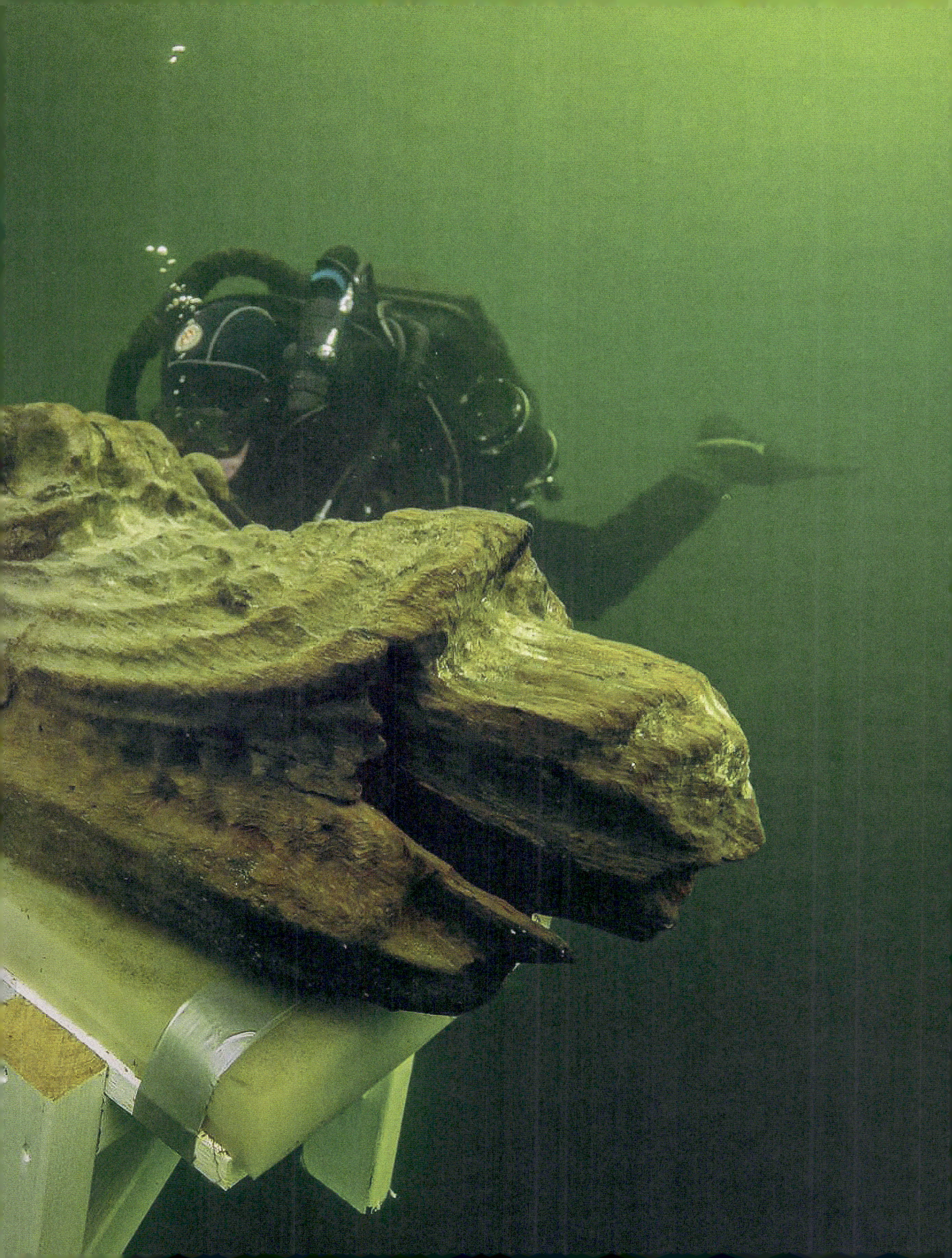

han den selbē rubin nit gesehen/ vnd meint man es sei in allerwelt kein so gůtter/ so schöner oder so grosser· wann er ist so gar rot vnd glestet dz man in gar ferre sicht· Vnd meinet man er sey des landes gröster schatze· Vnnd so das land ein künig erwelet so geben sy im den stein zů getzierde· vnd fürent in durch die hauptstat des lands· vnd so dz volck den stein ob im sicht so synt sy im gehorsam vnd haltent in für ein künig· Den steine treit er ann einer kronen· Vnd der gros Can/ der da meint herre zesyn über alle welt der hette gern den selben stein gekauft· aber er mocht jm nit werdē weď vm̄ trawung/ vm̄ gůt/ noch vm̄ bette· Auch ist der künig von macamerō ein rechter richter in synem lande· vnd laszet nyeman kein vnrecht thůn/ vn̄ züicht man sicher durch syn land· vnd hat syn land by tusent mylē weit

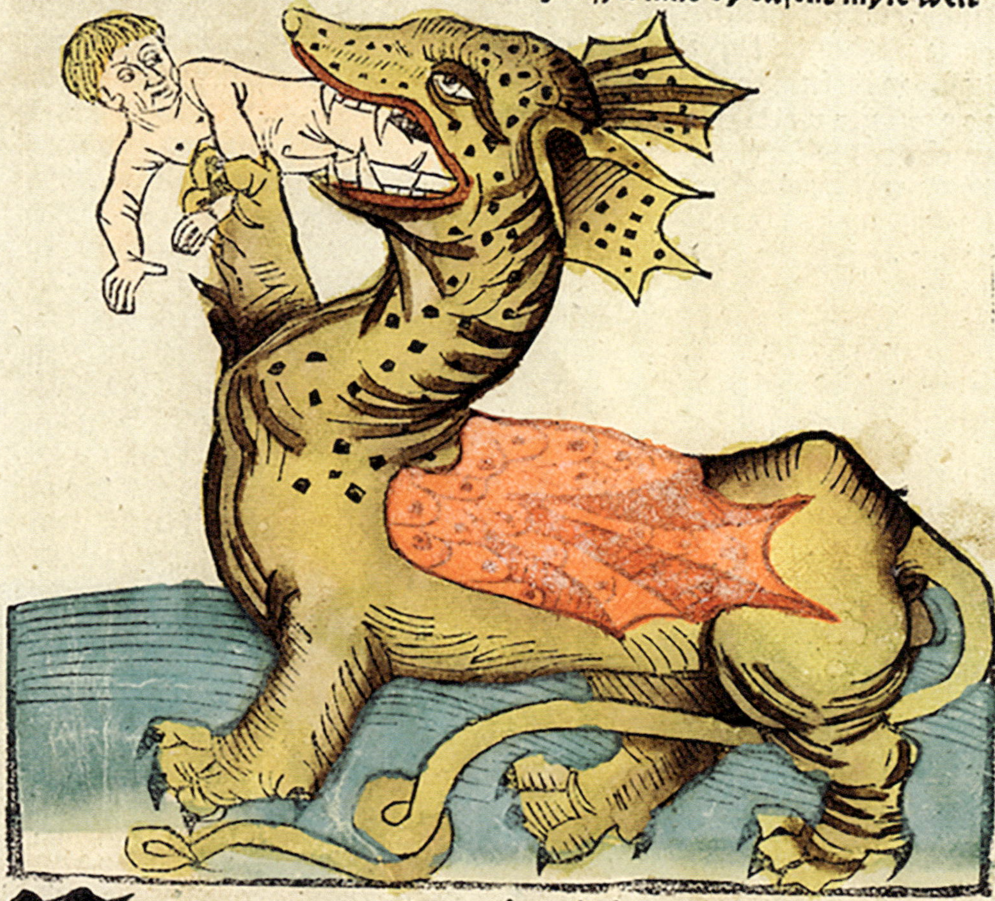

Das·xvj·Capitel

Dfylen ist auch ein wyt land· vnd hat wol acht hundert myl begriffen· vnd züicht man von macameron dar Aber fremde leüt mügent nit wol durch dasselbe land komē on der lantleüt hilff

and they have without a doubt played a major role for people over the course of history and in widely differing contexts. One way of clarifying the boat's role as a joint player – a kind of individual in its own right – is to name it, and another is to give it a 'face' in the form of a figure on the bow. A related variant is giving the actual boat eyes – something already practised in the Mediterranean area in ancient times.

It is interesting that this human approach to things can also be seen in the light of a theoretical and analytical perspective over the course of history, emphasising the interaction between man and material objects. It is, after all, we who create our various objects, but we ourselves are then also affected by them. This material dialectic does not, however, mean that the interaction needs to be seen as a symmetrical relationship. Objects and technology are part of society and have major consequences as regards the way it is organised, developed and changed. But it is people and groups of people in various historically given situations who are the actors. This relationship also to a great extent applies to the issue of warships' importance to societal changes at the beginning of the early modern period.[2]

The shipwreck from which the extraordinary monster figure has been recovered lies off the island of Stora Ekön near Ronneby in Blekinge. Bearing in mind that the water is relatively shallow there, and that the location has long been known as a good place of anchorage, the local population and fishermen have probably been aware over the centuries that an old vessel has been lying there on the sea bed. It was not, however, until 1971 that the wreck was discovered – by Doppingarna Diving Club from Ronneby. Since then it has been investigated on several occasions.[3] Using written sources, the wreck has also been identified as King John of Denmark's carvel-built ship *Griffen* or *Gribshunden* (The Griffin Hound), which was built in the early 1480s and sank in 1495 at this very location following a fire on board when the ship was at anchor.[4]

The late 15th century was a time of change as regards both the power situation in Europe and vessel design. A new feature of shipbuilding technique at this time was the practice of laying planks edge to edge using the so-called carvel planking system, instead of the overlapping clinker construction.[5] Changes were also made to the rigging, including use of several masts, as well as combining square sails on the bow masts with a lateen sail on the

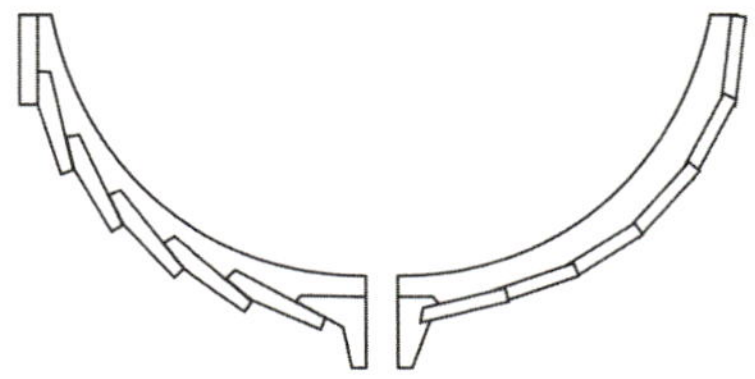

Above: Carvel-built or carvel planking is a method of boat building where hull planks are laid edge to edge and fastened to the boat's frame. The planks do not overlap, as they do with clinker-built boats (to the left in the illustration).

Left: A 1484 edition of the book *The Travels of Sir John Mandeville* includes an illustration a griffin devouring a human. The picture shows that in the late medieval period a griffin was not necessarily depicted with the head of an eagle.

Below: The Greek Siren Vase in the British Museum dates from around 480 BC. Odysseus' ship is depicted as having eyes, which was common in antiquity.

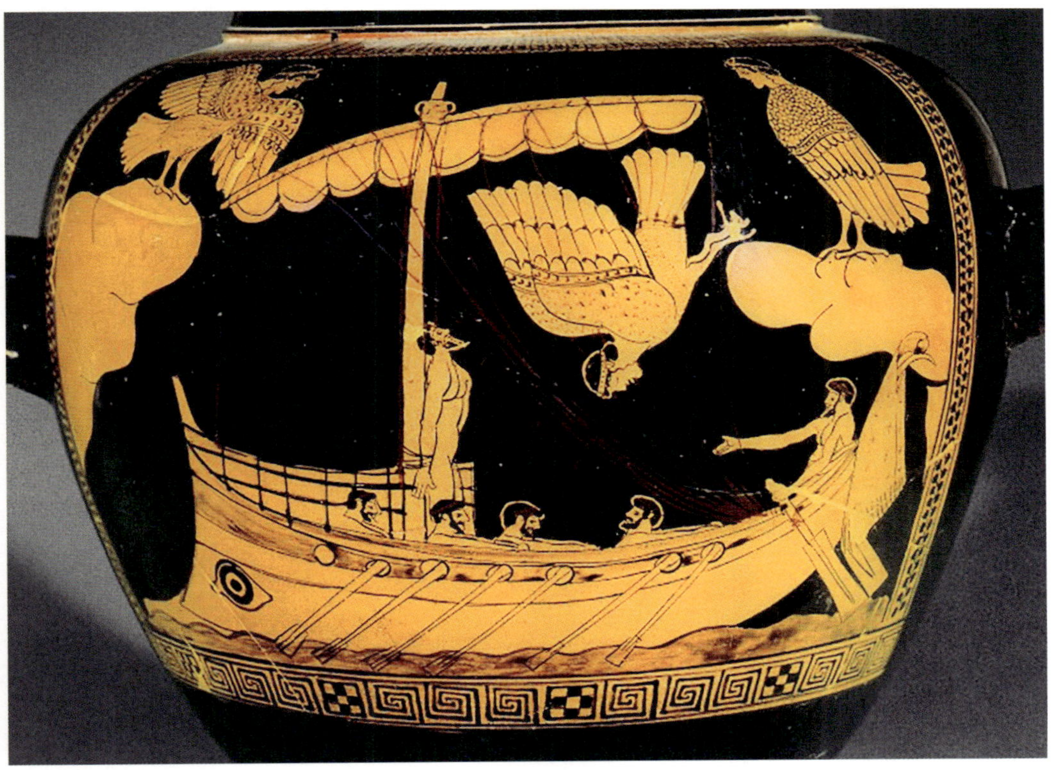

mizzenmast at the stern. Another new feature on board was the increasingly frequent use of gunpowder and cannon. During the 15th century on-board artillery still just comprised relatively small cannon that were fired from the gunwale or the topmasts. At the end of the century, however, one finds the first evidence of gunports at the side of the ship. Whether *Griffen* benefited from this new martial feature is uncertain.

The development of the new ships is closely linked to a general change in Europe involving the establishment of powerful new monarchies. These states needed large ships for defence and conquests, but also because of the struggle for resources and the possibility of bringing home riches from a world that through the geographical discoveries of the period was becoming much bigger.[6] The new carvel-built ships were a tool for the realisation of these ambitions, but developments in the art of shipbuilding were also a driver of change. The new shipbuilding and the need for ships, cannon and fleets required resources and a new organisation, and this led to a kind of arms-race spiral.[7]

The latter half of the 15th century was also a bewildering and fairly confused period of Scandinavian history. The Kalmar Union, through which Denmark and Sweden were unified, was still a reality, even though for long periods domestic regents ruled the Swedish area. The loss of *Griffen* in 1495 is directly linked to this power struggle, as in the early summer of that year King John was on his way to Kalmar to negotiate with Sten Sture the Elder in an attempt to re-establish the union. The Swedish magnate had greatly delayed John's accession to the throne in Sweden, which had long since been decided and agreed on. Because of poor weather the king and his fleet had sought protection at a known anchorage near Ronneby, but fire broke out on board the flagship *Griffen*, and the sources state that many noblemen, gold, silver and the king's 'best depository' were lost when John's big carvel-built ship went down.[8]

At the end of the Middle Ages, apart from goods and farms, powerful magnates and bishops often also owned a number of vessels, both large and small. Those who had the biggest resources in the Nordic countries during the latter half of the 15th century were the Danish Union kings of the House of Oldenburg, and it was they who had the most up-to-date naval vessels. King John's father Christian I created his own fleet in the middle of this century, and John was then able to expand it. The Iberian peninsula is deemed to have been a core area for the development of ship technology towards the end of the Middle Ages. It was here,

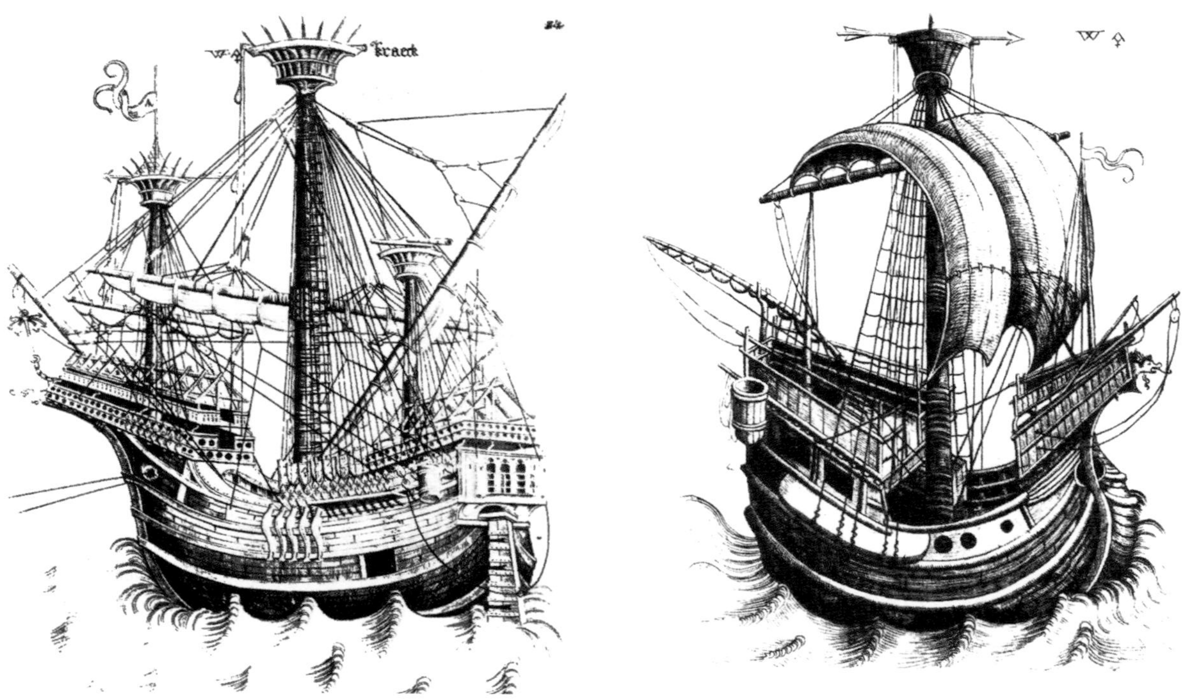

Two drawings by the anonymous Flemish master W.A. from around 1470 – a large ship with small cannons including on the top platform and a slightly rounder smaller vessel. Both have wooden sculptures at the front of the forecastle.

partly in conjunction with Prince Henry the Navigator's known voyages of discovery, that efficient new ships were developed that combined southern and northern building traditions. The new carvel-planking technique was part of this development. An interesting circumstance in this context is that in the mid-15th century Denmark is said to have had relatively strong contacts with Portugal, and that Danes in various contexts are said to have taken part in a number of maritime expeditions to distant destinations.[9] The state-of-the-art maritime technology of the time must thus have been well known in Denmark, and King John must have been familiar with it when he was developing his father's fleet.

The wreck off Ekön is the world's best-kept example of the new type of vessel, which was to be developed shortly afterwards, resulting in big Northern European naval ships, famous examples being *The Mary Rose* (1545), *Mars* (1564), *Vasa* (1628) and *Kronan* (1676). The vessel is contemporaneous with Christopher Columbus's renowned ship *La Santa María* (The Saint Mary) of 1492, though it is bigger. In design terms it also has much in common with what is to be seen of the well-known Mataró model – as it was known – from mid- 15th-century Catalonia.

The Mataró model has a sculpture similar to Griffen's right at the front, and similar figures are also depicted at the bows of many contemporary representations of ships. The wooden sculpture on the wreck off Ekön projected below the forecastle. It was part of the psychological warfare that was a feature of the boarding combat that still dominated naval warfare at this time. In an interesting way this also links up symbolically and iconograpically with the 'monster pictures' to be seen in medievall churches. Monsters in various terrifying guises often crop up in medieval Christian rhetoric as an allusion to the evil in our existence. The motif of falling into an infernal orifice and being swallowed alive was an indicator of what awaited sinners – though the monsters could also allude to courage, to the association between heaven and earth and even to Christ. On the church walls both the sea monster Leviathan and the 'whale' that swallows Jonas are depicted. Imaginative depictions of sea monsters are also to be found on the maps of unknown new waters that were beginning to be drawn in the late Middle Ages.[10]

As regards the ship's name, one might perhaps have expected a clear representation of a griffin with a snake's head right at the front of the ship. But this was obviously not the case.

Below, left: French ship with a prominent monster at its bow. From a translation of *The Aeneid* by Octavien de Saint-Gelais from 1500. The wood used to build *Griffen* comes from northern France, and it cannot be ruled out that the ship was also built on the French Atlantic coast.

Below, right: Ship with a dog-like monster at the bow from a description of a pilgrimage to Jerusalem made by Conrad Grünenberg in 1486.

Following pages: Bernt Notke's sculpture of Saint George and the Dragon in Stockholm's Storkyrkan was erected by Sten Sture to commemorate his victory over King John's father, Christian I, at Brunkeberg in 1471. While the Stockholm dragon is depicted with more typically Renaissance features, it belongs to the same historical context as the sculpture of *Griffen.*

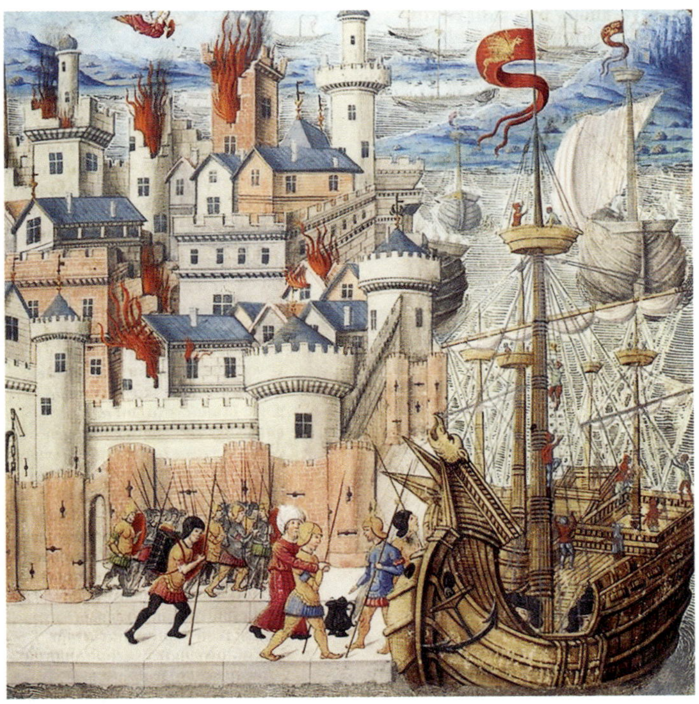

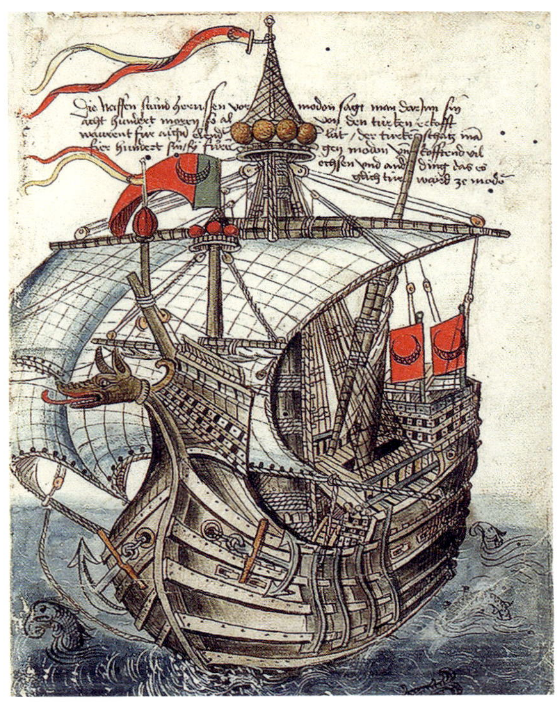

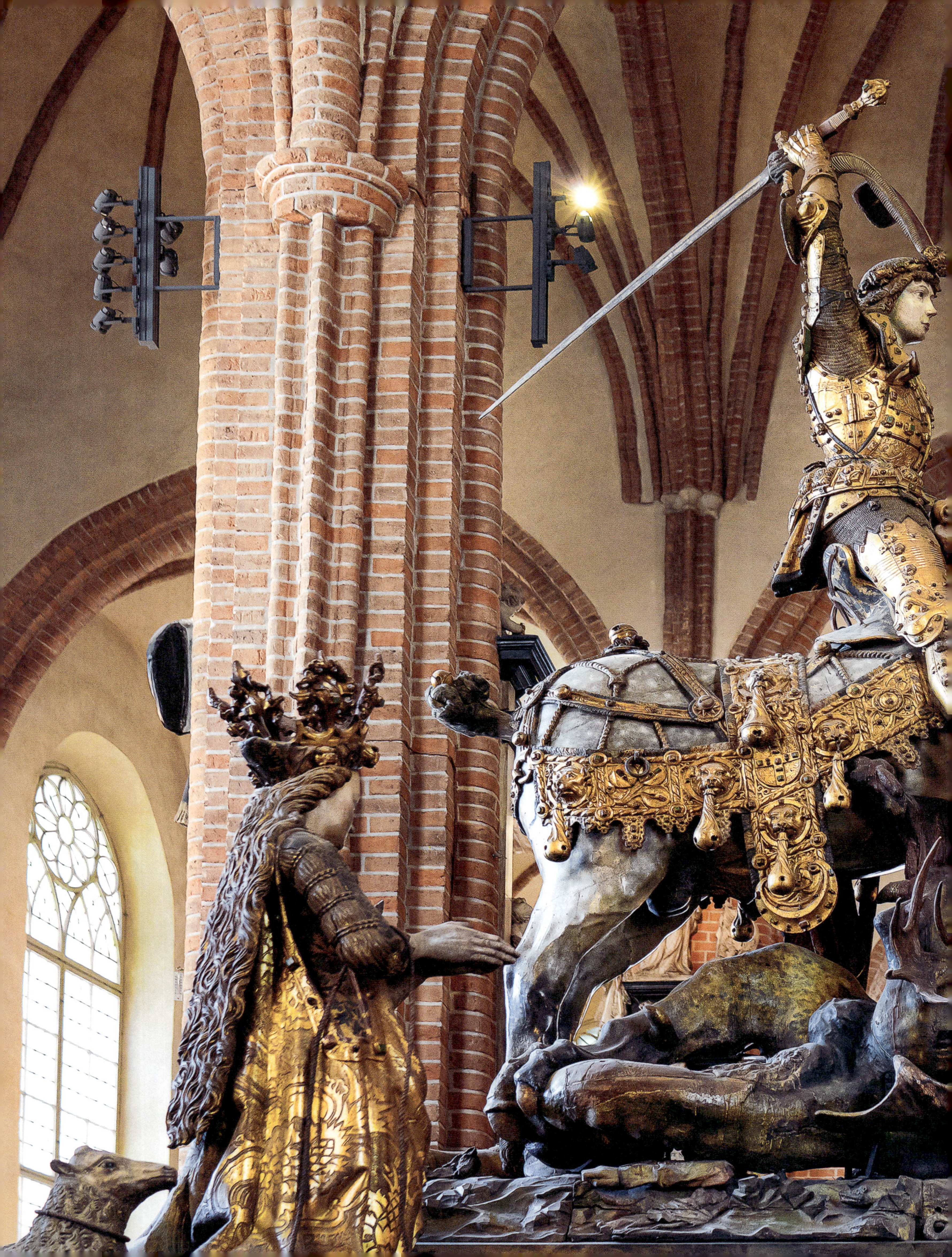

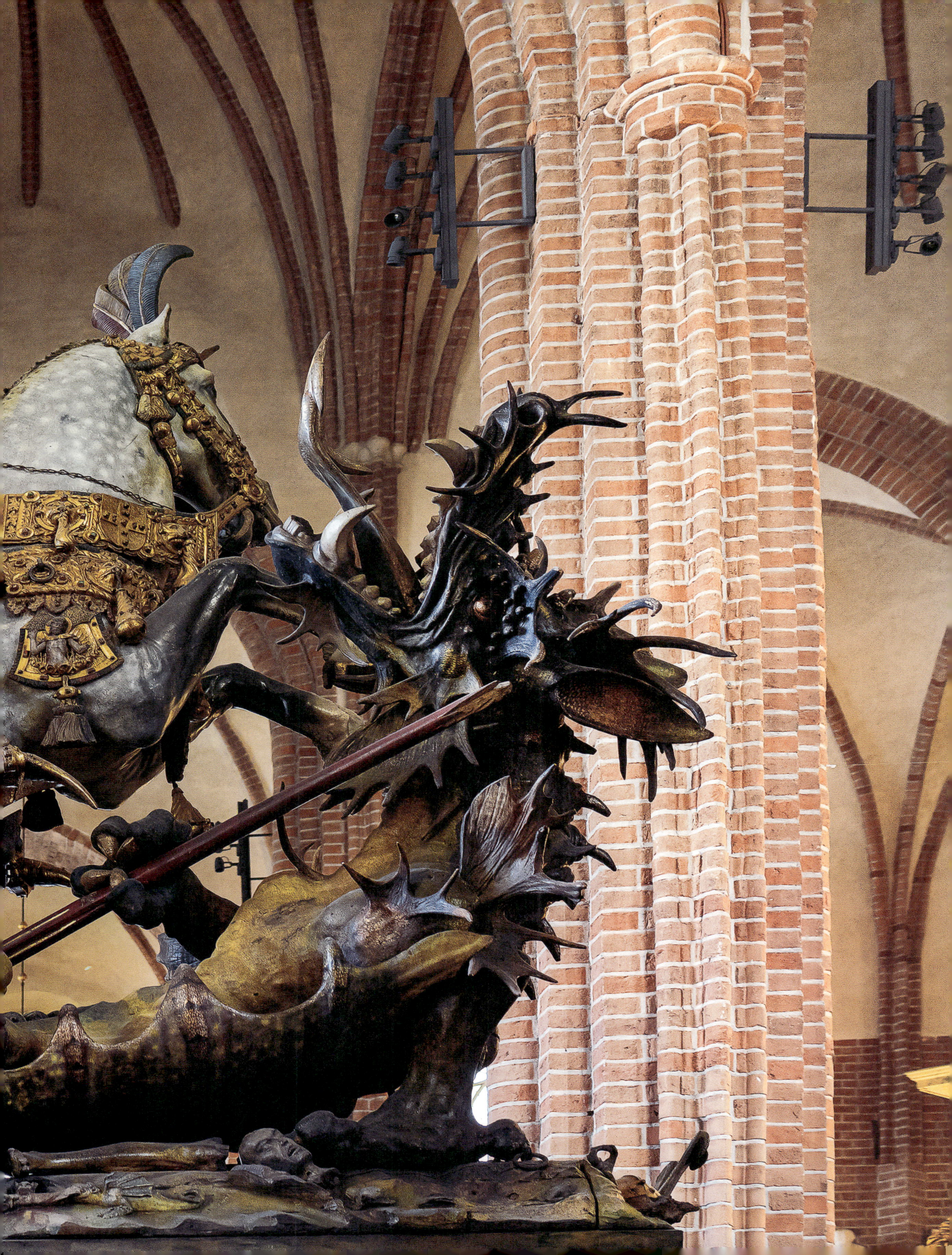

However, one should maybe not expect representations of griffins in the late Middle Ages to follow a strict pattern. An interesting example of this is a depiction of a griffin swallowing a man in an edition of *The Travels of Sir John Mandeville* of 1484.[11] The book was very popular in the late Middle Ages – Columbus, for example, took a copy with him on his travels. It is also known that the work was translated into Danish in the 1440s (see page 202).[12] It is safe to assume that there were copies of the book at the Danish court when *Griffen* was built in the early 1480s, and the illustrations in it may have inspired both King John and a sculptor! The picture in Mandeville also clearly shows that a mediaeval depiction of a griffin did not necessarily have a snake's head. One should maybe simply interpret the second name given to the ship in the sources – *Gribshunden* (The Griffin Hound) – as indicating that the figure was not particularly griffin-like and on distance actually have a rather doglike apperence? A alternative explanation could be that 'hound' here shall be seen more like 'best' or 'monster'.

The long, grimacing face is otherwise probably most like that of a dragon. The dragon was a popular motif in several contexts in the Middle Ages – perhaps above all, though, in representations of St George and the Dragon. With the Fall of Constantinople in 1452 the crusade ideology in Europe intensified. Nearly all European princes subsequently used a kind of crusading rhetoric in order to assert various claims. This also applied to the Danish kings, who saw an opportunity to direct these ambitions not only eastwards against the Turks and troublesome Swedes but also against sinners and infidels to the west. Icelandic and Greenlandic waters were plied by Danish ships, and maybe decades before Columbus they went even further west than he did.[13] And it was moreover to St George the Dragon Slayer that one prayed before going to war, and it was his cross that was often carried as a banner during battle.

Another context in which both dragons and griffins crop up during this period is on coats of arms. The shield of King John's father, Christian I, in a 16th-century portrait also shows amongst the other panels a Wendish monster, a heraldic relative to the griffen, a "lindworm". A coat of arms for John's son Christian II in St Bavo's Cathedral in Ghent is of the same type and composition, though the fantasy monster here has been given a head that is undeniably rather hound-like!

The 'monster figure' on *Griffen/Gribshunden* is contemporaneous with Bernt Notke's sculpture *St George and the Dragon* in Stockholm Cathedral (Storkyrkan). Despite the more Renaissance-like modernity of the Stockholm dragon's execution, there are obvious similarities in terms of appearance.[14] The monument in Stockholm Cathedral was erected by Sten Sture the Elder at the end of the 1480s to commemorate his victory over King John's father Christian I at Brunkeberg in 1471. One assumes that the bestial dragon represents the Dane. It is thus the very same Sten Sture who had the statue installed in Stockholm Cathedral that King John was to encounter in Kalmar at the time of the shipwreck off Stora Ekön.

John's eventual coronation under the Kalmar Union in 1497 took place in Stockholm Cathedral next to the Swedish victory sculpture. His own monster figure had then already lain on the sea bed off Ronneby for two years. But it is now no longer there!

Right: King Christian I and Queen Dorothea in a 16th-century portrait at the National Museum of Denmark in Copenhagen. The kings coats of arms show the Danish "leopard lion," the Norwegian lion with an axe, the three crowns and a Wendish lindworm. The one on the middle shows the yellow-and-red escutcheon of the House of Oldenburg.

Below, right: Christian II's coat of arms in St Bavo's Cathedral in Ghent.

Below: Ship carved in the plaster of Säby Church in Denmark. It is presumably contemporary with *Griffen* and could also be directly inspired by the ship.

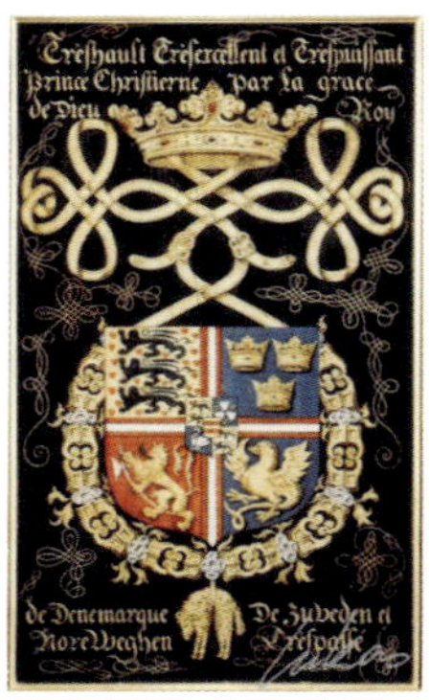

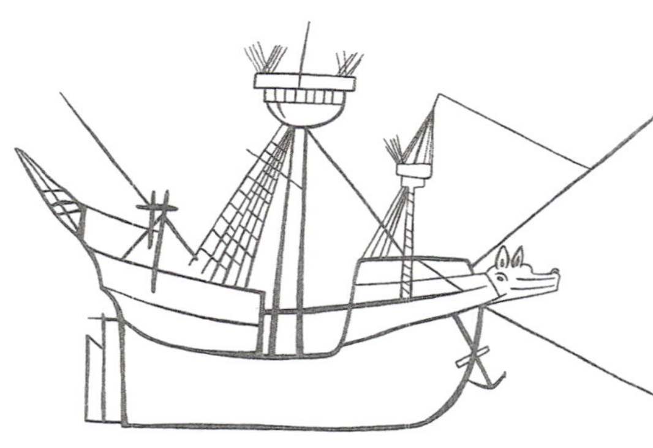

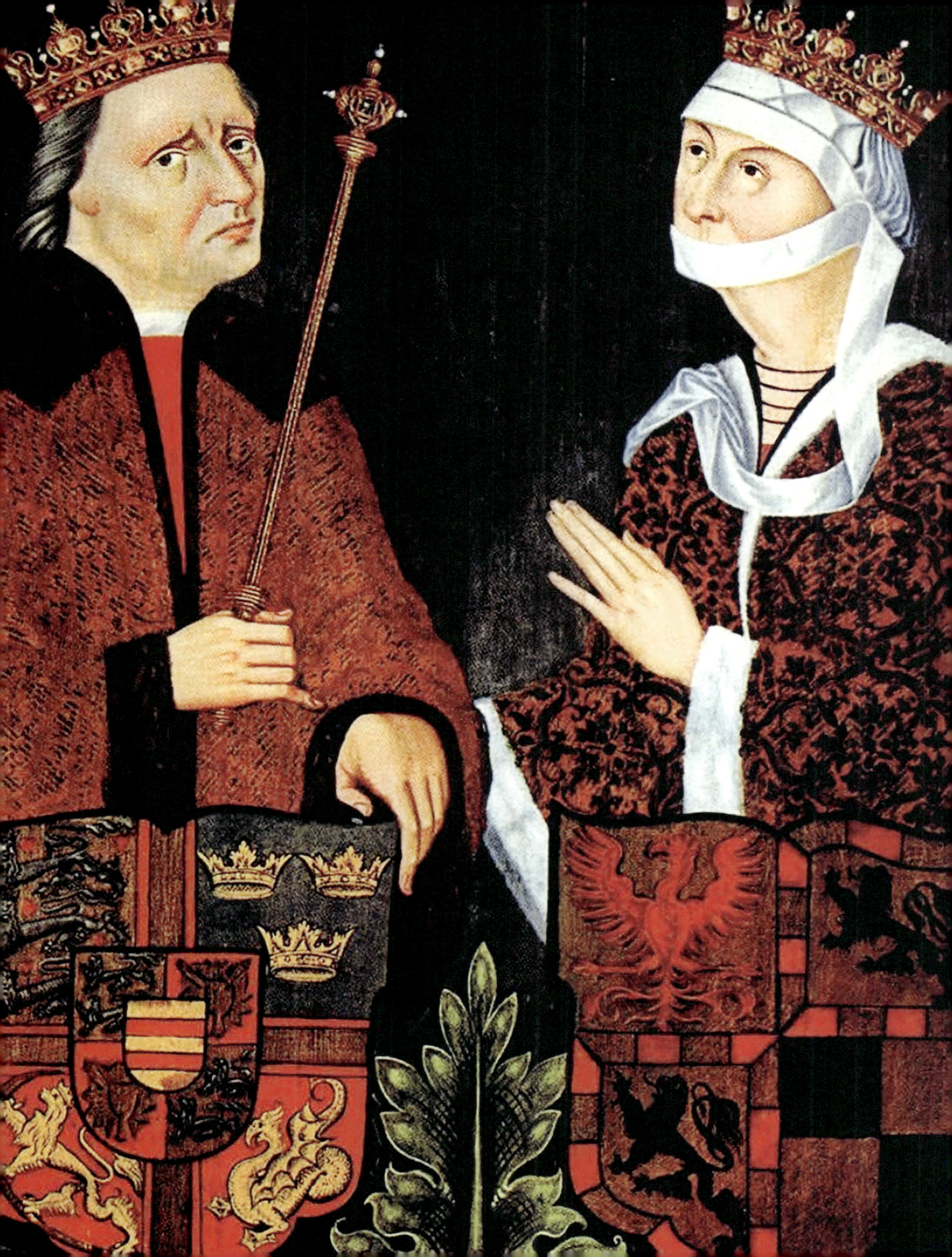

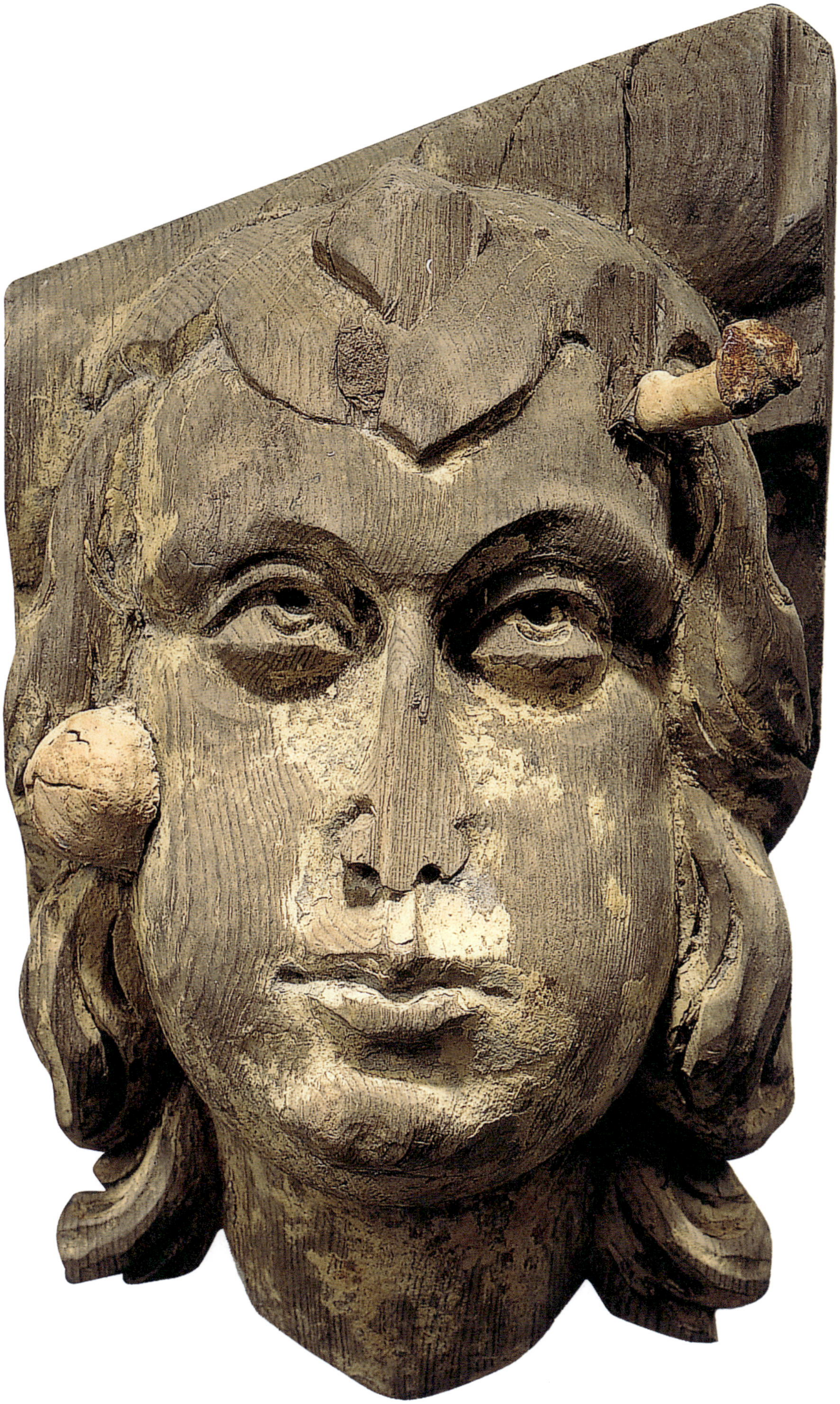

THE ROYAL SHIP KRONAN – PALACE AND WAR MACHINE

Lars Einarsson

Face carved in wood found in 1987 in the admiral's cabin on the quarter deck, located on the upper deck of the man-of-war *Kronan*. A large number of items carved in wood were found in this area, which originally decorated the Lord High Admiral's cabin on board.

AT NOON ON 1ST JUNE 1676 the Royal Ship *Kronan* (The Crown) exploded and sank during a battle between the Swedish navy and the allied Danish-Dutch fleet. The site of the battle was six kilometres off the village of Hulterstad on the southeast coast of the island of Öland. The ship capsized in the strong southeast wind when it was attempting to turn towards the enemy, probably because it was under too much sail. The loss of *Kronan* is one of the biggest naval disasters in Swedish history. Of the 850-strong crew, only forty or so survived.[1]

The keel of the Royal Ship *Kronan* was laid on 27th October 1665 on the island of Skeppsholmen in Stockholm, and it was launched on 31st July 1668. Because of funding problems the ship did not go into service until 1672. The designer was the Englishman Francis Sheldon the Elder, who built the vessel in the English style with three complete gun decks. When *Kronan* was built she was one of the world's largest vessels, at 53 metres long (*Vasa* 47.5 metres) and approx. 13 metres wide (*Vasa* 11.7 metres). It has been calculated that the displacement (weight without equipment) was approx. 2,300 tonnes (*Vasa* 1,210 tonnes), which according to contemporary information at the time of the launch was only exceeded by one Portuguese ship and two English ones.[2] It was probably armed with 110–114 bronze cannon (*Vasa* 64), which together weighed approx. 185 tonnes (*Vasa* 72 tonnes). The number of cannon makes *Kronan* the heaviest-armed vessel we know from this period. The intention was for her to be manned by 550 mariners and 300 soldiers.

In 1980 the wreck of the *Kronan* was discovered at a depth of 27 metres, six kilometres east of Öland, and since 1981 it has been the subject of annual marine-archaeology investigations under the auspices of Kalmar County Museum. Approximately 35,000 objects have been recovered from the wreck thus far.

During the 17th century decoration of naval ships reached its absolute apogee. Symbolism and aesthetics were the guiding forces behind the design, and were two mutually inseparable components. In particular the royal ships, i.e. those ships named after regalia, were intended to glorify royal power and leadership of the kingdom by virtue of their magnificent embellishments. In its entirety the ship demonstrated power, strength and ideals. In this context it is important to emphasise that the ships' powerful, carved symbols of their names – *Kronan* (The Crown), *Svärdet* (The Sword), *Äpplet* (The Apple), *Fenix* (Phoenix), *Neptunus* (Neptune) etc. – were not intended to compensate for the beholder's possibly deficient reading skills. The aim was that they should instead reinforce the feeling of power and continuity – a physical expression of a meaningful context that in its most extreme form was aimed at the limited group able to interpret the objects' significance.

Even though idiom and symbolism remained fundamentally the same, gradual changes

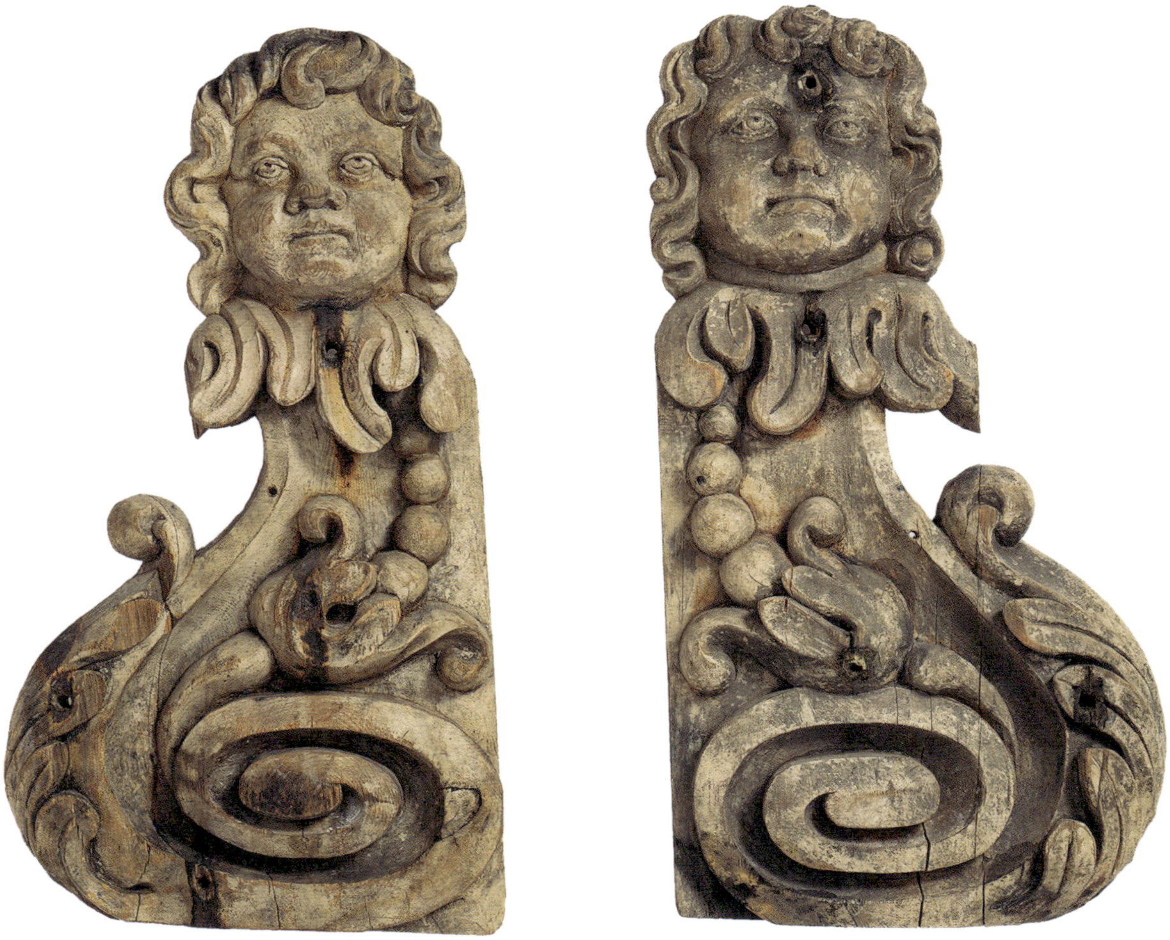

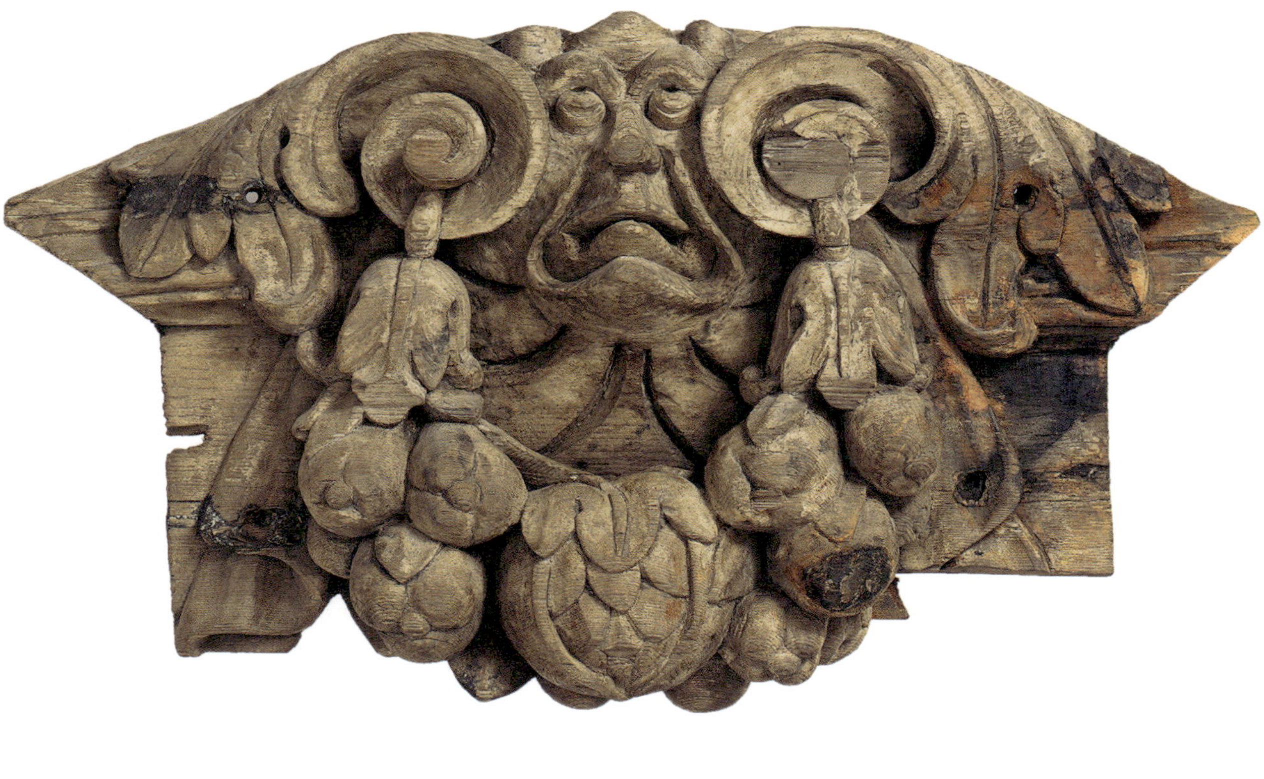

took place during the latter half of the 17th century. In pace with the development of prosperity, the social elite and Sweden's position as a major power, the importance of art as a means of expression also became more important. The increased contacts with foreign countries resulted in new ideas and technologies, and imported labour with new knowledge affected the design of artworks.

The sculptures were an integral part of the vessel, and the ship's master builder was the one who by and large decided on both design and decoration. Those who produced ships' decorations were divided into sculptors and woodcarvers. The sculptors were higher up the pecking order than the woodcarvers. Below them they had a large number of journeymen, apprentices and pure craftsmen. The most eminent sculptors and woodcarvers were not employed solely on decoration of vessels. The greater part of their work was performed for clients on land, and it was in the main rich aristocratic families who made use of their skills.[3]

Left: Two putti and a mascaron found in the admiral's cabin on *Kronan.*

Below: Some of the interior sculpture from *Kronan.* Flower ornament with traces of gold leaf.

Chief responsibility for decoration of the Royal Ship *Kronan* probably lay with the master sculptor Jost Schütz. His name crops up in the Admiralty's accounts for nearly forty years, until his death in 1674.[4]

An extract from the accounts concerning Schütz's work on the ship *Fenix* in 1651 does not state any precise details, but gives an indication of what he was able to produce. Listed here are decorations such as beakheads, crowns, lions, fish, ornamentation, grapes, edgings, gunwales, pictures and 'a small weapon'.[5]

More than 500 sculpted figures have been recovered from *Kronan.* The best preserved ones include caryatids, herms, putti, mascarons, festoons, garlands, corbels, pilasters, capitals, shields, edgings and cartouches (see image to the left and below). There are also a large number of fragments of decorations of a varying nature. The finds also include full-scale sculpted figures representing humans. The gaping masks, elegant acanthus loops and godlike figures of the Baroque constantly recur in the material. Despite their initially being hard to detect, traces of gilding and strong colours are still detectable after several centuries on the sea bed (see image below).

The decorations were originally organised in stylish arrangements based on the ship's design, symmetry and stability. Externally they were largely concentrated on the ship's bow and stern. On naval vessels, design amidships generally involved both sparing decoration

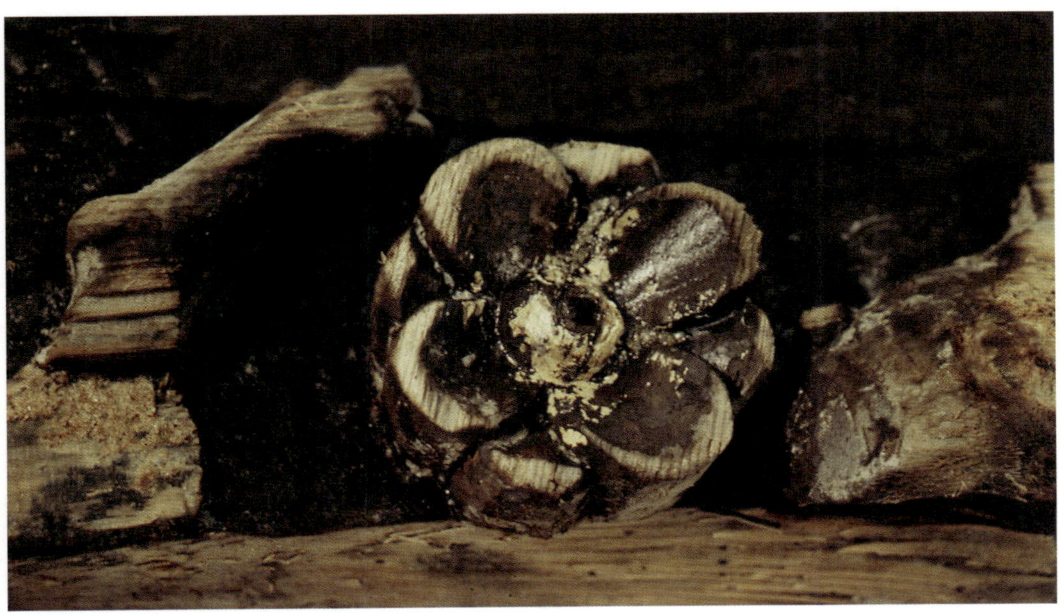

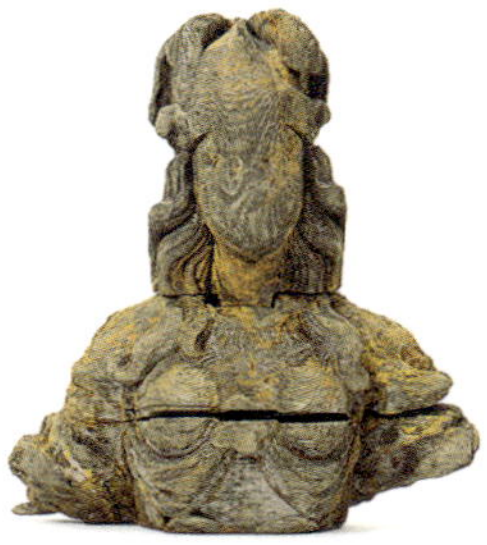

Right: Life-size wooden sculpture, most likely intended to depict Charles X Gustav in a regal contrapposto stance, dressed in a feathered helmet and Roman *triumphator* garb. Compare with the figure above. The carved wooden bust was found at the bottom of the sea 35 metres north of *Kronan's* stern post. The figure is also fitted with a feathered helmet and what appears to be a cuirass – symbolic attributes that could be linked to a godlike royal personage.

Below: Model "Ö 3" in the Maritime Museum's collections in Stockholm depicting the man-of-war *Äpplet* (1661). Most likely belonged originally to Francis Sheldon the Elder.

and negligible coloration.[6] The transom on English three-deckers during the latter part of the 17th century was characterised by its verticality and by the rows of windows on the upper and middle decks, which provided light for the admiral's and officers' cabins. The quarter galleries – the bay-like extensions to be found on the sides of the ship at the stern – were relatively narrow and, as with the transom, decorated with rows of sculpted figures that reinforced the vertical impression of the ship's stern.[7]

The bow sections of the ships built in the English style (*Kronan* being an example), during the 17th century were gradually given a more upright design. The previously more horizontal design had made things easier in terms of positioning of cannon aimed forwards on the upper decks and for use of the beakhead for toilets for the crew. On the more upright aft section the figurehead is bigger and heavier, and it was also given a more prominent position. On the biggest ships the lion as a figurehead was now replaced by horsemen.[8] This change has in symbolic terms been linked to the centralised power needing to give concrete form to its position of increased might, especially after England's restoration of the monarchy in 1660. No less than eight big English ships (80–102 cannon), of which seven were First Rate and one Second Rate – *Prince Royal, St Andrew, The Prince, Royal James, Royal Charles* (1673), *Royal Charles* (1680), *Britannia* and *Royal Catherine* – were decorated during the latter half of the 17th century with horsemen on the beakhead.[9] The horseman is highly symbolic. The horse rears up under its master in a so-called *pesade*. Controlled by the rider, the horse bends its neck down towards its throat and front legs, so that the bottoms of its hoofs are nearly at right angles to the ground. This pose signifies someone who has control over his horse – and his kingdom. A model of a vessel from the 17th century in the collections of the Swedish Maritime Museum, Model Ö 3 (see image below), has just such a horseman as its figurehead. The model was formerly deemed to represent the English ship *Naseby*, renamed *Royal Charles* after the English Regent Charles II's accession in 1660. It is probably instead the Royal Ship *Äpplet*, built by Francis Sheldon the Elder in Gothenburg and launched in 1661, that the model should be associated with.[10]

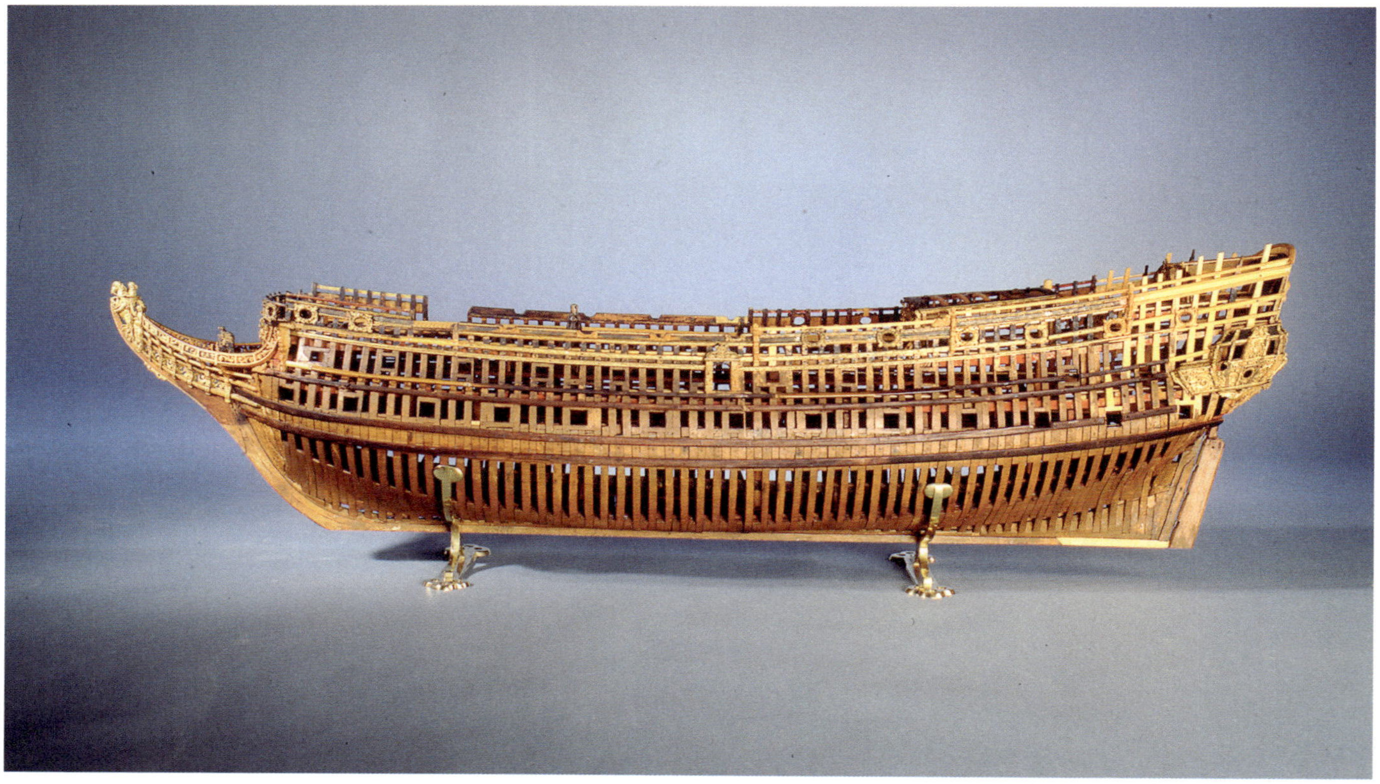

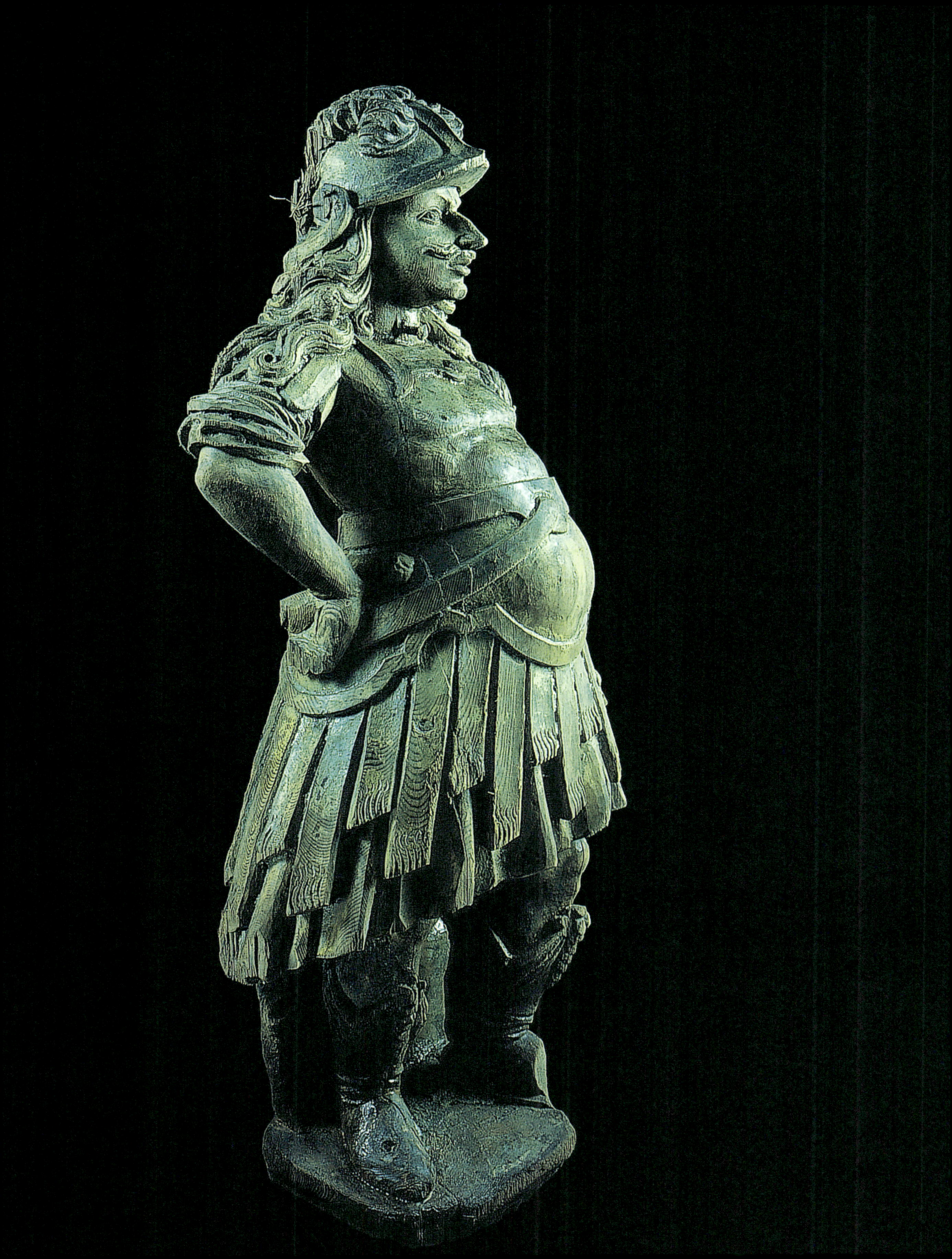

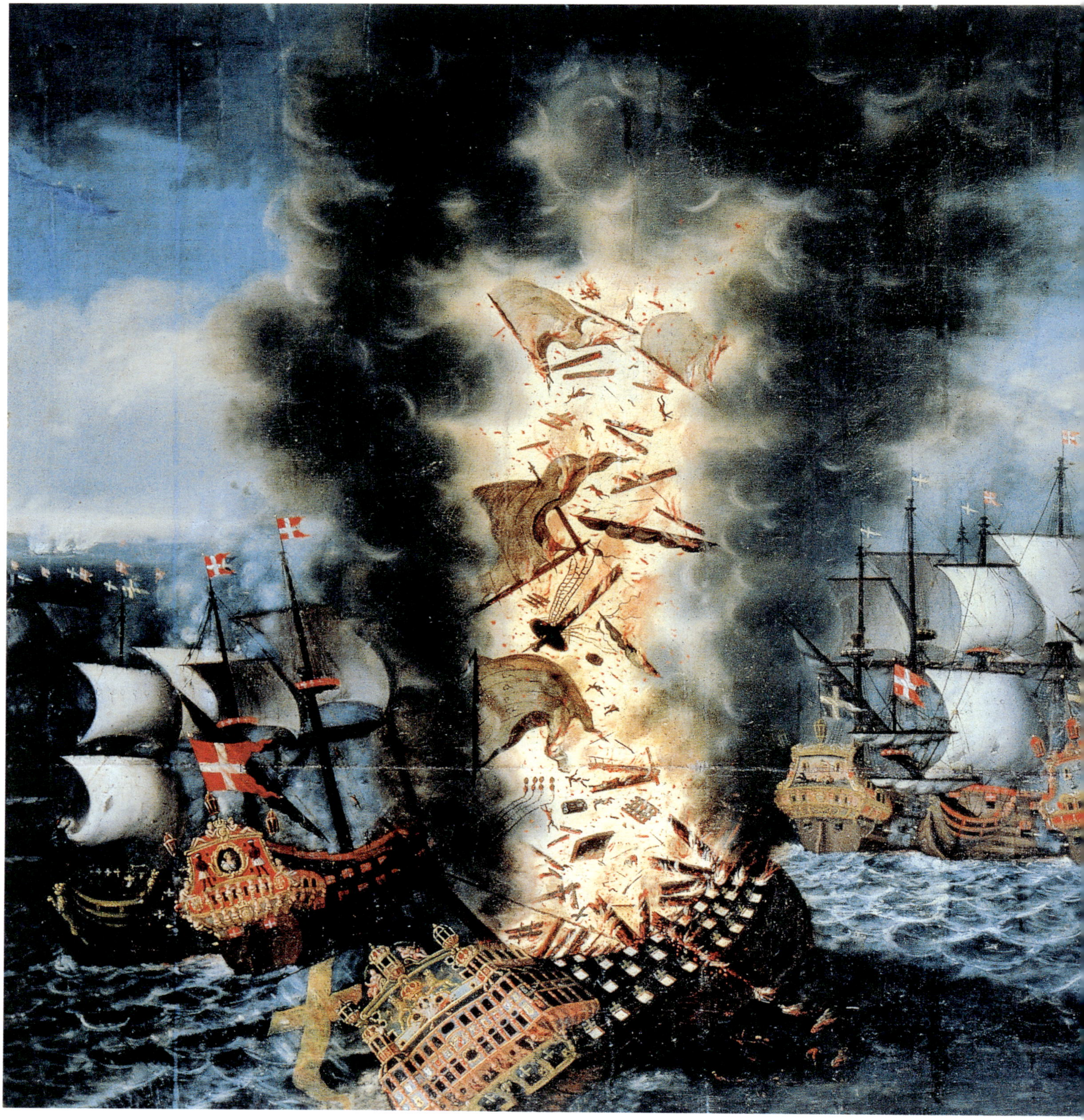

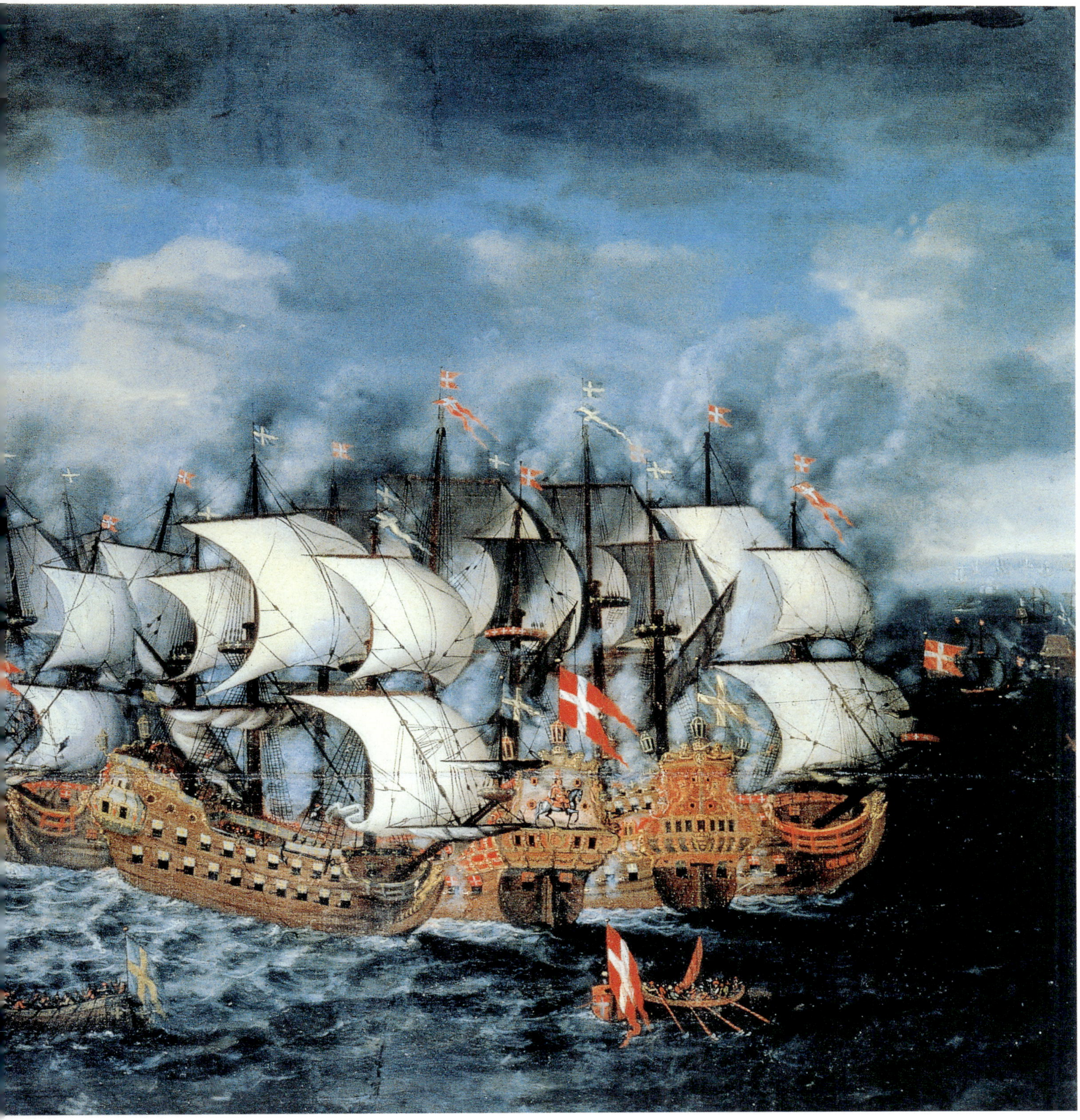

The internal wooden sculptures recovered from *Kronan* were mainly found in the stern section of the ship's upper decks. On the poop deck was the captain's cabin, on the upper deck the admiral's cabin and on the middle deck the officers' mess. These areas were embellished with magnificent decoration in the spirit of the time, to create a stylish setting for the persons of high rank who had their quarters there. Most of the sculptural findings towards the inside of the hull were in their original positions. This provides a clear picture not only of individual sculptures but also of the exact positioning and design of entire sculptural units at the time the ship sank.[11]

Preceding pages: The Battle of Öland on June 1, 1676. *Kronan* explodes. Oil painting by Claus Møinichen.

Below: Charles II's visit to the English fleet on the Thames in September 1672. The 100-cannon ship *The Prince,* to the right in the foreground, can be compared in size and appearance to *Kronan.* Oil painting by Willem van de Velde the Younger.

Apart from a large number of caryatids and the two magnificent royal figures (see image on page 214 and 215) – full-length and half-length respectively – adorning the transom, the external embellishment of *Kronan* is still relatively unknown. It remains pressed into the oxygen-free clay under the port side at a depth of 27 metres. In all the small exploratory shafts dug under the port side over the years, however, a wealth of remarkably well-preserved sculptures have been observed in their original positions towards the outside of the ship. The excellent preservation conditions in the protective clay mean that *Kronan's* exterior on the port side should be in pretty much the same condition as when the ship sank. An interesting comparison can be made with the Royal Ship *Vasa's* opulent ornamentation. The large majority of *Vasa's* sculptures had fallen off and were found around the wreckage, as during the 300-plus years on the sea bed they had for natural reasons come loose from their original but exposed positions on the outside of the ship.[12] Sound detective work and puzzle-solving facilitated reconstruction of the ship's appearance and placement of the sculptures in their original positions. In *Kronan's* case, however, well-protected and unique treasures of art history remain slumbering under its port side in the depths of the Baltic Sea.

The Royal Ship *Vasa's* decorations are probably the best-preserved sculptural contexts from 17th-century floating palaces,[13] though the opulent embellishments from the Royal Ship *Kronan* constitute interesting material for comparison. Unlike the wooden sculptures from *Vasa*, those recovered from *Kronan* represent the stylistic ideal of the Caroline High Baroque, which in terms of design is more driven and more affected by foreign influence.

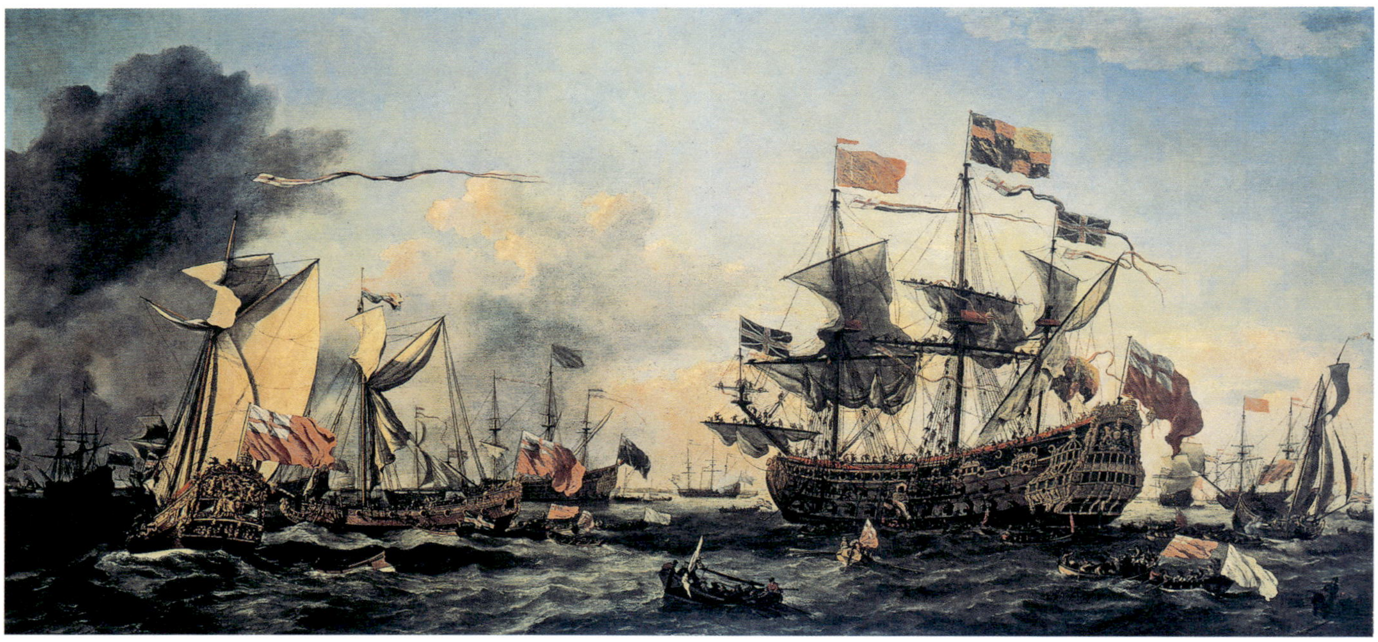

Kronan also displays a wealth of internal decorations, different coloration and a higher degree of preservation. Colour analyses of *Vasa's* sculptures indicate that they were painted in a wide variety of resplendent colours.[14] Based on studies of the sculptures recovered as well as comparisons with contemporary vessel models and paintings, *Kronan* instead appears to have been dominated by red, black and a significantly large amount of gilding/gold paint. In terms of their type, the sculptures from both *Vasa* and *Kronan* basically do not deviate from those to be found in palaces on land of the same period. The ships can thus justifiably lay claim to the epithet 'floating palaces'.

Above all in England, outstanding source material in connection with the decoration of English ships is available. It principally comprises vessel models and maritime paintings (see image to the left) that can be studied in museums but has often also been published.[15] Artists such as Willem van de Velde the Elder, Willem van de Velde the Younger and Jan Karel Beecq have left a wealth of contemporary depictions of naval ships, both during construction and in battle. They have often been accorded a high documented value, but the fact that several of the battle paintings, especially the spectacular ones, were sometimes completed several decades after the actual event and were occasionally by a different artist, constitutes a problem in terms of source criticism, in particular with regard to the details. The lack of contemporaneity of event and finished painting is thus demonstrable, as is the interdependence between different depictions of the same object and event.

The magnificent battle painting by the Dane Claus Møinichen (ca 1660–1726) is the most spectacular depiction of *Kronan* and the Danish-Dutch fleet's victory over the Swedish at Öland on 1st June 1676 (see image preceding pages). As no drawings of *Kronan* have been found, this is the closest we can get to a contemporary representation of the ship. The painting was completed in 1686, ten years after the battle, as a tribute to the Danish king Christian V and as part of the décor for the audience hall at Frederiksborg Castle in northern Zealand. It is hanging there to this day, together with several other paintings of Danish victories. The composition is dramatic. On the left of the picture the capsizing *Kronan* can be seen. In a blood-red cascade, sails, ship's timbers, swords, muskets and maimed crew are being hurled skywards. In the right-hand part of the picture the Swedish fleet's second-biggest vessel, the Royal Ship *Svärdet*, is to be seen, surrounded by several enemy ships.

There is no doubt that the aim of the painting was to glorify the Danish victory by artistic means. The propaganda dominates the composition, which is compressed both temporally and spatially. *Kronan* is exploding at the same time as *Svärdet* is seen surrounded by two enemy ships, despite there having been several hours and a considerable geographical distance between these events. In the middle of the picture the artist has appropriately enough inserted the two conquered Swedish ships *Neptunus* and *Järnvågen* (The Iron Scales), which Danes and Dutchmen took to Copenhagen in triumph after the battle. As a symbol of the Swedish defeat one sees *Kronan's* faded blue-and-yellow stern flag disappear into the depths. For the sake of clarity the closed gunports on the ill-fated Swedish ships are painted with blue-and-yellow flags, and those of the Danish victors with red-and-white ones. This must be seen as yet more propaganda, as neither Swedish nor Danish ships are known for having decorated their gunports in this way. The painting also clearly shows that the artist knew that *Kronan* was built in the English style, with a rounded stern above the waterline and a small transom, and that *Svärdet* had been constructed in the Dutch manner, with a flat stern right down below the waterline. Otherwise, the ships' hull structures are far from precisely reproduced, and neither was this the aim of the assignment. What was crucial was clarification of the ship's identity, nationality and fate.

Let us first state that as regards source criticism the painting leaves much to be desired, but what from a sculptural standpoint is of the greatest documentary interest about the painting is the appearance and decoration of the transoms. This is where the painting's purpose is reflected. The ships of particular importance – the sunken ones (*Kronan* and *Svärdet*), the conquered ones (*Neptunus* and *Järnvågen*) and the big Danish flagships *Christianus Qvartus (Kristian IV)* and *Christianus Qvintus* (*Kristian V*) – all have their transoms facing the beholder. The transom was the ship's face, as it revealed its identity. The ship's prestige and status were given concrete form in the design and content of the decoration. Thanks to his painstaking work and the wealth of detail, Møinichen gives us the opportunity to carefully study the ships' features. Not for a single moment does he allow the beholder to doubt the ships' identity and rank. On *Kronan's* and *Svärdet's* transoms the golden heraldic lions flank the regalia – crown and sword. On *Neptunus*' transom one sees the sea god being drawn in his chariot by a seahorse, holding his trident in his hand. *Järnvågen* is symbolised by two scale pans in which the iron that was so important to national prosperity was weighed before being shipped out.

The stern of model "Ö 24" in the Maritime Museum's collections in Stockholm. The model is believed to represent the ship *Postiljon*, also called *English Frigate*. The ship, launched in 1664 in Gothenburg, was built under the direction of *Kronan's* master shipwright Francis Sheldon the Elder. The carved wooden sculptures on both corners of the stern, in the form of a man in Roman *triumphator* garb, are strongly reminiscent of the wooden sculpture recovered from *Kronan* (see page 215).

On *Kronan's* transom the Swedish heraldic lions hold the royal crown above the head of a bust representing the Swedish regent, who in the heat of battle with the superior Danish forces is now heading for ruin, together with his subjects on board. A similar sculpted wooden bust has been found on the sea bed a little way from the wreck's sternpost (see page 214). There is a striking similarity between the figure in the painting and the wooden sculpture from the sea bed. The site of the finding below the wreck's sternpost and the sculpture's symmetrical shape indicate that it was centrally positioned on the ship's transom. Even though it has been smoothed down by sand and water for hundreds of years, one can nevertheless discern a figure with his arms by his sides, a plumed helmet, a cuirass and long, curly hair that extends down towards his armpits. The style is well in keeping with the image of the king by the grace of God, as depicted in sculptures and paintings. The artist's assumed close knowledge of the ship's transom can be temporally and spatially linked to *Kronan's* master builder Francis Sheldon the Elder. Six months before completion of the painting Sheldon took up his appointment in the service of the Danish king at the Holmen naval shipyard in Copenhagen.[16] It thus cannot be excluded that the 25-year-old artist had access to the sketches of the ship that the master shipbuilder saw as being his personal property and possibly took with him to Denmark when he took up his new post there.

What is perhaps the most eye-catching sculpture recovered from *Kronan* is the 60-kilo, 155-centimetre wooden sculpture of a man that was found in the opening to the bay-like quarter gallery in the admiral's cabin on the upper deck (see image page 215). In all likelihood it originally belonged to the external decoration on the port corner of the transom. Paint residues on the sculpture show that it was to a large extent painted yellow, probably using orpiment. The foot position representing the so-called contrapposto pose, the attire, the crooked nose, the plumed helmet and the substantial paunch are all symbolic attributes associated with a prince. The sculpture represents Charles X Gustav in Roman warrior's attire, posing in full scale on his biggest warship. In shipbuilding of the late 17th century it was customary for vessels to be embellished with sculpted full-scale figures known as counterpieces at each corner of the stern between the ship's side and the transom. A figure nearly identical to that recovered from *Kronan* is to be found on a contemporary model in the collections of the Swedish Maritime Museum. The model is said to represent the ship *Postiljon*, also called *Engelska fregatten* (The English Frigate), which was designed by *Kronan's* master builder Francis Sheldon the Elder, built in Gothenburg and launched in 1664.

A few more years of marine archaeological investigations of *Kronan* remain in order to complete the overall investigation. A synthesis of the information gleaned from the sculptural findings etc. will then be undertaken. The flagship *Kronan's* rich, varied and well-preserved findings offer a unique opportunity to create an overall picture of 17th-century society, both at sea and on land.

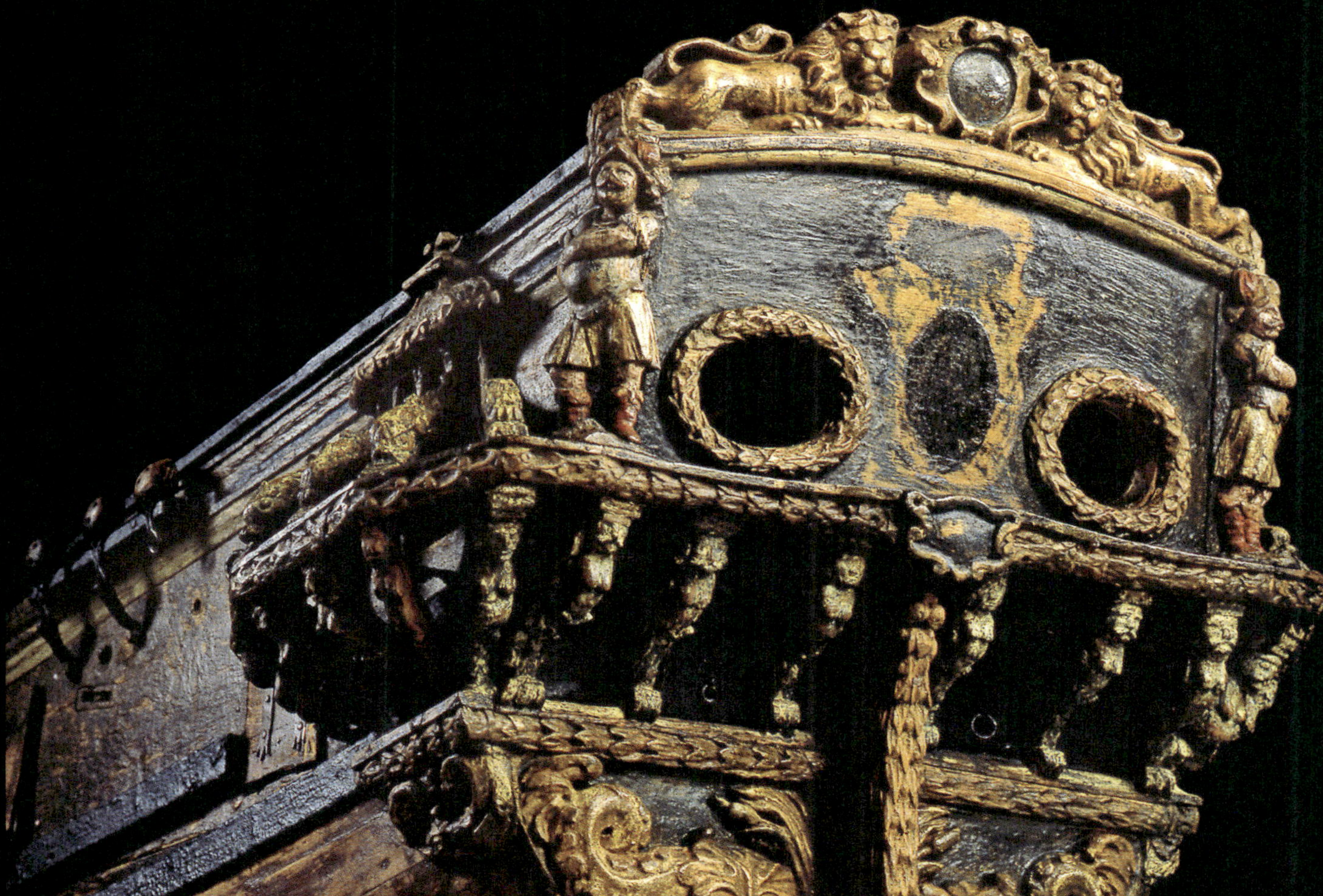

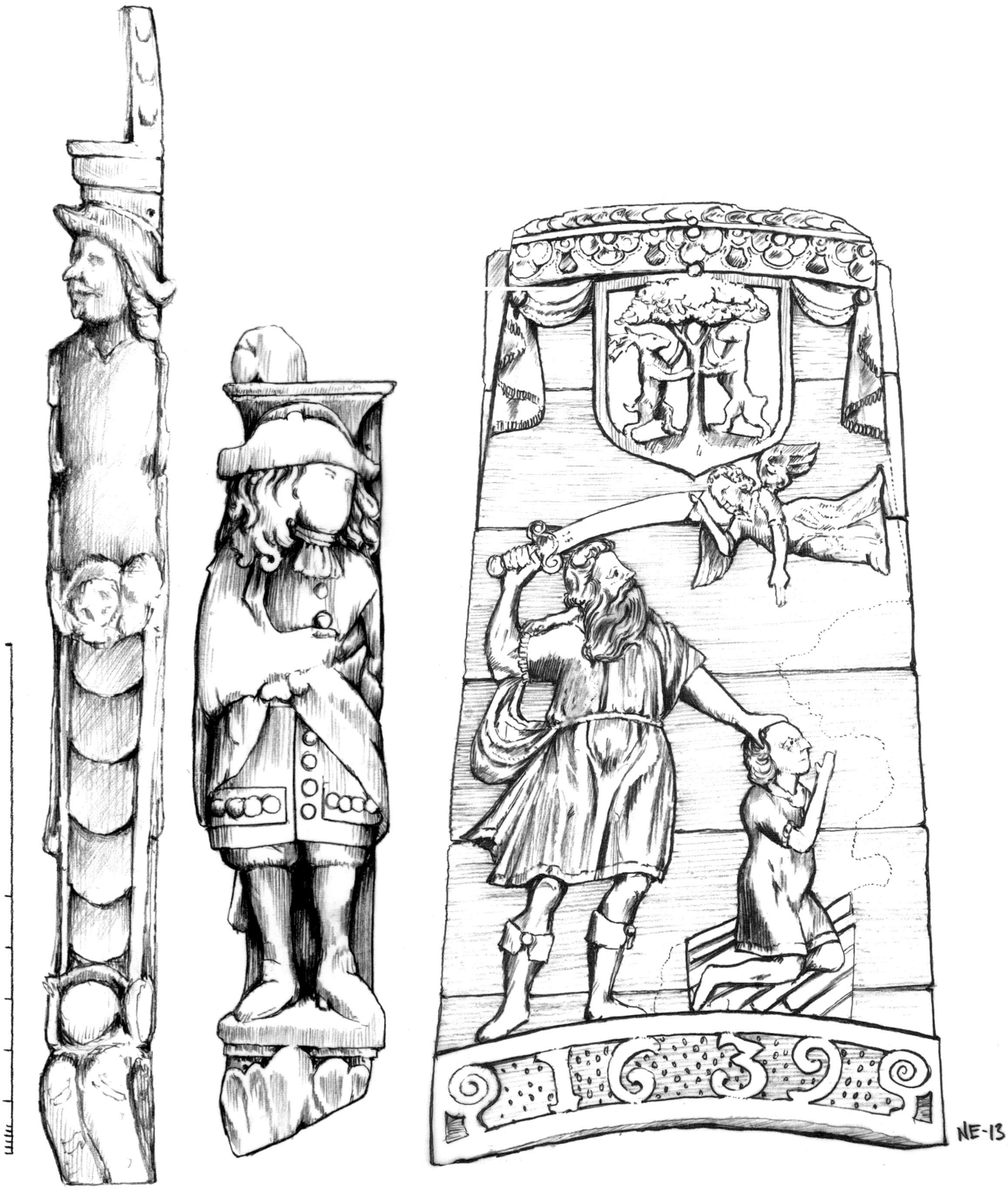
1639
NE-13

BETWEEN LION FIGUREHEADS AND LIONS OF FASHION

Niklas Eriksson

DURING ITS PERIOD as a great power Sweden expanded over an ever increasing area. The country was thinly populated – principally by farmers. Apart from agriculture, the economy was based on export of raw materials. In the Netherlands the situation was pretty much the reverse. The country's limited area accommodated a large population, and the Dutch economy flourished thanks to global trade. The Dutch were interested in Swedish raw materials, and the Swedes were interested in finding a market for them. The two countries' differing circumstances and needs made for a well-functioning and lucrative collaboration. The Dutch influences on 17th-century Sweden also came to be significant, through migrant families who established both trading houses and industries. A lasting result of this influence is the many 17th-century buildings erected in Sweden by Dutch master builders in the Dutch style.[1]

There was a steady flow of merchant vessels of a type known as fluyt sailing back and forth between cities such as Stockholm and Amsterdam. They are said to have been introduced in 1595 in the town of Hoorn in North Holland, where a merchant by the name of Pieter Janszoon Liorne realised his vision of the ideal merchant vessel. From the side, a fluyt looks like just any old three-masted ship. It is actually not until you look at the hull from above that you see the characteristic shape – like an oblong rectangle with slightly rounded corners. The stern is almost pear-shaped and is topped by a strikingly slender, elevated transom.

Thanks to standardisation as well as certain technical innovations such as wind-powered sawmills, the fluyts were cheap and easy to build. The fact that they could also be sailed with relatively small crews made transportation cost-effective. This type of vessel is thus deemed to have contributed significantly to the expansion and economic success of Dutch shipping during the 17th century.

Like most ships of the period, fluyts were built without using any design drawings. The shape of the hull was determined by rule of thumb, and by copying vessels that had proven to work well – a situation that hardly encouraged change and innovation. Fluyts thus appear to be an unusually uniform type of vessel. They were built in their thousands and became a familiar feature at all European ports, as well as ports on other continents that in one way or another had contact with Europe. The fluyts are thus also to be found in the colonies and the New World, and they became so common as to be almost invisible.

Even though fluyts were the commonest merchant vessels of early modern Europe, there were for a long time no detailed descriptions of what they actually looked like. They were certainly depicted in large numbers in harbour scenes, or in the background of paintings of East Indiamen or men-of-war, but they rarely constituted the main subject matter and were not represented in detail. In recent years, however, archaeological investigations of wrecked fluyts have contributed to new information on this legendary type of vessel coming to light.[2]

On the left are the "corner men" carved figures from the unidentified fluyts *Lejonvraket* (The lion wreck) and *Spökskeppet* (The ghost ship). The stern to the right comes from a stranded fluyt and hangs in Søndre Harritslev Church in Denmark. The fluyt *Abrahams Offer* (Abraham's sacrifice), presumably built in 1639, and the coat of arms at the top reveal that the ship had its home port in Hindelopen, located in Friesland.

Following pages: Wolfgang Hartmann's etching provides a lively view of Stockholm in the mid-17th century. It shows clearly here what a key role large ships played in the city's public space. Warships were primarily anchored at Skeppsholmen in the foreground, but merchant vessels can be seen along Skeppsbron in the Old Town. These include a number of fluyts.

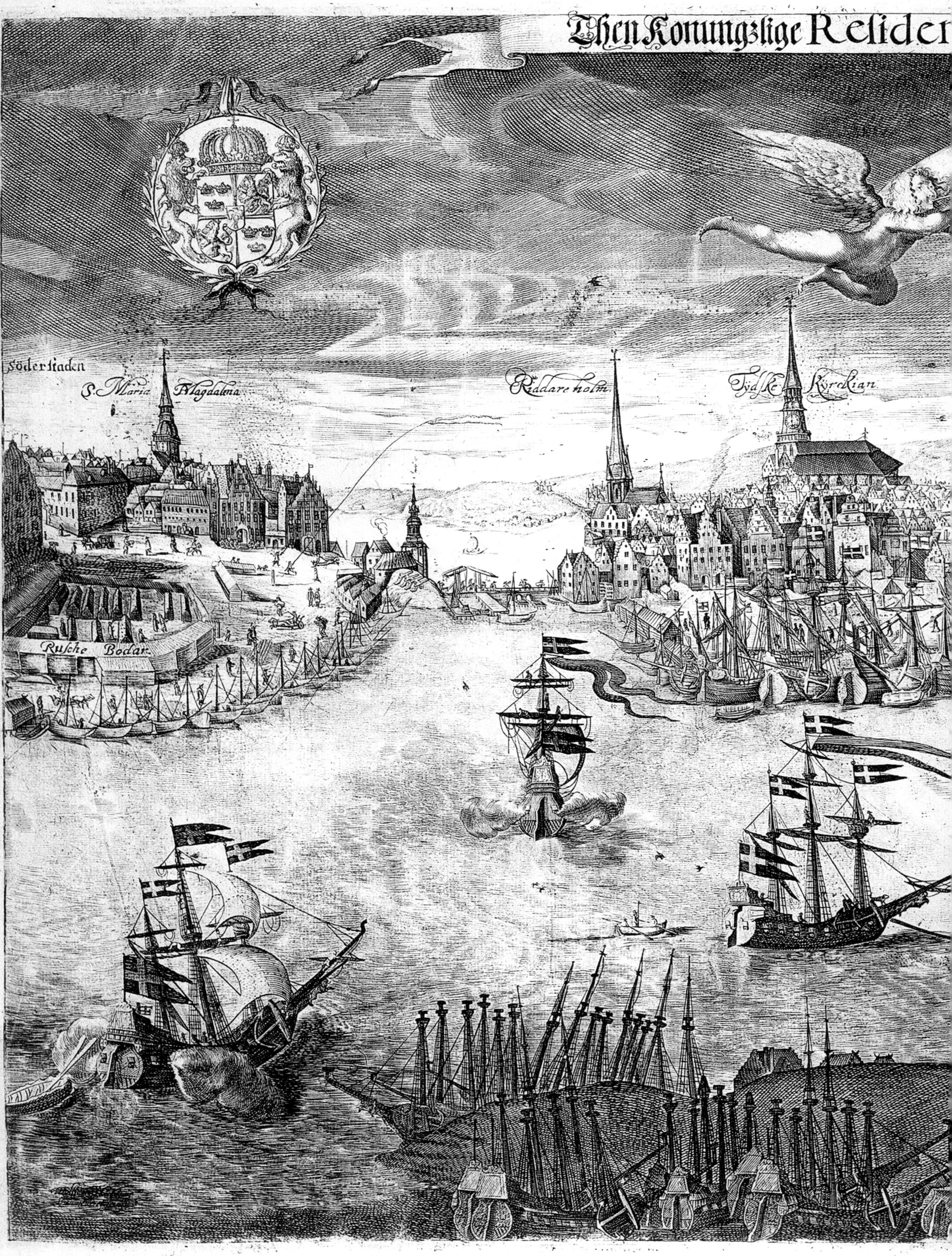
Then Konungzlige Reliden
Söderstaden
S. Maria Magdalena
Riddare holm.
Tydske Kyrckian.
Rusche Bodar.

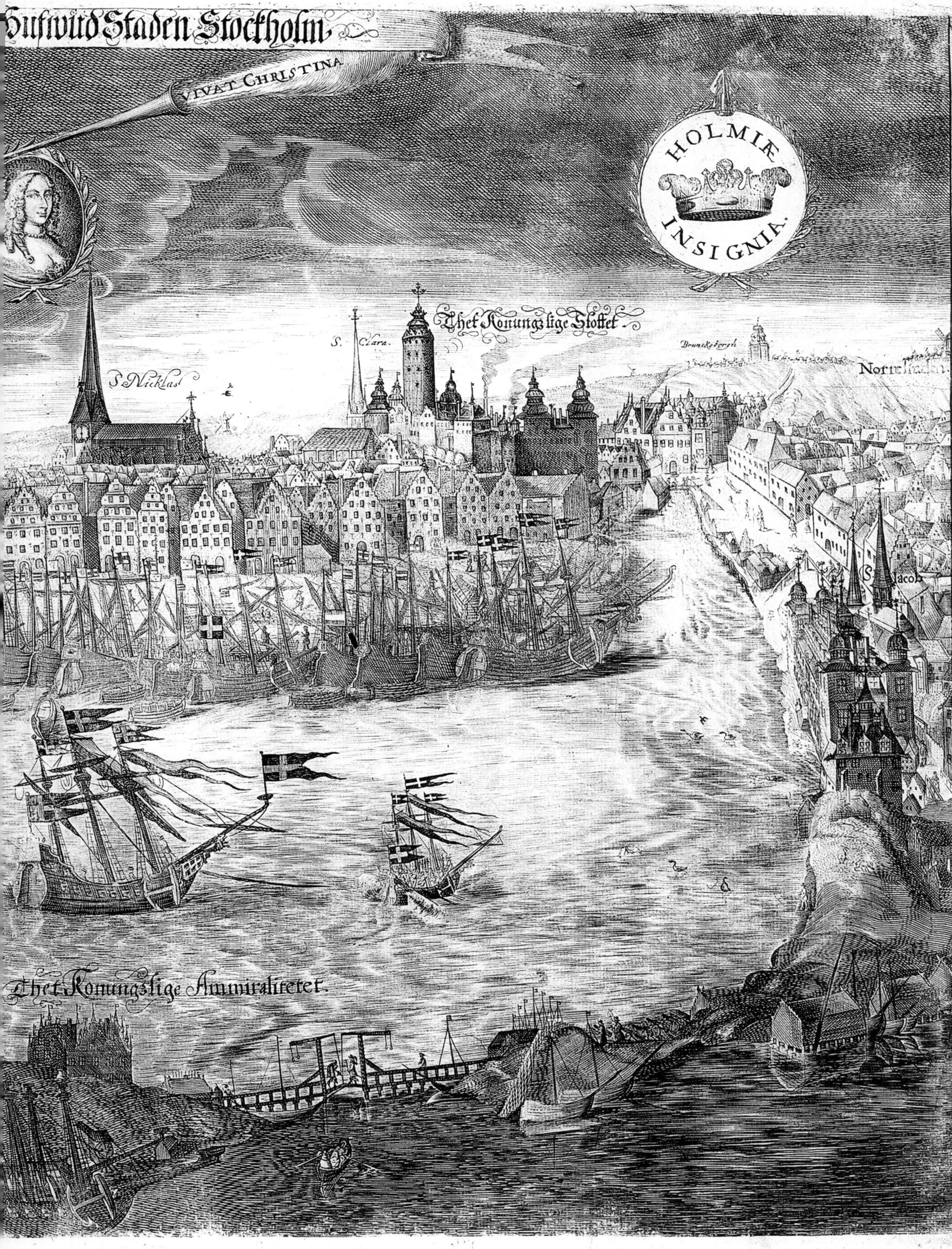
Hufwud Staden Stockholm.
VIVAT CHRISTINA
HOLMIÆ INSIGNIA.
Thet Konungzlige Slottet.
S. Clara.
S. Nicklas
S. Iacob
Thet Konungslige Ammiralitetet.

The fluyts plied routes between various ports, and their sculptural decoration and architectural expression were adapted to this very context. As with urban buildings, the fluyts were attired with various items of sculpted and painted ornamentation, most of which were to be found on the narrow transom, at the top of the pear-shaped stern. Transoms were the magnificent facades of early modern ships, and those of fluyts bear many similarities with house gables in Dutch towns.

The so-called Dutch Golden Age has been described by various authors as the origin of capitalism and the modern economy.[3] The merchant families who built their fortunes on trade with markets such as the Baltic countries and Sweden expressed and manifested their affluence through the facades of their houses. Up until the 17th century the tall, narrow buildings were provided with serrated stepped gables, but after that they came to be replaced by the elegantly curved neck-gable facades that remain a common feature in towns and cities in the Netherlands and many other European countries. Seen against the light, the contours of these facades are not dissimilar to those of fluyts' sterns – one indeed talks of a flowing urban environment.[4]

At a time when most people could barely spell their own name, imagery and symbols played a major role. The fluyts' names are expressed by means of painted or sculpted motifs on the transom. For such communication to work, the beholder must be able to recognise and decode the motif. The fluyts thus often have fairly prosaic names such as *Halvmånen* (The Half Moon), *Rosen* (The Rose), *Svanen* (The Swan), *Solen* (The Sun) or *Vintunnan* (The Wine Barrel), which could easily be communicated with the aid of pictures.

Another option was naming fluyts after well-known episodes from the Bible, e.g. *David and Goliath*, *Jonah and the Whale* and *Noah's Ark*. Several vessels bore names that can be linked to the story of Abraham and the sacrifice of his son Isaac. At Sønder Harritslev

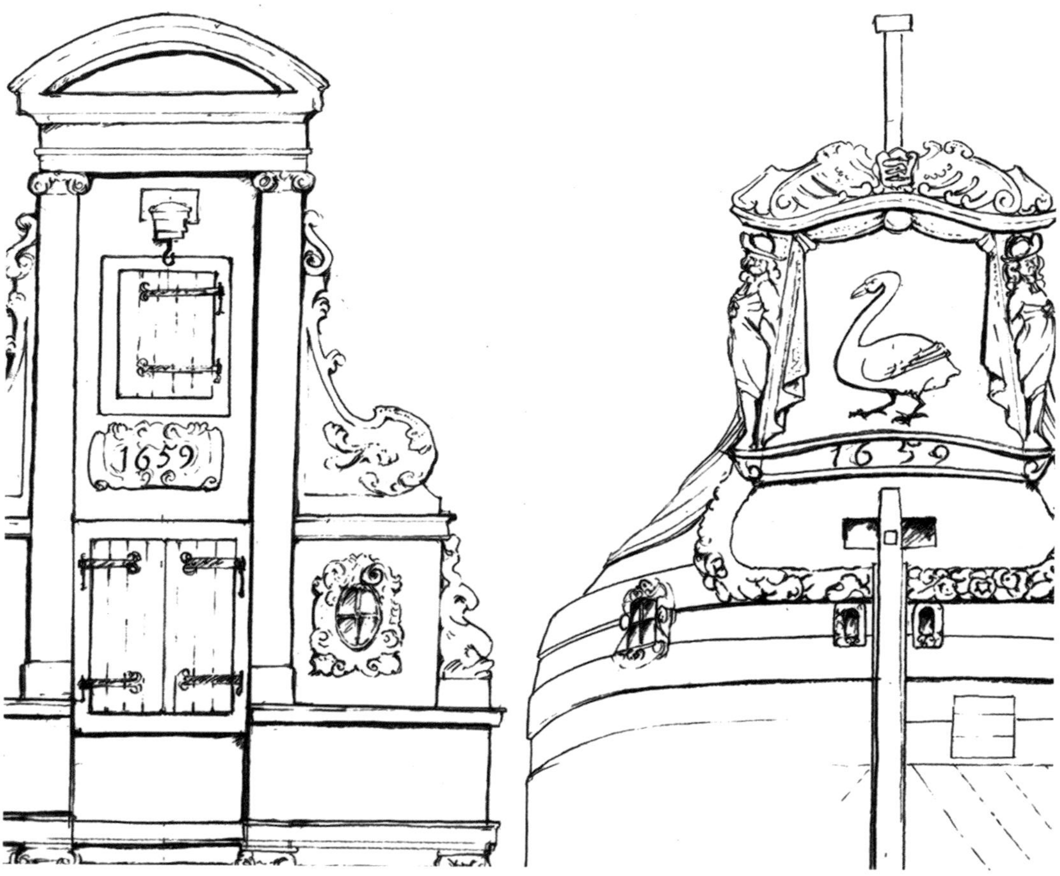

A neck-gabled house placed next to the stern of a reconstructed wreck with the working name *Spökskeppet* (The ghost ship).

Church in Denmark there is an almost complete transom from a fluyt depicting the scene where Abraham raises his sword to cut off his son Isaac's head, but is stopped by the angel.

To emphasise that the scene is from long ago, the biblical names were often represented by figures dressed in archaic Roman-inspired clothing. A fragment of a figure in Roman armour has been recovered from the socalled 'Jutholm wreck' in Dalarö harbour, and may have been part of a biblical scene. The advantage of expressing names with symbols is that they work independently of language. The vessel *Rosen* becomes *The Rose* when it sails to England, but the problem with the symbols is that they allow rather differing interpretations. The ship *Gotlandia*, for example, is called *Lammet* (The Lamb) in certain sources – something that is explained by the fact that a lamb forms part of the province of Gotland's coat of arms.[5]

The symbols and the range of names were limited, and several vessels consequently had the same name, but the ornamentation included other differentiating characteristics. At the top of the transom, above the area bearing the name, one often found a coat of arms showing the home port, and at the lower edge of the transom the year of construction was given. Both coats of arms and years of construction are often to be found on the facades of houses in towns, but in this case they are the owning families' coats of arms.

A more or less obligatory feature of fluyts was the sculpted men to be found on either side of the transom. In the Netherlands such a figure is known as a *hoekman* – literally 'corner man'. Archaeological investigations have shown that they depict merchants, dressed in the very latest style of clothing. The appearance of these 17th-century urban dandies is reminiscent of the shipowners who held shares in the fluyts. Sculpted wooden merchants were pretty much obligatory on the vessels, but they do not seem to appear at all on houses with neck gables. Through the fluyts' corner men, the merchants were a constant presence and

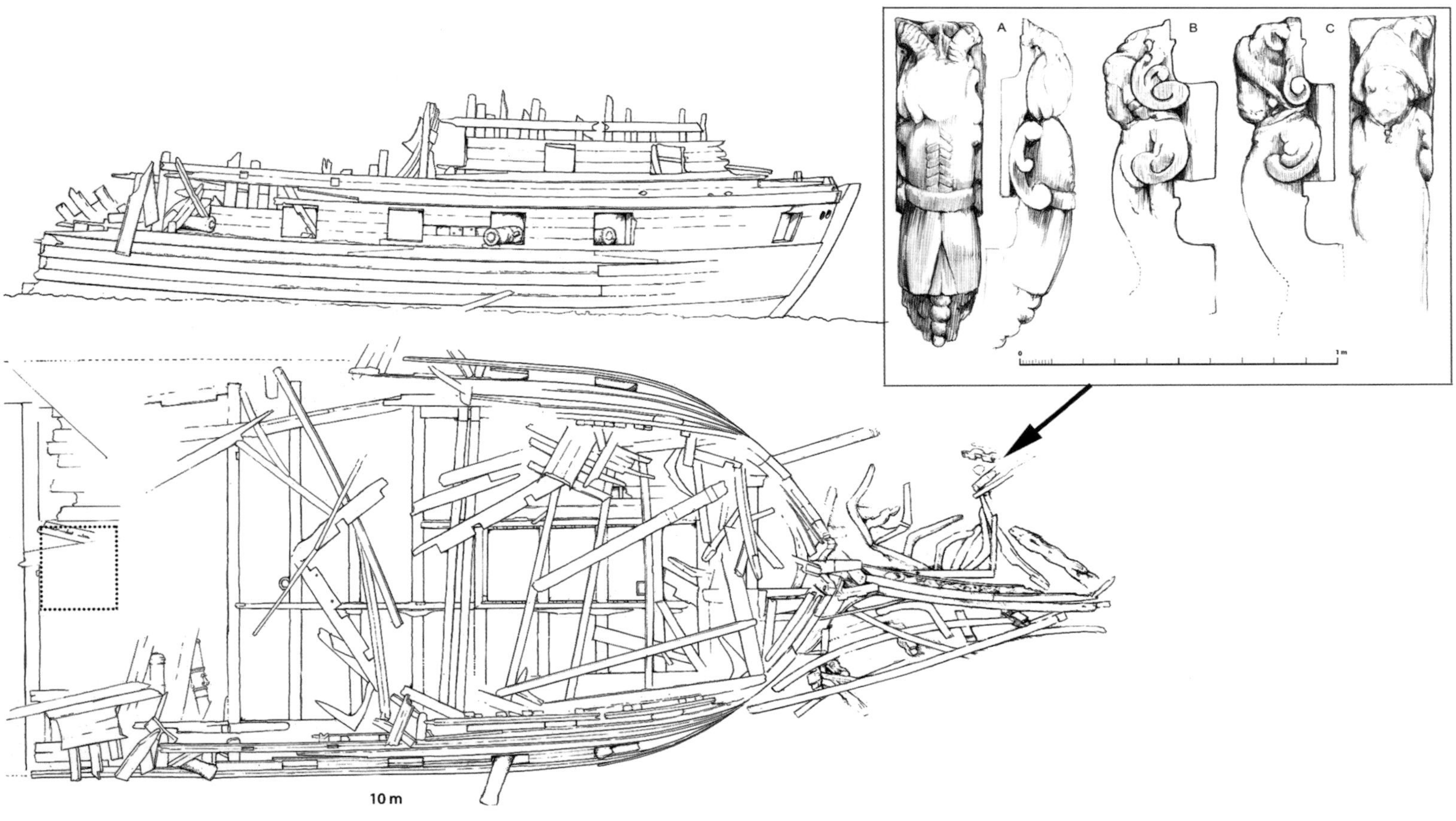

The deteriorated wreck of the royal ship *Svärdet* (The sword) was discovered in 2011. The richly decorated stern was blown to pieces during the Battle of Öland in 1676. At the bottom of the sea in front of the surprisingly intact forebody lie several sculptures that had been attached to the figurehead. The inset drawing shows a few of them. The figures in armour were carved by the Swedish admiralty's sculptor Jost Schütz.

were literally nailed to the vessels – an almost over-explicit way of showing that the fluyts are primarily to be associated with trade.[6]

But the Dutch collaboration did not just apply to trade. During the first half of the 17th century the fleet recruited master shipbuilders in the Netherlands. They built ships they were familiar with, or in the 'Dutch style', as it was known. A well-known example is the royal ship *Vasa*, which sank in 1628. But there have been many more. In 1660 the master shipbuilder Jacob De Voss, who was recruited in Amsterdam, laid the keel of a big ship at Skeppsholmen in Stockholm. The vessel was launched in September 1662, and was named *Svärdet* (The Sword). Its size was only to be exceeded by *Kronan* (The Crown), which was launched in 1668.

In all likelihood, this was the model that the English master shipbuilder Francis Sheldon showed to the sculptors in Gothenburg when *Riksäpplet* (The orb) was being built. During repair work, one of the upper windows was lowered. The field at the top of the stern was originally smaller. The caryatids, coats of arms and the grotesques at the bottom of the stern can be recognised from the sculptures salvaged from the wreck.

Svärdet first entered into service in a big way in conjunction with the Scanian War, which broke out in 1675. The year after that, the ship took part in the fateful Battle of Öland, which began with *Kronan* capsizing and exploding. In the subsequent chaos and confusion *Svärdet* was surrounded by enemy ships. After an hour and a half's intense bombardment it was set alight by a fireship – a vessel filled with combustible material that was ignited and then sailed off towards the enemy. The fire reached the gunpowder store, and only around fifty of the 650 men on board survived the violent explosion.

The wreck was discovered in 2011, and it still shows signs of the violent combat. The June 1676 battle is evident by virtue of the cannon still projecting through the gunports and the charred sides of the hull, penetrated in several places by Danish and Dutch cannonballs. Abaft the mainmast the explosion destroyed the hull, and thus also the ornaments on the transom and the quarter galleries.

But there are also certain clues as to what the stern looked like. When Erik Dahlbergh was out at Skeppsholmen doing sketches in preparation for the monumental collection of engravings *Suecia Antiqua et Hodierna*, he caught sight of *Svärdet*. The simple sketch of 1664 is the only one known to have been made of *Svärdet* while the ship was still afloat. Two lions are suggested on the transom, holding a large sword between them.[7]

In the bow the most striking detail was the cutwater – the opulently decorated platform projecting in front of the stem. In the mid-17th century, cutwaters were made somewhat shorter than, for example, at the time of the *Vasa*. The Admiral of the Realm Carl Gustav Wrangel, however, preferred the older, slightly longer type of cutwater, and wrote to the Admiralty that *Svärdet* should be equipped with one 29–30 feet long.

Lions were without a doubt the commonest figureheads during the 17th century, and they sometimes involved a symbol indicating the ship's name. Extant letters reveal that the ship *Riksnyckeln* (The Key to the Realm), which was contemporaneous with *Svärdet*, was embellished with a lion holding a key in one of its front paws. The practice is also recognisable on the *Vasa*, where the lion is holding a shield featuring the Vasa sheaf. *Svärdet's* cutwater was to be provided with 'a lion with a sword in its front right paw'.[8] This lion is now to be found on the sea bed, its nose and front paws buried in the sediments.

Under *Svärdet's* cutwater there is a frieze over seven metres long depicting a bird, possibly a fish and a human-like face in profile. A similar frieze, featuring sea creatures, birds and heads as well as snakes, is to be found on *Vasa's* cutwater. The art historian Hans Soop has traced the motif to Ovid's so-called metamorphoses, including the Greek myth of Peleus's wooing of Thetis, goddess of the sea.[9]

The armour- and helmet-clad male figures that had been placed along *Svärdet's* sides were sculpted by the admiralty sculptor Jost Schütz (1610-1674). His name occurs in the admiralty's accounts as early as in 1634, and it is perhaps no surprise that the motifs on the cutwaters of *Svärdet* and *Vasa* – built a good thirty years earlier – appear so similar.[10] The Dutch dominance in the field of Swedish shipbuilding was, however, finally broken. This had in practice already happened when *Svärdet* was constructed at Skeppsholmen in Stockholm. In Gothenburg a newly recruited English master shipbuilder had been hired, and there was a political background to this.

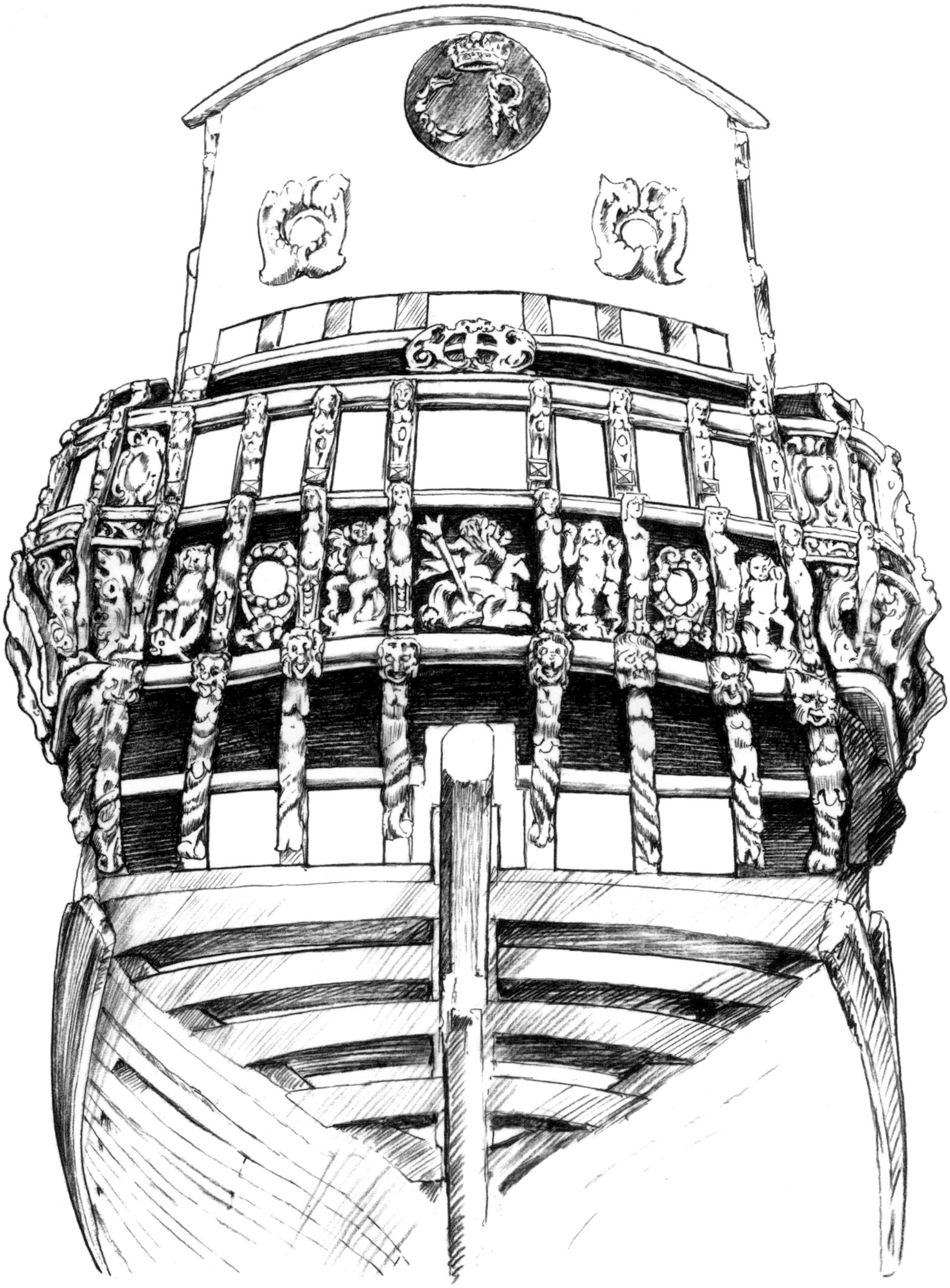

At the time, as now, the entrance to the Baltic Sea was via the Sound (Öresund). On either side of the Sound were the rival states Denmark and Sweden. But if either state were to become too dominant it would be able to dictate terms for Baltic trade on its own. The Netherlands thus intervened in the Danish/Swedish conflicts on several occasions, and by consistently supporting the weaker party it was possible to ensure that the balance of power was retained. At the beginning of the 17th century, for example, the Netherlands were aligned with Sweden, but by the middle of the same century they were instead collaborating with Denmark.

In parallel with this a tug o' war for dominance over global trade was in progress, above all between England and the Netherlands. Both countries also had interests in the Baltic Sea. In the Netherlands the Baltic route was sometimes known as *Moedernegotie* (mother of all trades), whilst in England one spoke of the Baltic Sea as "the lost trades". When the Netherlands aligned themselves with Denmark it was a natural step for Sweden to eye up England with a view to possible collaboration. And three master shipbuilders were signed up. The best-known was Francis Sheldon, and his first big ship in Gothenburg was named *Riksäpplet* (The Orb).

English ships differed from Dutch ones in several respects – they were, amongst other things, narrower and had a greater draught. *Riksäpplet* was about as long as *Svärdet* but considerably more slender, the bow was sharper, and the stern featured the elegant rounding that was characteristic of 17th-century English shipbuilding.

English and Dutch shipbuilding also differed as regards the sculptures, and with a trained eye the difference was easily discernible. The decoration on big 17th-century English ships is characterised by clear vertical and horizontal features that form a net draped around the transom, the quarter galleries, the cutwater and sometimes also parts of the sides of the hull.

Some of the sculptures salvaged through the years from *Riksäpplet* (The Orb). The bust of a man with outstretched arms was probably placed at the centre of the stern. That is also probably true of the upper section of the coat of arms shown beneath it. The coat of arms with the cross was probably placed between the large windows at *Riksäpplet's* stern. The grotesque figure with the cloven hoof was placed at the bottom of the stern along with a number of sisters. These sculptures are characteristic of English ships from the 17th century.

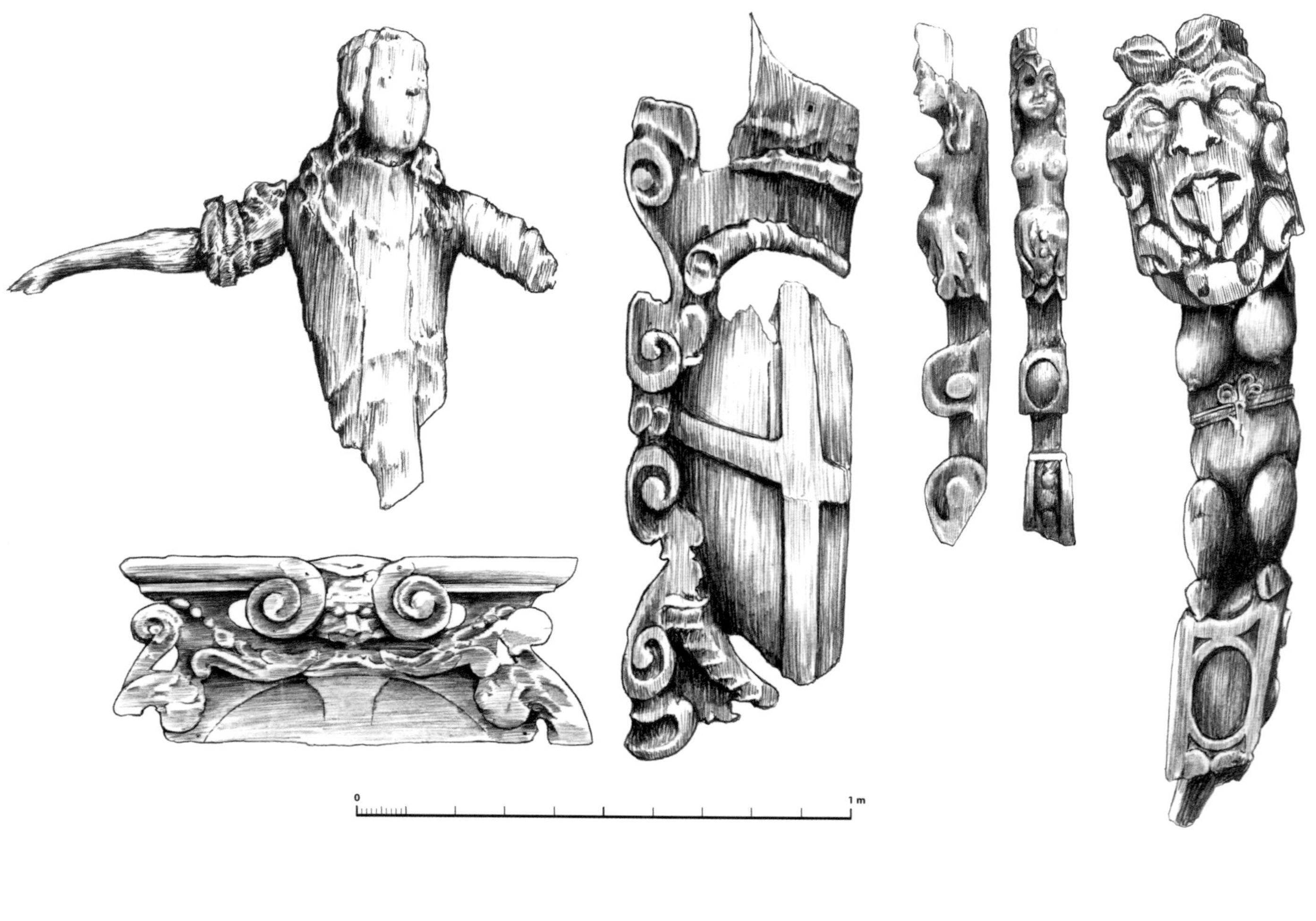

The horizontal elements comprised broad profiled edgings, whilst the vertical ones were made up of sculptures stacked on top of each other. The sculptures were sometimes gilt, but were usually painted monochromatically in yellow paint, in imitation of gold, against a dark background – usually black. Dutch and Swedish ships were instead painted polychromatically in natural colours.

Whilst building *Riksäpplet*, Sheldon discussed with the sculptors in Gothenburg whether they could create sculptural elements for his new ship. As a basis he showed them a model – probably the model with the inventory number 'Ö 3' kept at the Maritime Museum in Stockholm. The sculptors replied that they could envisage doing the work, provided they had a master sculptor. This can be interpreted as their lacking the expertise for such a big undertaking, but the more probable explanation is that they were not conversant with the pictorial repertoire required. Discussions with the Admiralty about also recruiting a sculptor from England were thus entered into.[11]

During the Scanian War the Swedish royal ships went down in order of size – first *Kronan* and then *Svärdet*. Upon the fleet's return to Dalarö in the Stockholm archipelago, *Riksäpplet* came adrift, crashed into a cliff and sank, only five days after the fateful Battle of Öland. Over the past 150-plus years, extensive salvage work has been carried out on the wreck. Many sculptures have been found, and are now held in museum storerooms or are in the possession of private collectors. The sculptures include the typical caryatids that made up the vertical, upright element of the sculptural net on the transom, as well as the burlesque, cloven-hooved female figure crouching in the vault right at the bottom of it. They are recognisable both on Sheldon's model and on other models of English ships of the period.

One difference between English transoms and contemporary Dutch ones was the fact that the former were to a greater extent punctuated by windows. The fact that double rows – big cabin windows – were a new feature in Sweden when *Riksäpplet* was built is revealed by the discussions arising when it came to designing the national coat of arms, which told you that the ship was Swedish, and the sculpted orb, which told you the ship's name. Because of the transom being punctuated by so many windows it was hard to find room for these sculptures.

The design gave rise to many discussions between the shipyard and the Admiralty. Space was ultimately found for all the obligatory motifs, and what are probably the eroded remains of the actual orb motif were found during the salvage work in the 1920s.

For many people of the early-modern period, wooden vessels featuring rich and varied imagery were natural components of everyday life. They were able to understand and decode the meaning and import both of lion figureheads and of sculpted lions of fashion in a way that is not always directly possible for latter-day observers. A great deal of research thus remains to be done into the background and meanings of all the sculpted motifs that were found on ships and are now held in public and private collections or are lying on the sea bed.

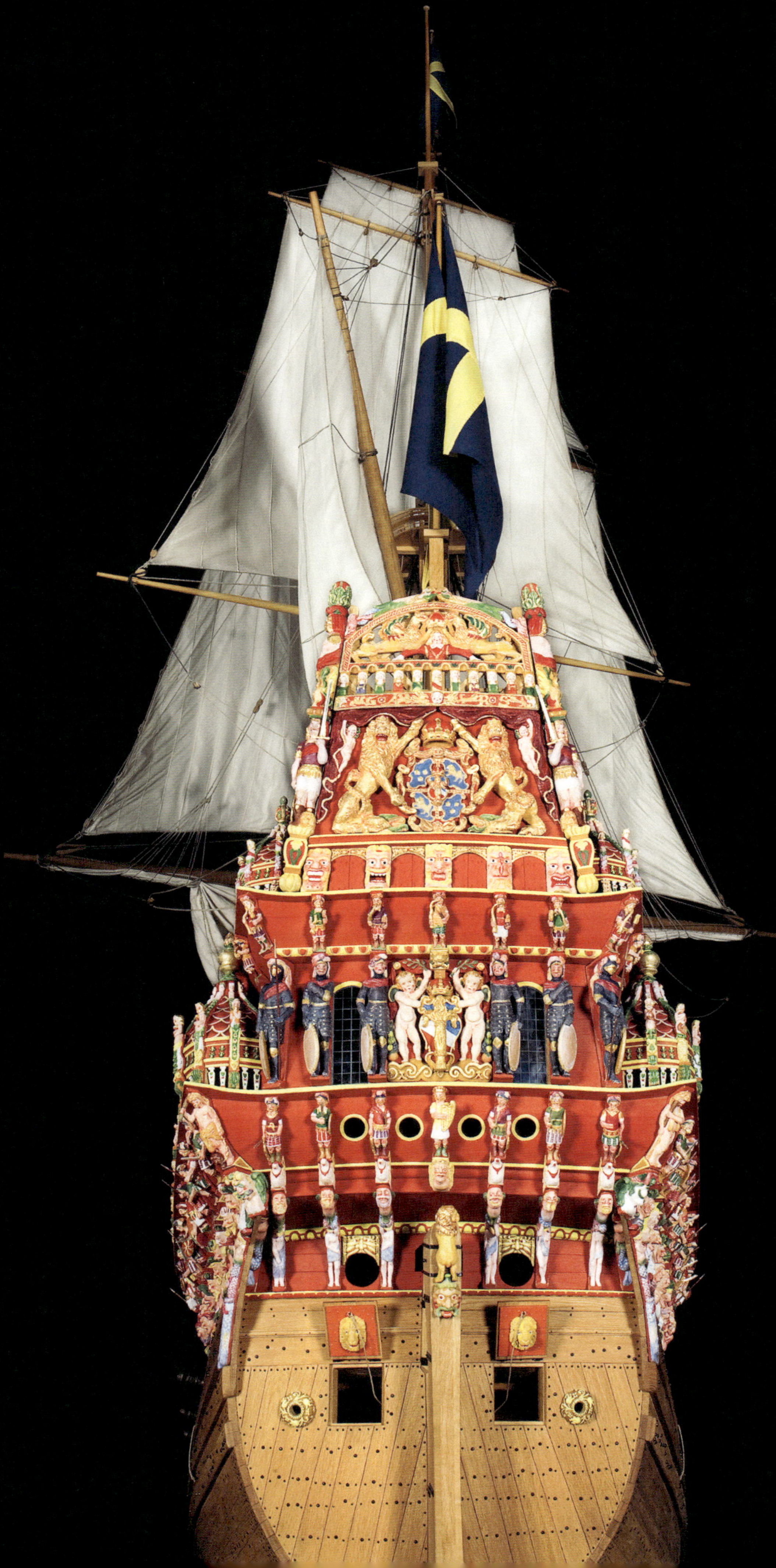

VASA – THE METAPHYSICAL WARSHIP

Fred Hocker

All of the carvings that ornamented *Vasa* served more than a decorative function. They all played an important role in reinforcing the ship's purpose as a warship.

WHEN VISITORS ENTER THE Vasa Museum, they are awed by the monumental ship apparently floating in the floor. Its size and presence are overwhelming in the confined space of a building. Its surfaces are covered in hundreds of carved figures: armoured knights, Roman warriors and emperors, festoons of fruit, garlands of flowers, snarling lions, lounging tritons and mermaids, and fantastical creatures of all descriptions. In its original paint, the ship was a riot of colour and movement, exuberant and even vulgar. To the modern eye, especially one used to the clean lines and surfaces of post-modern architecture, or the unrelieved grey of modern warships, this level of ornament seems tasteless and wasteful. Many visitors wonder why so much cost and effort was spent on decoration, and even if so many heavy sculptures contributed to the ship's loss.

While we might see all of these carvings as ornament, and wonder about the expense, this is not how a person of the 17th century, especially the patron of such work, would have seen them. These are not simply decorative details added to dress up a brutally functional object, but an essential part of the ship's purpose. The carving and joinery in the ship account for 8.6 per cent of the total cost of the ship, and over 20 per cent of the labour cost, which represents a substantial investment (see table below). This is not simply a matter of prestige or conspicuous consumption, of "showing off", but was a commitment of resources to the ship's strategic mission.

Vasa	*Dalers*
Salary for Dutch carpenters	10400
Swedish carpenters	5300
Sawyers and borers	2400
Sculptors and joiners	4600
All types of iron bolts	8700
All types of iron hardware and anchors	2300
All types of nails	5500
Oak planks 1500 pieces at 3 dalers each	4500
All types of oak timber	9600
Total	53300

At the front of the figurehead, a lion holds the Vasa coat of arms – featuring a sheaf of corn. It also displays the name of the ship. In the 17th century, the ship was usually known as *Wasen*.

Every warship ever built, from Athenian *triereis* to nuclear-power ballistic missile submarines, has two related functions. The first and most obvious is its physical function: a floating gun platform, mobile airport or hidden missile silo. Few warships ever fulfill this purpose, serving for years or decades without ever encountering the enemy before sinking or being retired and scrapped. The second function is metaphysical, to be a potent messenger of the owner's self-image, policy and ambition. It fulfills this role in war and peace, whenever someone sees it or thinks about it. In modern terms, a ship's existence and appearance are components of its role in power projection. Just keeping a fleet in being, even if it is never used, is an essential part of its strategic effectiveness as a deterrent to aggressive enemy action and a promoter of stability.

In modern navies, warships are usually very plain or practical in their appearance, especially in the age of stealth technology. Most navies paint their ships in a range of grey schemes, which were originally developed as camouflage. Warships intended for inland or close inshore are sometimes painted in terrestrial camouflage, as the Swedish navy has sometimes done with warships tasked with the defence of the archipelago. In tactical terms, however, most current target acquisition is by radar beyond visual range, so the colour scheme is less and less relevant to battlefield survivability. In a larger sense than the purely tactical, those grey colours have come to have the meaning of "warship", and it is rare to see a merchant ship painted in this way. In the 19th century, some merchant ships were painted in schemes to mimic warships, with false gunports along the sides, in the hope of deterring pirates, but such merchant camouflage is no longer effective or necessary.

In the 17th century, this metaphysical function was carried by the richly carved and painted figures attached to the exterior and interior. They represented ideas in a language common to much of Europe, and which was also used in churches, palaces and various public buildings to communicate key values which the owners wished to promote. Some of the messaging is quite straightforward and might be understood by many, while other symbols required a solid classical education to interpret and some would not have been understandable without a guide to explain them. Some laud the owner and the home nation, while others denigrate or mock the enemy. As a whole, they communicate ideas central to the owner's identity and how he or she wished to be perceived. Taken together, they could also make statements about war aims and national policy.

From 1621 to 1629, Sweden was at war with the Polish-Lithuanian Commonwealth. The causes of the war were complex, with economic issues mixed with religious tensions and Swedish imperialistic ambitions, but one aspect was purely dynastic. The king of Poland and grand duke of Lithuania, Zygmunt III Waza, was the son of King John III of Sweden and his queen, Katarzyna Jagiellonka, the daughter of King Zygmunt I of Poland. He was thus eligible for both thrones. He was elected to the Polish throne in 1587, and succeeded his father in Sweden in 1592. As has been shown repeatedly throughout history, it is difficult to be the ruler of two countries simultaneously, especially when they differ in language, form of government, religion and economic interests. In the 1590s, Zygmunt's paternal uncle, Duke Karl of Södermanland, used fears of the Catholic Zygmunt bringing the Counter-reformation to Sweden as well as frustrations with Zygmunt's divided attentions to encourage a revolt. Zygmunt was deposed in 1599, and Karl named first as protector and then king, as Charles IX. When Karl died in 1611, his son Gustav succeeded him as Gustav II Adolf.

Zygmunt did not accept his deposition and throughout the early part of the 17th century sent agents into Sweden to sound out potential allies for a counter-revolt. Gustav Adolf could not afford an attempt to undermine his legitimacy, and in 1621 invaded Polish territory in Livonia (modern-day Latvia), capturing the city of Riga. The war continued until a stalemate was reached and a truce was agreed in 1629. The imagery on *Vasa*, built at the

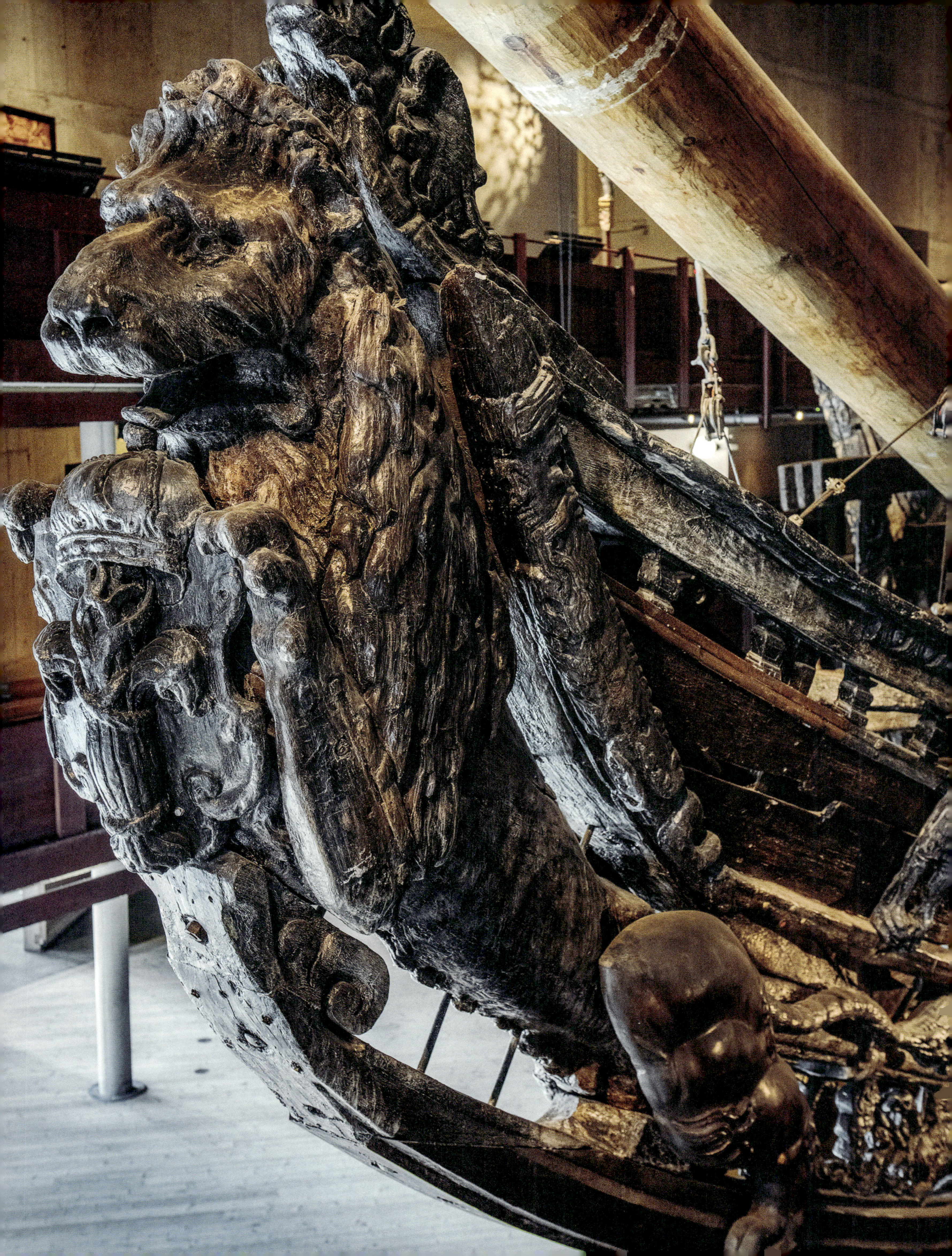

height of the war in 1626-1627, combines more general themes of classical imperial ambition with specific references to the dynastic struggle and mockery of the Poles to reinforce a central message establishing Gustav Adolf's legitimacy.

The overarching theme of much of the sculptural programme is to reinforce Gustav Adolf's claim to the Swedish throne. In a sense, this could be considered a statement of war aims. In his public statements at the outset of the war, he demanded that Zygmunt renounce his claim to the Swedish throne. This message appears in several forms, either by analogy or by setting the king in the company of other great warriors and rulers.

Below: The upper section of the stern, with the young Gustaf Adolf being crowned by two griffins who symbolise his father, Charles IX.

Right: Caricature of a Polish nobleman with a shaggy moustache, fringe and traditional župan tunic crouching under a table. The figure makes fun of Sweden's enemy in the 1620s.

At the bow, ten full figures of Roman emperors are mounted on each side of the beakhead. They are in chronological order, from Tiberius to Septimius Severus or Didius Julianus. Augustus, the first and greatest of the emperors, is missing. This is probably to show that Gustav Adolf is the new Augustus, and he was compared to Augustus in panegyrics and other propaganda. The Latin form of his name, Gustavus, was even an anagram of Augustus. By extension, this would mean that Sweden is the new Rome. At the foremost point of the beakhead, the gilded figurehead of a leaping lion (a *lion salient* in heraldic terms) holds a shield with the arms of the royal house, a *fascine* (bundle of sticks). This is also the ship's nameboard, as the Swedish for this heraldic device is *vase;* in the 17th century the ship was usually known as *Wasen*, the bundle of sticks. In a sense, it is the king's family name, and its choice for the most powerfully armed ship in the navy is an agressive statement about royal power and the king's personal connection to the kingdom. By choosing this name, the king emphasised his role as not only sovereign but head of the dynasty.

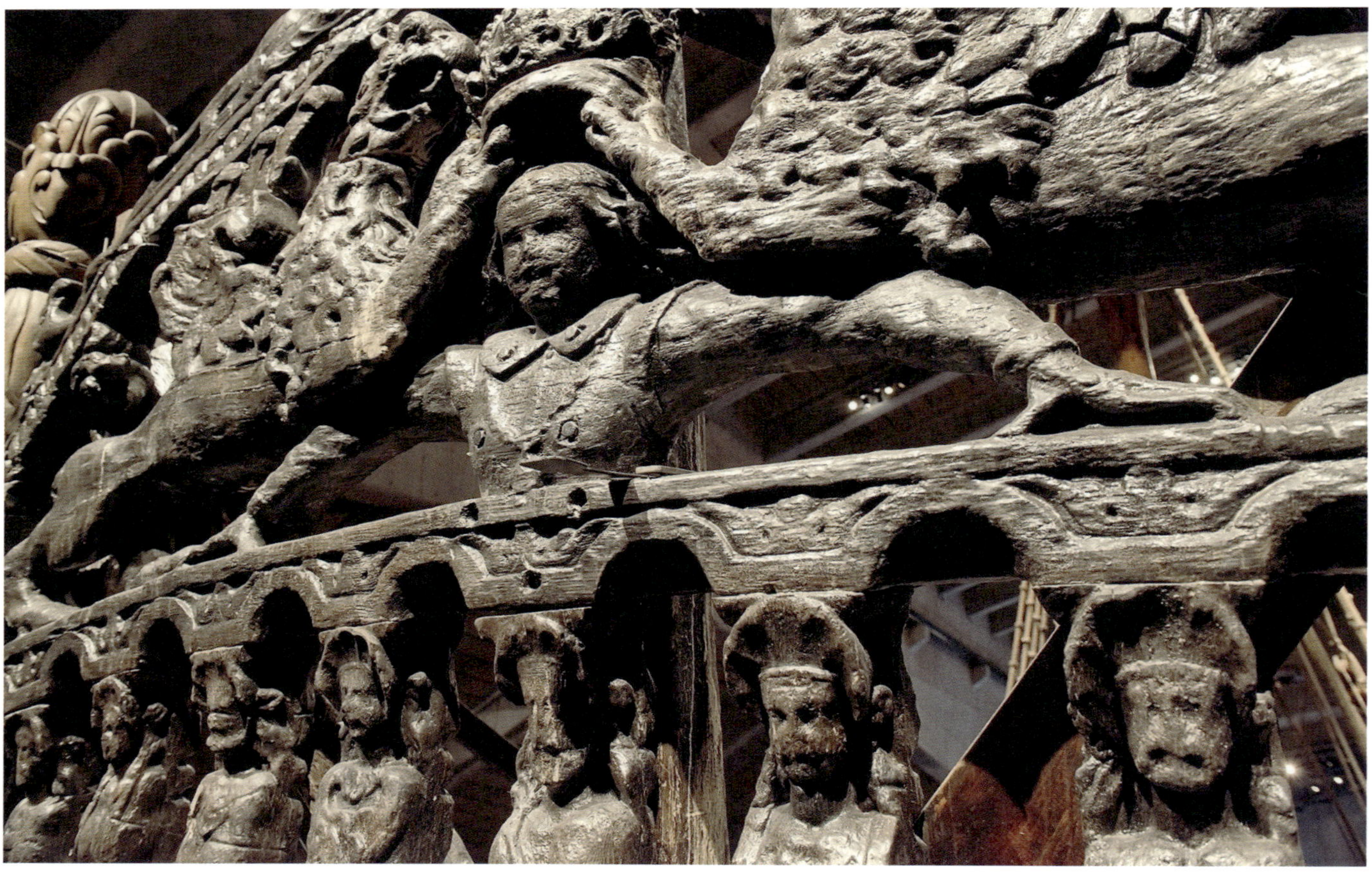

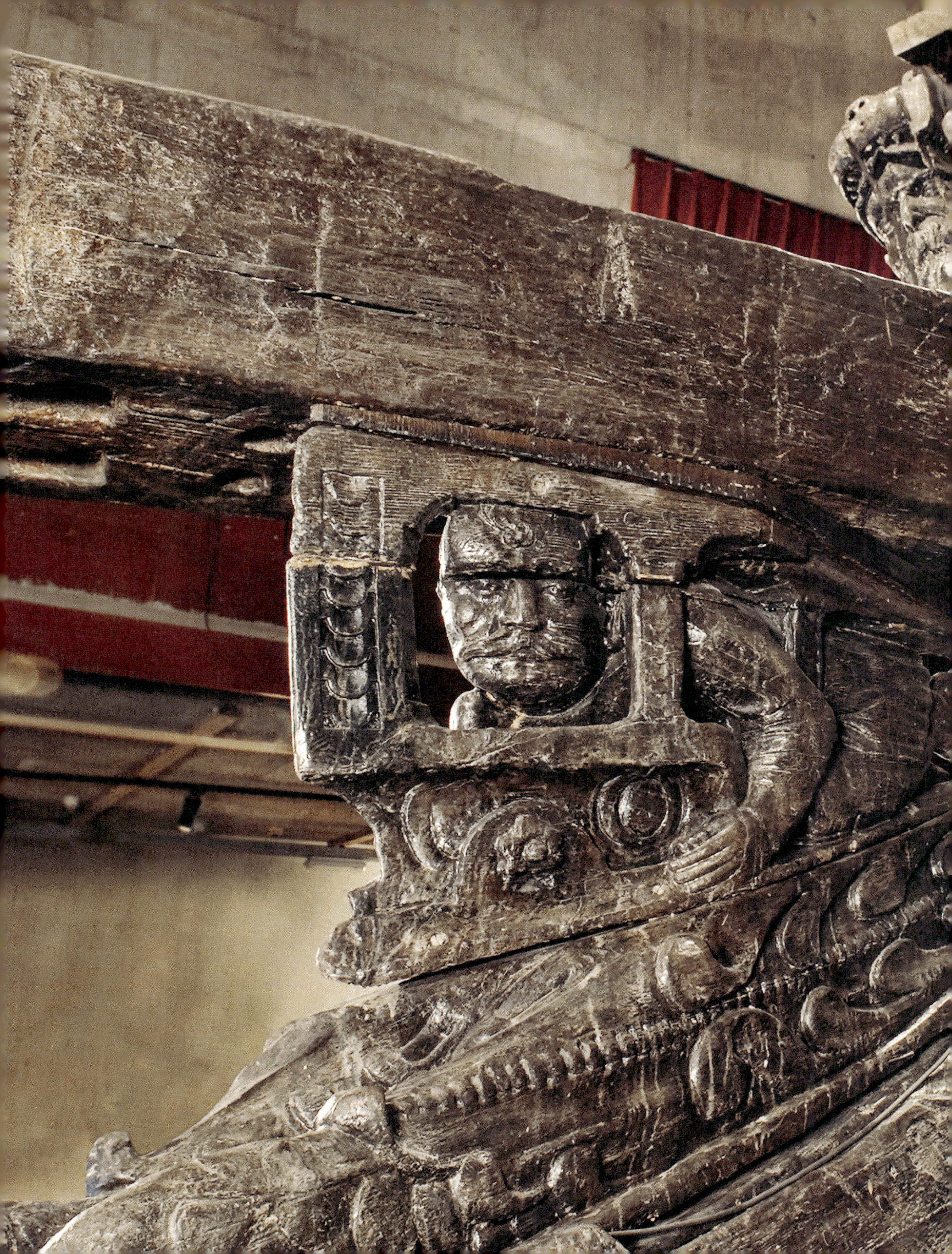

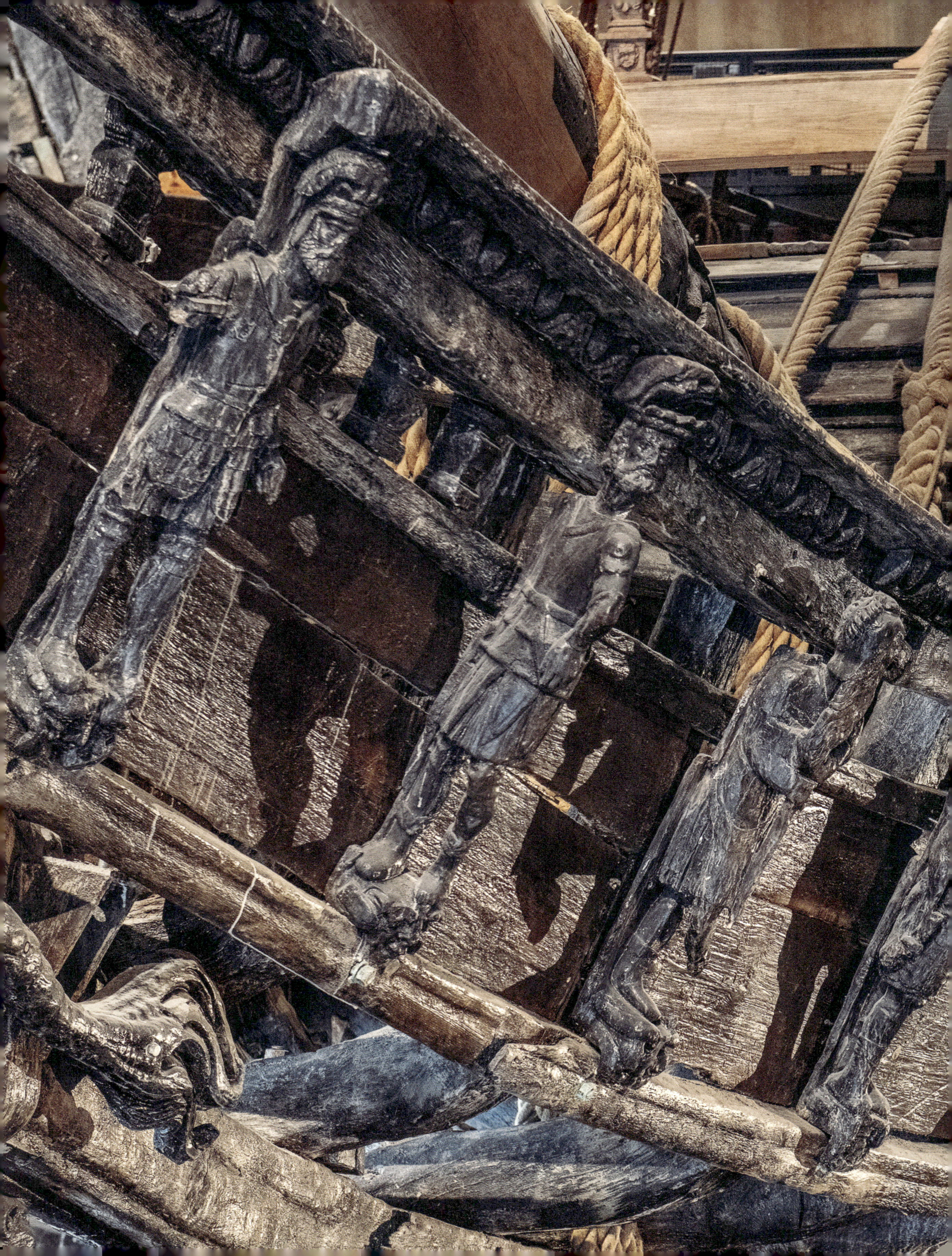

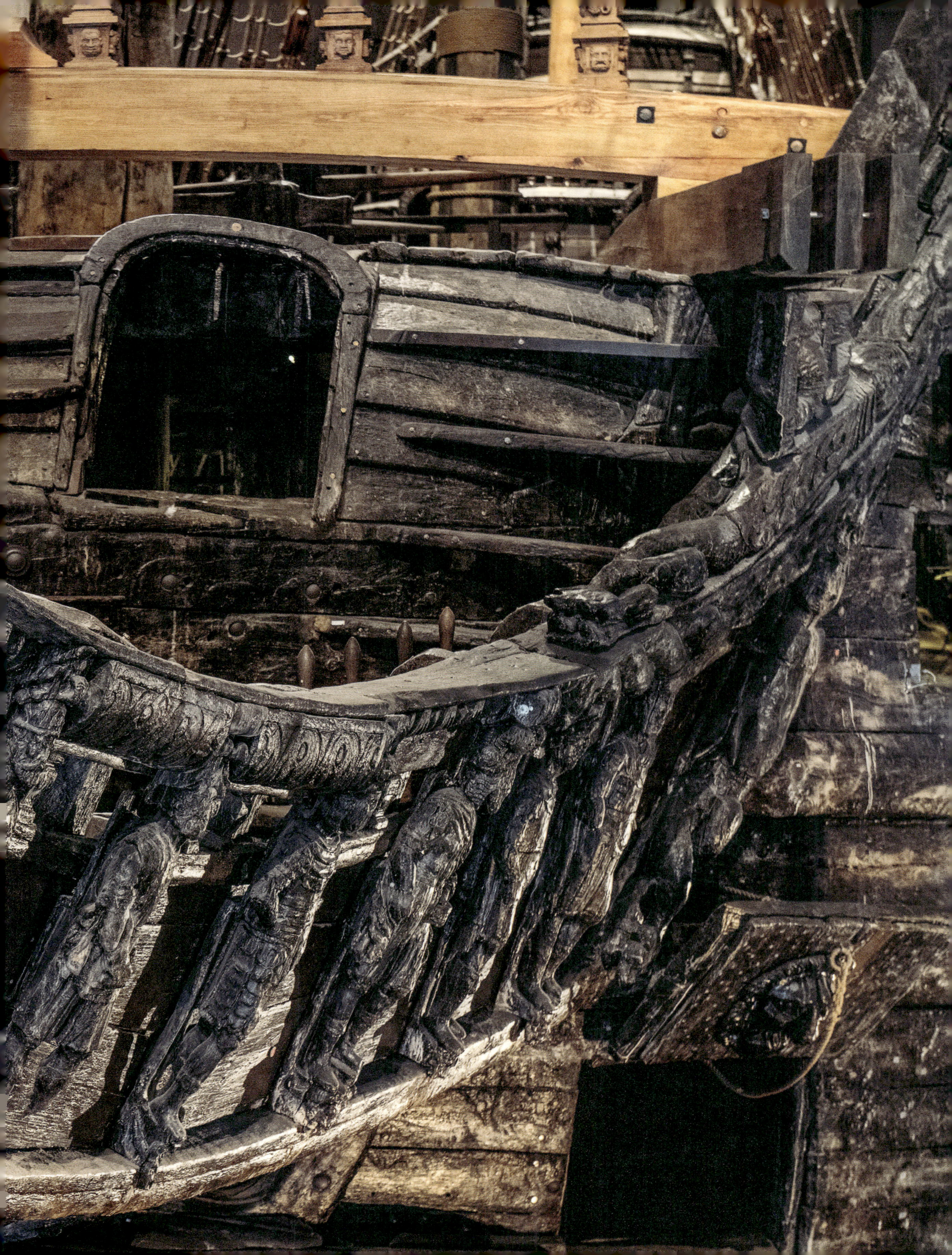

At the stern, the towering transom provided a large, flat panel which was often seen as the ship's "face", a canvas on which the most important images could be composed. This tells a story in several registers. At the base, the royal arms appear again, held by a pair of putti with olive branches, symbolizing the king's proclaimed commitment to peace, backed by festoons of fruit, a symbol of prosperity. Gustav Adolf, although he seemed to enjoy war, always maintained that he went to war in order to establish a just and lasting peace in which the people of Sweden would be free to prosper and to worship in the true (Lutheran) faith. The ship's name is flanked by armoured knights, who may symbolise the king's forbears and establish Gustav Adolf as the latest in a long line of warrior kings.

The central theme of the transom is the identity of the owner, the kingdom of Sweden. The national arms, the three crowns of medieval Sweden quartered with the lion of the Folkunga dynasty (the dynasty who united Sweden in the Middle Ages), are held by two completely gilded lions wearing royal crowns and is surmounted by a royal crown. This indicates that Sweden is an independent kingdom with a long history. The use of so many lions throughout the ship is another reference to royal power, as this animal has a nearly exclusive relationship to royal houses. It also has strong Judeo-Christian symbolism and recalls the king's commitment to the faith.

Between these two registers are a row of warriors carrying torches and trumpets in the form of goat's or ram's horns, the *shofar* of the ancient Hebrews. These are a reference to the Biblical story of Gideon, who defeated a much larger force of Midianites thanks to his faith in God. This was a story with which the king identified, and he presented himself as a modern Gideon, the leader of a small country able to defeat greater forces because of God's support. Not only is Sweden the new Rome, it is also the new Israel.

At the top of the transom is a complicated allegory. Over a banner with the letters GARS are small busts of six women and five men. A life-size figure of a young man holds his arms outstretched over them, while two griffons hold a crown above his head. GARS is an abbreviation for *Gustavus Adolphus Rex Suecorum* (Gustav Adolf King of the Swedes). The eleven people are the Swedes themselves, with different figures dressed as peasants, soldiers, burghers, nobles and priests, the estates into which the people were formally divided. The young man is Gustav Adolf, who came to the throne as a teenager. The griffons represent Charles IX, his father, whose heraldic symbol as duke of Södermanland was the griffon. The young king thus receives the crown from his father, a bold statement to Zygmunt in Poland, "I am the rightful king of Sweden, and you never will be."

This type of imagery was part of a larger tradition in Renaissance Europe, in which princes made use of a wide range of classical and biblical examples to represent their power and ambition on buildings, ships, and furnishings. Roman themes were common, and are among the defining ideas of the age, which gloried in the rediscovery of the ancient past. Political leaders were especially fond of Roman associations, as they recalled the grandeur and reach of a pan-European empire. From the Bible, Old Testament military heroes, such as David and Gideon, were also popular, while New Testament examples are largely missing from warships. In Scandinavia, kings claimed descent from the Goths of antiquity, and so Gothic imagery was also important. A prince should be shown to have his roots in the distant past and to be recovering or recreating the glory of the ancient world.

Not all of the symbolism in the sculptural programme is high-minded allegory. Some of it is dirty jokes. A figure of a crouching Polish nobleman, trapped in a cage, forms the support block for the cathead, the crane used to lift the ship's anchors. This is a triple humiliation for the enemy. In Polish noble society, the punishment for slander was to crawl under the table and bark like a dog, to admit that one had the manners of a dog. Every time the anchor was raised, the enemy was crushed under a ton and a half of iron. Finally, this sculpture does not face outward, but inward towards the beakhead. It is a reminder to the crew of how to think of the enemy, and the best place to look that figure in the face is from the ship's toilet.

There are also some elements which have no deeper meaning. The aesthetic tradition of the period abhorred empty surface, and so the spaces between main elements was often filled with purely decorative or whimsical carvings. These include grotesque faces, foliage, drapery, geometric figures and a wealth of other details. Many of these were the equivalent of stock images, often taken from pattern books which circulated widely in northern Europe. It was a part of the sculptor's art to combine specific messaged imagery with these ornaments to create a harmonious and effective composition.

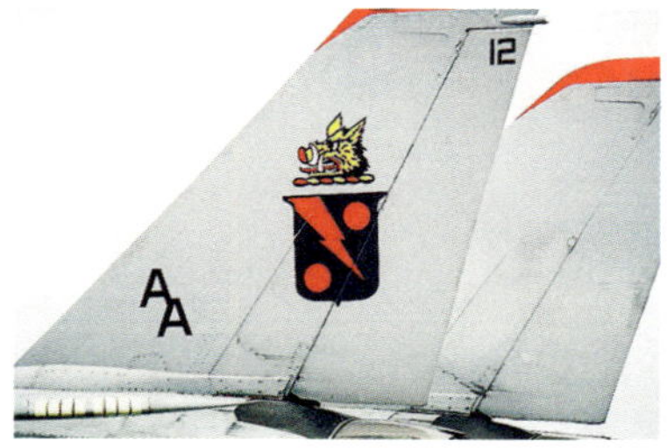

Decoration on the tail of an F-14 type fighter plane flown by the Red Rippers (then VFA-II) in the US fleet.

Preceding pages: Roman emperors lining the sides of the galleon. Each was originally identified by a banner with his name held in the mouth of a lion on which the emperor stood. Some names are still legible.

Although modern warships seem far removed from the extravagant carvings of *Vasa*, there is one place where the same kind of thinking can still be seen. On board an aircraft carrier, where the ship and its aircraft are an unrelieved symphony of greys, a few aeroplanes stand out. On US carriers, the squadron commanders' aircraft often have brightly painted tails. These identify the squadron, usually by both official designation and a rich vocabulary of symbols and badges. Squadrons are given relatively wide latitude in chosing their badges, mottoes and names, and so the tail art seen represents the specific self-image of the squadron, much as *Vasa* is meant to portray Gustav Adolf as he wished the world to see him. Some squadron badges make use of traditional heraldry, while others are more cartoonish, referring to specific events in the squadron's history.

VFA-II is one of the oldest squadrons in the US Navy. Called the Red Rippers, they use a heraldic badge, although with the brashness of military aviators. The central element is a dark blue shield divided in half by a red lightning bolt from upper left to lower right, with a red circle in each half. In heraldic tradition, the lightning bolt is called a *bend sinister*, and indicated that the owner was the natural (meaning illegitimate) son of his father. The meaning of the badge is thus that the pilots are a bunch of bastards with big balls, and are to be feared. This is not far removed from some of the more low-brow imagery on *Vasa*, and performs the same function. It tells the world who owns the aircraft (or ship) and how they want to be seen.

Note:
The sculptures on *Vasa* were the subject of Hans Soop's dissertation, published as *The Power and the Glory*; this remains the essential reference on the carvings. The sculptures are also the subject of a new book being prepared by Anna Maria Forssberg, due to appear in 2020. This will address the context of the sculptural programme, the fabrication of the carvings, the research into the original colours, and the meaning of specific compositions.

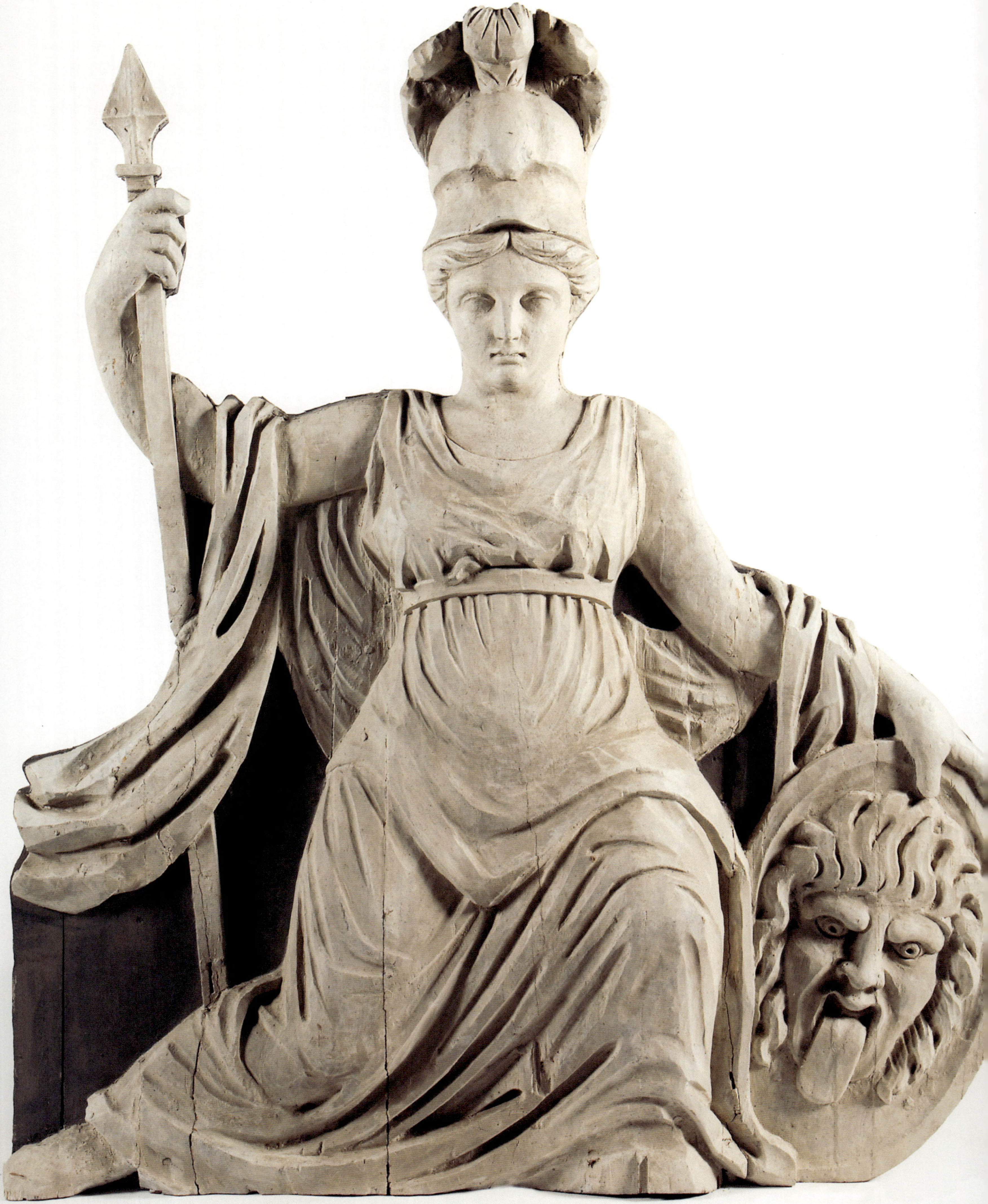

VESSEL ORNAMENTS IN STOCKHOLM'S MARITIME MUSEUM COLLECTIONS

Ingrid Ulfstedt

Stern decoration carved by Emanuel Törnström for the frigate *Af Chapman*, launched in 1830.

STOCKHOLM'S MARITIME MUSEUM has an extensive group of vessel ornaments in its collections that originated from Sweden's merchant and naval fleets. The emphasis is on specimens from the late 18th and 19th century. A few ornaments from before that time are included in the museum's marine archaeological material.

The museum's collections were formed from two older collections that were merged when the current museum opened in 1938. One has its roots in the collections of museum objects and models housed at the naval yards in Karlskrona and Stockholm. Some of this material was showcased in the Naval Pavilion at the Stockholm Art and Industrial Exhibition of 1897 and then brought together again several years later in a maritime museum located at various Stockholm addresses over the years. The second collection consisted of specimens and keepsakes from merchant shipping, which people at the turn of the 20th century realised were disappearing and which had been collected from across Sweden, including through appeals in the press. The Association for Sweden's Maritime Museum in Stockholm, which was formed in 1913 and whose members included ships' officers and merchant shipowners, launched a drive to collect funds for this purpose.

Ornaments from the Karlskrona naval yard mostly ended up in that city's Naval Museum, but a few hundred smaller specimens and a number of larger ones were transferred to the Maritime Museum in Stockholm[1] as early as 1912. For many of the objects, there is information about which vessels they come from, and in such cases we also know more about who designed them and who carved them.

Less background information is available about the ornaments collected from the Stockholm naval yard. There is currently no explanation for this difference. At present, nothing can be said about who designed or carved them, although it is highly likely they share the same origins as their Karlskrona counterparts. Since 2002, the Naval Museum in Karlskrona and the Maritime Museum in Stockholm have been part of the same government agency. With this merger, parts of the two naval yard collections that have been separated can now more clearly complement one another and can perhaps be allowed to tell their shared story.

The ornaments from the merchant fleet have a more varied background, and in most cases there is insufficient documentation. The objects came, for example, from salvage companies, old naval yard environs, dredging operations and the sheds and other outbuildings of coastal residents. There was rarely any information provided apart from who donated them to the museum.

The two groups of ornaments have also been treated differently at our museums. For natural reasons, figureheads from naval yards could be placed in a context to a greater extent. They are often fraught with symbols, and since it is known what the symbols represent, the history of the vessel can be linked to the history of Swedish defence. Their iconography – identification of the allegories and symbols – and iconology – interpretation of the meaning of these symbols in their historical origins – have been described in detail.[2] This is especially true of the ornaments for Gustav III's fleet, with Greek mythology and idealised virtues frequently featured. The figureheads of the merchant fleet, in contrast, have seldom been incorporated into any account, analysed based on their depictions or explained in iconological terms. They often hang as decorations without a name and no explicit context. The exceptions are the stories of sailors and their relation to the sea, with figureheads described based on the symbolic meaning they held for these sailors.[3]

For several years now, the museums have been working with contextualisation, and in conjunction with this there has been discussion about what is known in the Nordic countries as *tingens metod* or 'the method of the object', which entails opening up museums and collections to the public and involving more people in telling stories about the objects. Throughout their history, these objects have had relationships with many people, all of whom have given them meaning. It can also be fruitful to highlight these people's perspectives and stories. In terms of vessel ornamentation, particularly interesting contributions could be provided not just by art historians but also by sculptors and by economic and business historians.

Naval vessels have been built in a number of Swedish cities, including Stockholm, Gothenburg and Kalmar. Starting in 1680, ship-building took place largely in Karlskrona. Great emphasis was placed on the ornamentation of naval vessels, and commissioned work was of high quality. Among those who produced models and influenced the design are well-known names such as Jean Bérain the Elder (1640–1711), Nicodemus Tessin the Younger (1654–1728), Jean Eric Rehn (1717–1793), Johan Tobias Sergel (1740–1814) and Carl August Ehrensvärd (1745–1800). But the shipbuilder Fredrik Henrik af Chapman (1721–1808) also designed ornaments, which shows how integral decoration was to the vessel's construction.[4] Jean Bérain, the renowned designer responsible for decorating vessels in the French

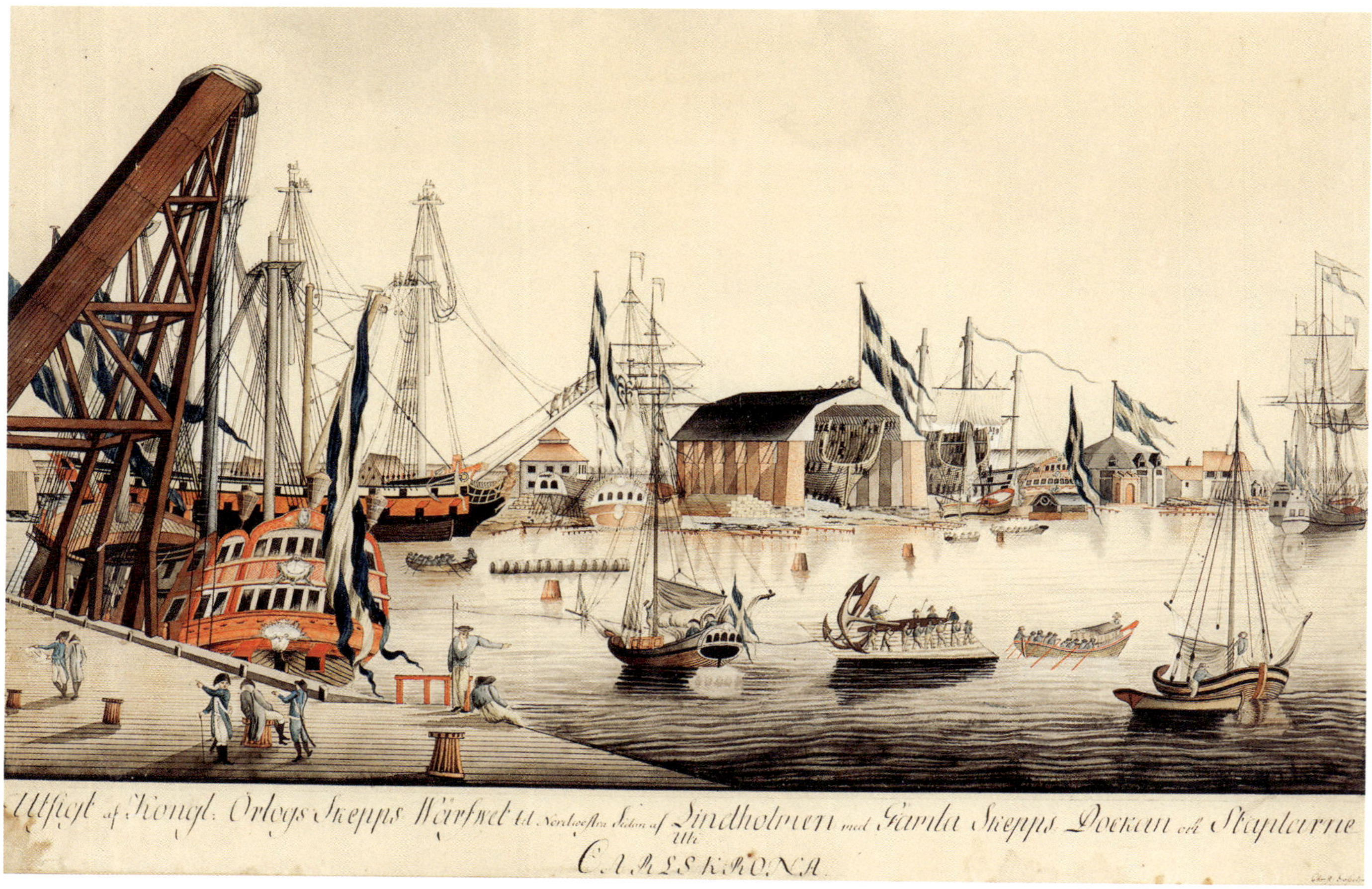

"Utsigt af Kongl. Örlogs Skepps Warfwet til Nordwestra Sidan af Lindholmen med Gamla Skepps- Dockan och Staplarne Uti Carlskrona." ('View of the Royal Naval Ship Yard on the Northwest Side of Lindholmen with the Old Ship Dock and Wharf Pilings in Karlskrona'). Pen-and-ink drawing with watercolour, signed Christ. Gosselman.

fleet, among other work, introduced ornamentation that was better adapted to the vessel's functions. Nicodemus Tessin the Elder brought this style to Sweden after having encountered his French colleague on a study trip to France in 1677–1678. It has also been assumed that Tessin the Younger, who oversaw the Swedish navy's decoration programme, contacted Bérain about proposed drawings for the stern of the ship-of-the-line *Konung Karl*, launched in 1694.[5]

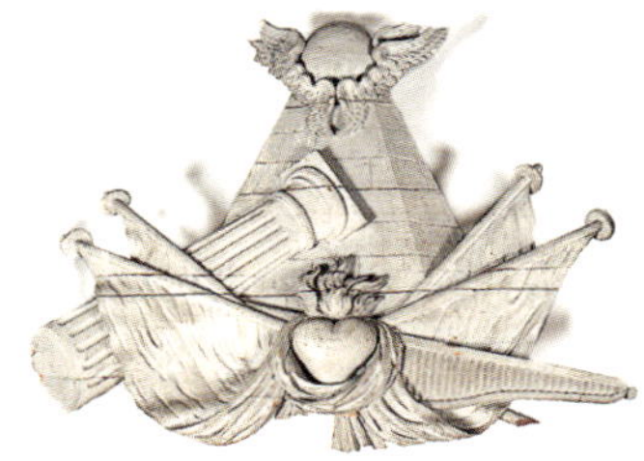

Stern decoration with strong features from the Pyramid of Cestius, which frequently recur in Jean Eric Rehn's work.

There were always sculptors linked to these projects. Work from the following people has been preserved and identified: Henrik Schütz (1650–1701), stern ornaments for the ship-of-the-line *Carolus XI*, built at Skeppsholmen in Stockholm in 1677–79; Johan Törnström (1744–1828), primarily the figureheads for the frigates and ships-of-the-line built in Karlskrona during the extensive shipbuilding programme of the 1780s; and his son Emanuel Törnström (1798–1833), including the figurehead and stern ornaments for the ship-of-the-line *Af Chapman*, launched in 1830. From the steamship era, the last great period of vessel ornamentation, there is work by Lars Nordin (1808–1873), including decorations for the steam corvette *Thor* from 1841 and the steam-powered ship-of-the-line *Stockholm* from 1856; by Henrik Nerpin (1834–1908), the steam-powered ship-of-the-line *Vanadis* from 1862; and by Carl August Sundvall (1855–1941), the steam corvette *Saga* from 1877. Of the second group, Lars Nordin was the last admiralty sculptor employed at the naval yard; the others worked in their own shops.

In the rich material from perhaps our best known master shipbuilder, Fredrik Henrik af Chapman, there are not only drawings and models for the development of technical solutions and new vessel types but also, as noted, sketches of ideas for vessel decoration, with ornaments for figureheads, transoms and galleries. They include pictures of vessel ornaments made during his study trips to England and France in the 1750s, but also sketches of ornaments for the vessels he designed for construction starting in the 1760s and all the way to the late 1790s.[6] We know that some of these ideas were carried out, and some specimens are also preserved.

The first major man-of-war that Chapman designed is the 70-cannon ship *Kung Adolph Fredrik*, launched in 1775. Both Chapman and Jean Eric Rehn, who designed a number of royal palaces, made drafts of the ornamentation on the stern.[7] In a pen-and-ink drawing with watercolour signed Christ. Gosselman, presumably from 1782, Karlskrona naval yard is depicted in great detail with shipyard buildings and vessels, both older ones and some under construction. All have been identified, including *Kung Adolph Fredrik,* seen lying in the Polhem Dock with its stern facing the viewer. Some details on the stern cannot really be distinguished, but the actual decoration is closer to the scheme suggested by Rehn than Chapman's own.

The collection includes a beautifully carved ornament of particular interest. It is probably a stern ornament, but any information about who made the model, who carved it and which ship it belonged to or was intended for is lacking. The ornament is sculpted in the shape of a pyramid made of finely dressed masonry and crowned with an orb encircled by Hermes' wings. At the base, a flaming heart is flanked by four crossed standards and a quiver of arrows. A Tuscan pillar tumbles from the centre of the pyramid.

No other ornament in the collection has such strong classical features as this one. One of the creators of the Swedish counterpart to classicism, the Gustavian style, is Jean Eric Rehn. He was an architect, designer and professor at the Royal Academy of Painting and Sculpture, the precursor of Sweden's Royal Academy of Arts, starting in 1757, as well as palace architect from 1763. Among the sketches of buildings from antiquity that he made when he visited Rome in 1755–56 are a number that feature a pyramid and column, including the Pyramid of Cestius.[8] As early as 1762, Rehn used this motif in a burial monument located at Järfälla which is a smaller version of the Roman pyramid. The affinity between this pyramid, the earliest on Swedish soil, and the pyramid in the stern ornamentation is unmistakable. It has never before been possible to show that an ornament had been produced from Rehn's drawing, but in light of this evidence (sketches and the stern decoration executed), it is interesting that this has typical Rehn features.

Between 1782 and 1785, a series of ten ships-of-the-line and ten frigates were built in Karlskrona following a streamlined method that Chapman introduced. Johan Tobias Sergel made sketches of figureheads for two of the ships, *Kronprins Gustaf Adolph*, launched in 1782, and *Fäderneslandet* (The Fatherland), launched in 1785, but neither was ever executed.[9]

As early as 1777, Gustav III issued a directive that the ornamentation of naval ships was to be executed by painters and sculptors with artistic training and that they should be members of the Royal Academy of Painting and Sculpture. The directive was issued because the newly launched ship-of-the-line *Konung Gustaf III* had been outfitted with a figurehead depicting the king which was clearly based on a model by Chapman but was very poorly executed. The king therefore decided that recruitment requirements for sculptors needed to be more stringent.[10] Some of the stern ornamentation for the same ship-of-the-line is preserved. It was carved by Johan Törnström, who was an admiralty sculptor in Karlskrona starting in 1781. The transom ornamentation was probably also rejected. Törnström, who was not a member of the academy, received his qualifications during his time as Masreliez's apprentice when he worked on the interior of the Stockholm Royal Palace and from his work at the Djurgården naval yard in Stockholm. Chapman presumably came to know Törnström during his time at the same naval yard. It is not fully clear how much freedom Törnström had in creating the individual ornaments. During his first ten years in Karlskrona, however, he worked with Carl August Ehrensvärd (1745–1800). There is no doubt about his drawing and execution skills, and he is thought to have had a collegial relationship with Sergel. In an unsigned cartoon, Törnström bemoans his lot in life, which must have been more circumscribed and less lucrative than his colleague's.

One of Törnström's first assignments was to carve the figurehead for the royal yacht *ESPLENDIAN*, built at the Karlskrona naval yard in 1782 and conserved in the Maritime Museum in Stockholm. In this man-of-war, we see the full range of Törnström's composition skills and craftsmanship. There is no information about who drew the model. Törnström executed the decorations for another three royal yachts, designed by Chapman, the state barges *Galten* (The Boar), *Delfinen* (The Dolphin) and *Draken* (The Dragon), all in the

Below: 'Törnström collects treasures. Sketch by Sculptor Törnström'. Unsigned and undated.

Top right: Figurehead from the royal yacht *Esplendian*, carved by Johan Törnström around 1782. Below is another figure carved by Törnström. This is for the frigate *Euridice*, launched in 1785.

Next pages: The brigantine *Bull*, which in 1842 became the second Swedish vessel to circumnavigate the globe.

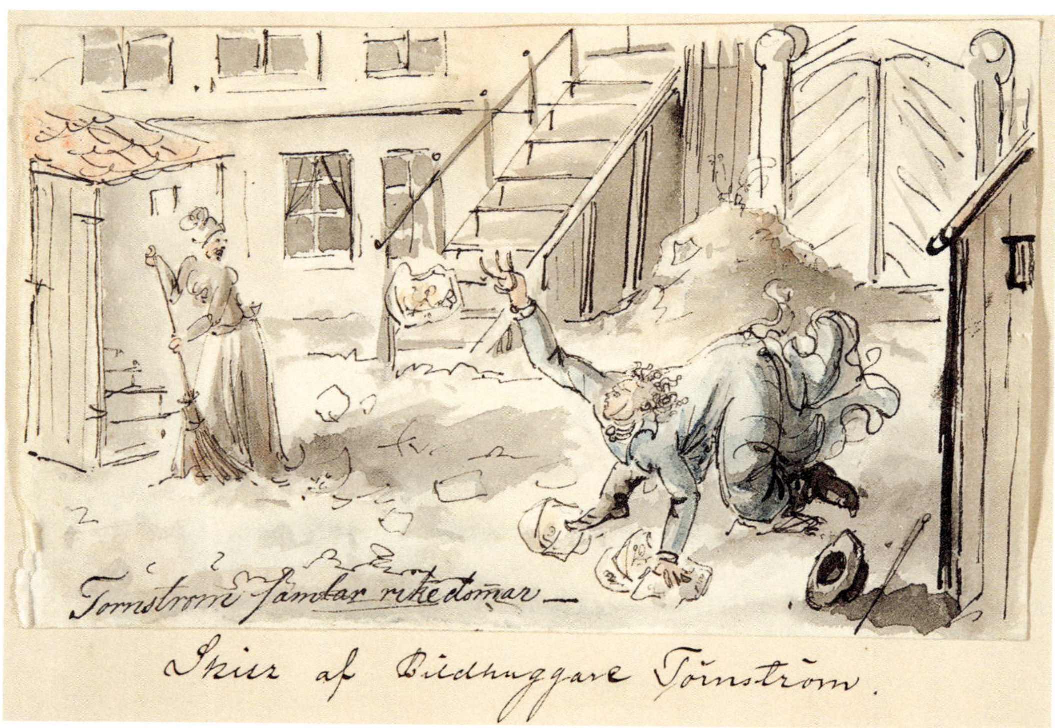

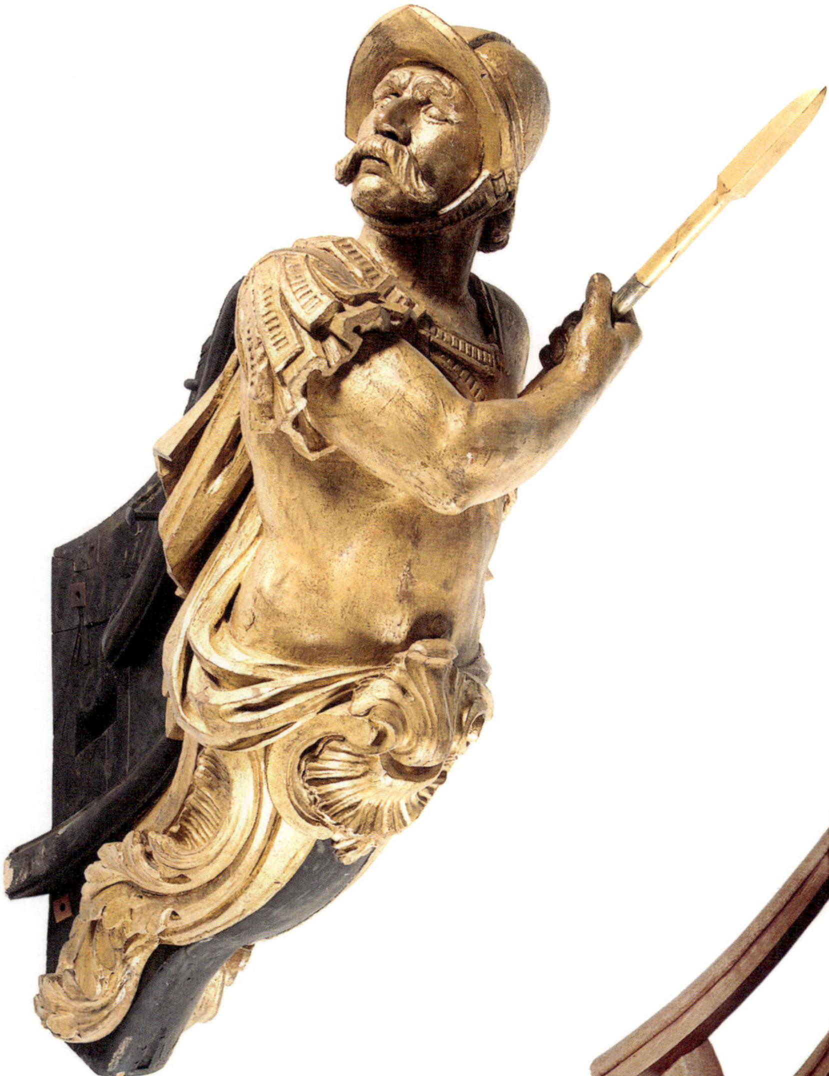

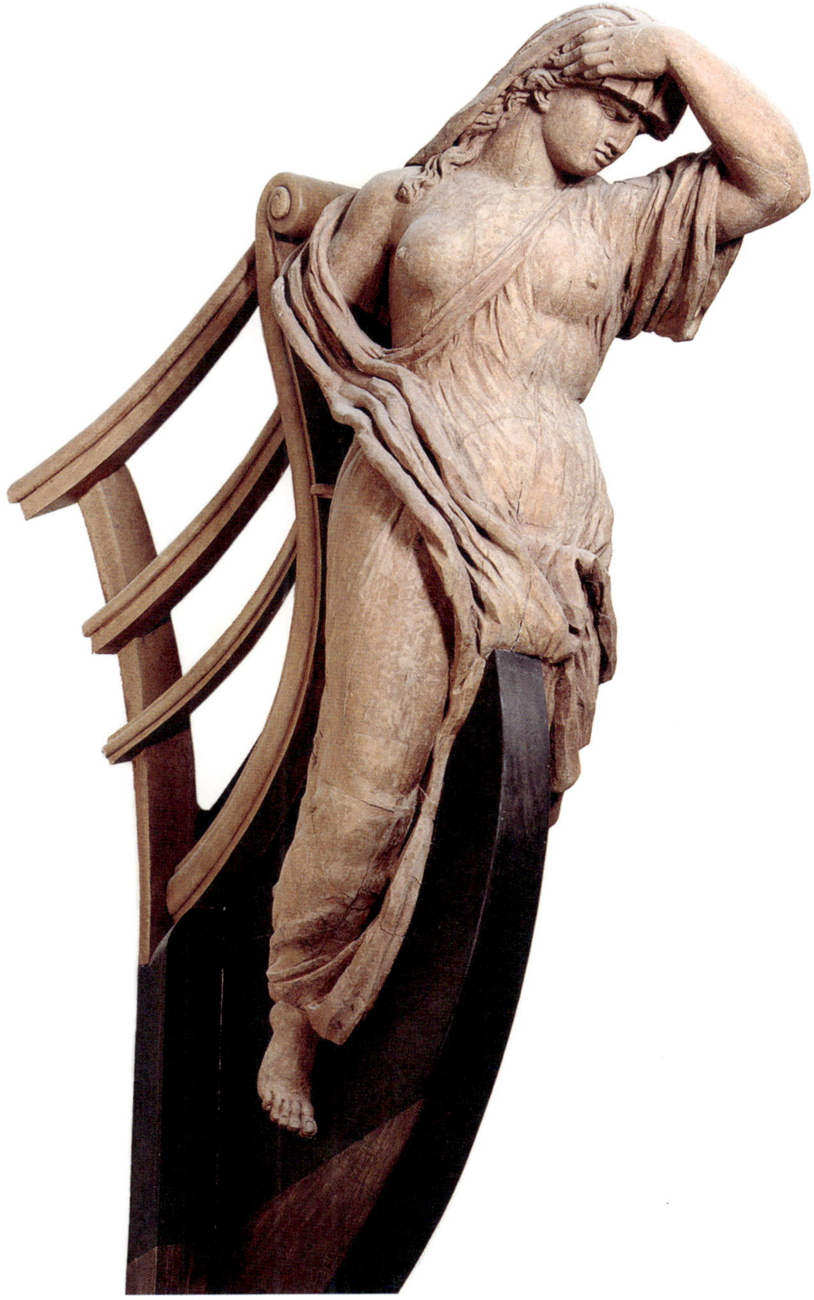

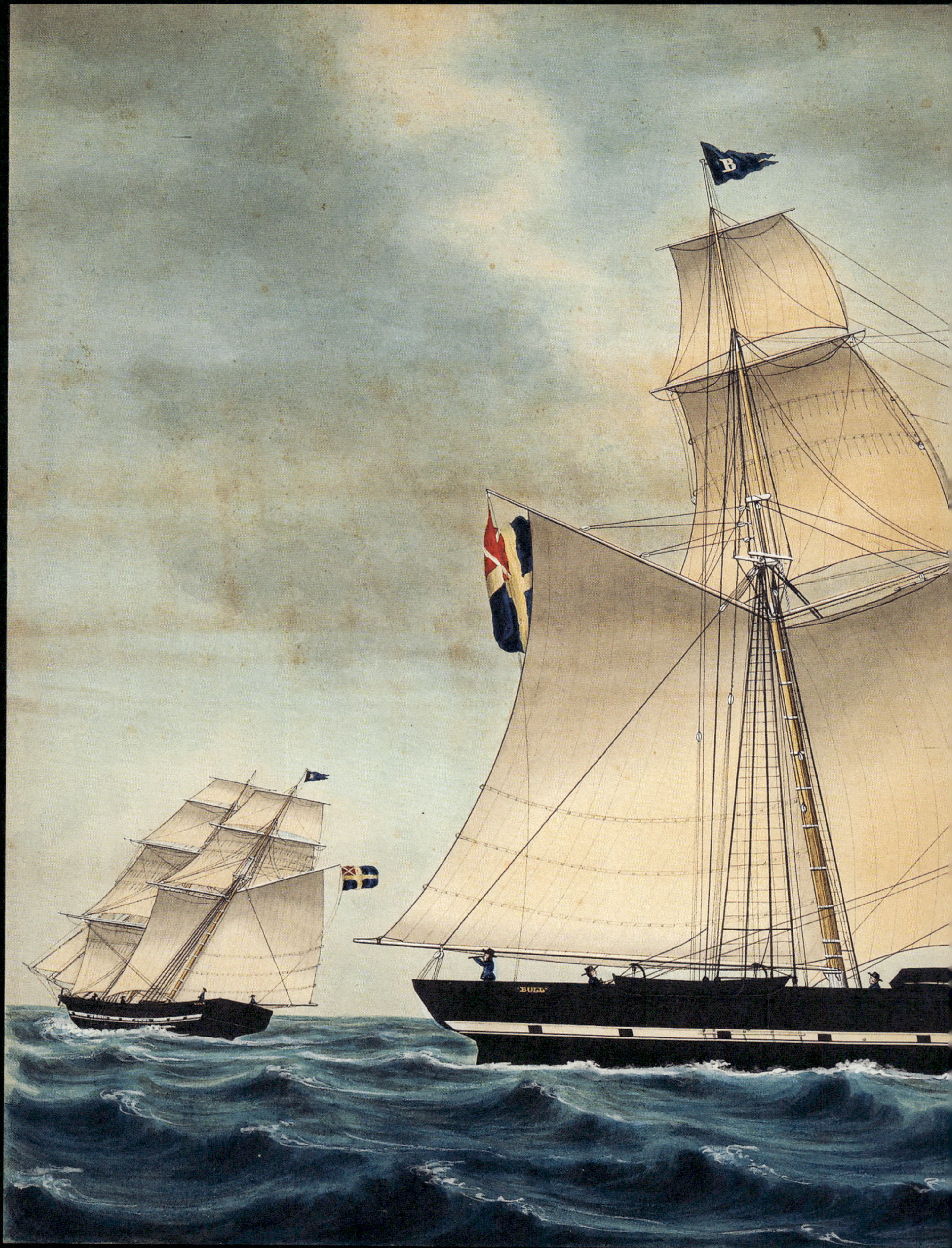
BULL

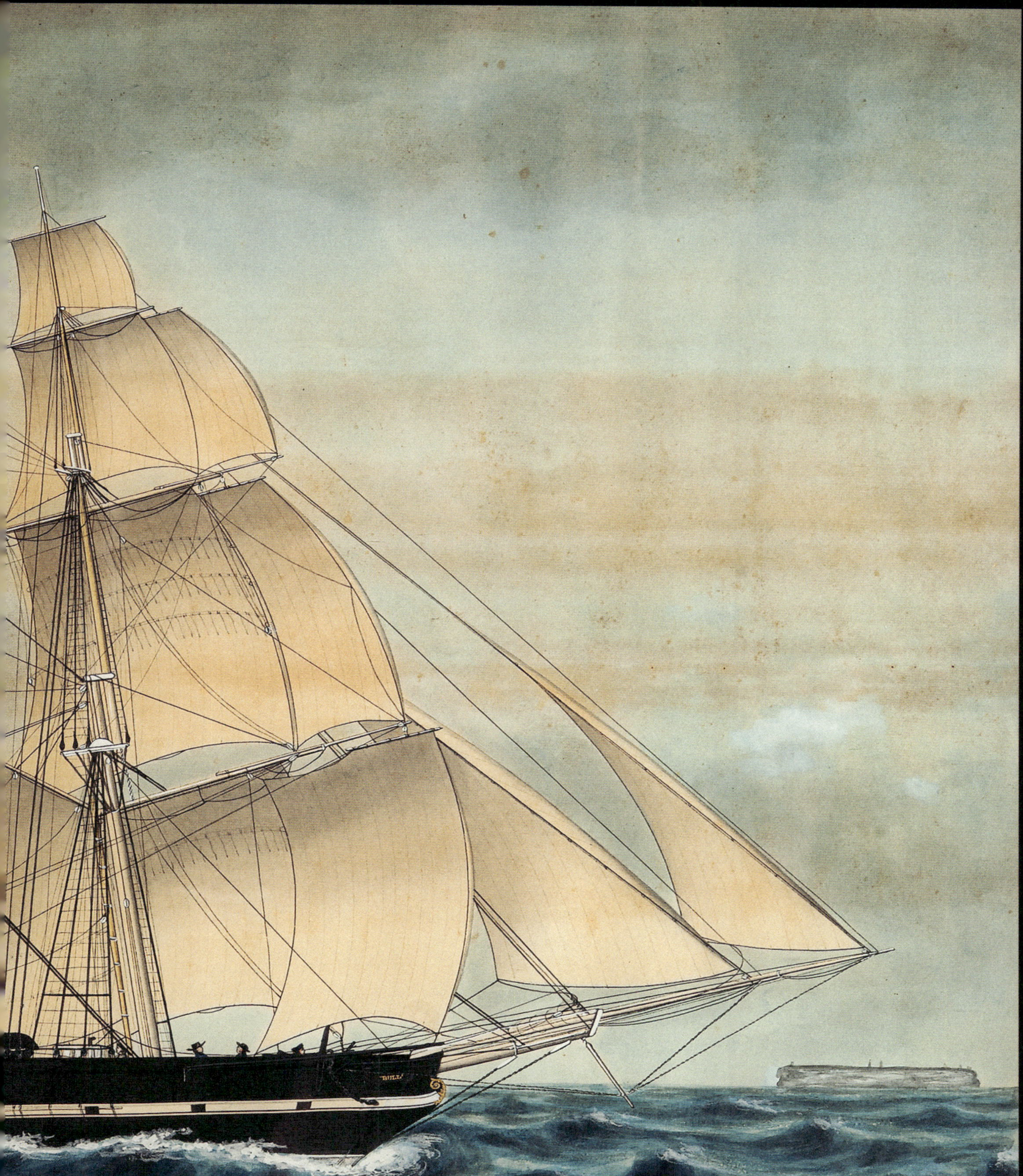

Maritime Museum's collection.[11] Many of the figureheads for the frigates and ships-of-the-line in this series are also conserved. The frigates were named after figures from classical antiquity while the ships-of-the-line were named after masculine virtues, which were embodied by the figurehead. However, the first ship-of-the-line built was given the name *Kronprins Gustaf Adolph*. Among the better known figureheads are that for the frigate *Euridice*, launched in 1785.

Törnström was succeeded in his work as ornamental sculptor at the Karlskrona naval yard by his son Emanuel (1798–1833). He served as an apprentice to his father but also studied with the portrait sculptor Gustaf Göthe (1779–1838). One of his pieces, the sculpture for the transom of the frigate *Af Chapman*, launched in 1830, shows his fine ability to achieve plasticity. Emanuel Törnström also carved the figurehead, a portrait bust of Fredrik Henrik af Chapman. A number of vessels built during Emanuel Törnström's time featured portrait-like ornamentation. He provided the ship-of-the-line *Carl XIV Johan*, for example, with a full-length figurehead depicting the king. The ornamentation of figureheads was otherwise reduced at this time to emblems and portrait busts.

Figurehead from the snow *Wynyard*, depicting the 4th Marquess of Londonderry, Frederick William Stewart, owner of the Wynyard estate.

The Törnström era was followed by the age of the steamship, which also made inroads in the navy. The conditions for ornamentation changed then since vessel sterns gradually took on a different shape. Among the ornaments carved by Lars Nordin, the last ornamental sculptor employed at the naval yard, were those for the fleet's only side-wheel steamer, the steam corvette *Thor*, which was made out of wood and launched in 1841. Many of the ornaments in the Naval Museum and Maritime Museum collections that belong to vessels built during Lars Nordin's time and which have no information about the maker can with some certainty be considered to have been carved by him. They consist mainly of portrait busts and shield emblems.

For a few decades, vessels continued to be made out of wood although they were equipped with steam engines. This was also true of the steam frigate *Vanadis*, launched in 1862, which is known primarily for circumnavigating the globe between 1883 and 1885. The stern was decorated with a figurehead depicting the Norse goddess Freya, carved by Henrik Nerpin. Nerpin represents a new category of sculptors. He provided ornamentation not just for naval but also merchant vessels. Sculptors, including Nerpin, now had technical training from one of Sweden's recently established technology institutes. Nerpin had also studied at the Paris Institute of Technology and taught ornamental drawing and modelling for decades at the Swedish Handicraft Association's Sunday and Evening School in Stockholm, later renamed the University of Arts, Crafts and Design.

Nerpin was in great demand, including as a decorator for the Rörstrand porcelain maker and as an interior designer at the Stockholm Royal Palace. He also designed objects for the applied arts and had his own workshop in Stockholm. Along with the Karlskrona naval yard, his clients included Stockholm shipowners. One was the Rydberg Foundation, which arranged the first formal training for seamen "to train skilled sailors for the practical profession of seamen". Nerpin carved the figurehead for the foundation's first training ship, the full rigger *Abraham Rydberg*, built in Karlshamn in 1879, which was designed as a portrait bust of the foundation's founder and benefactor, the shipping company owner and wholesaler Abraham Rydberg.

The ornamentation of naval vessels now declined in importance. The last large battleship made out of wood was the steam corvette *Saga*, launched in 1877. Like *Vanadis*, its name was taken from Norse mythology. Saga was the goddess of poetry and history and is depicted in Carl August Sundvall's figurehead with a vellum scroll in one hand – the symbol of written history. With the other hand, she lifts the veil of truth, seeing everything as it really is and, seemingly, into the future. The figurehead was finally completed in 1884 and was the last figurehead commissioned for the fleet. Like Nerpin, Sundvall was a teacher at the School of Handicraft.

While figureheads for the navy fleet decreased in size later in the 19th century, merchant vessels – brigs in particular – were decorated with increasingly larger figureheads, and new

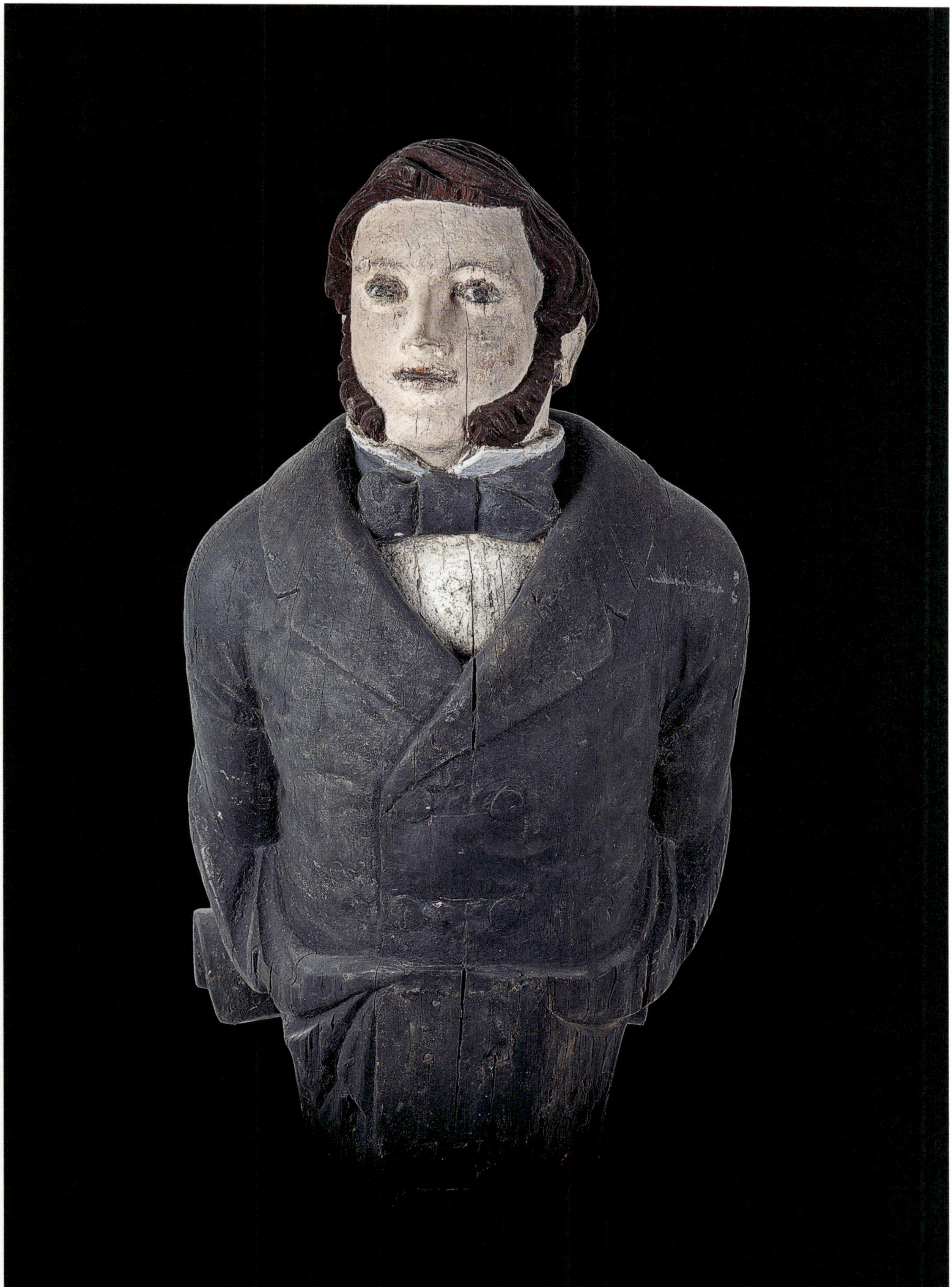

motifs also appeared. Before 1815, small ornaments were predominant. Now full-length female figures, male busts and new animal motifs were introduced.

However, smaller vessel types such as barges, ketches and brigantines lacked figurehead decorations for a long time and were fitted instead with a small shield with leaf ornamentation, known as a "stävknorr". One example is the brigantine *Bull*, built in 1841 at the Fredrik Liljevalch shipyard in Umeå. Another smaller vessel, the brigantine *Mary Ann*, was built at the same shipyard three years earlier. *Mary Ann* was the first Swedish vessel to circumnavigate the globe in 1840. Shortly after that, *Bull* made the second and third Swedish voyages around the world. All three sailings were under the command of Captain Nils Werngren.

Figurehead from the Flying Horse Line's iron-hulled barque, the clipper *Pegasus*, built in 1884 in Liverpool.

The collection includes figureheads depicting portraits of men, which was a typical decoration on Swedish merchant vessels in the mid-19th century. In only a few cases is there information available about which vessels these ornaments came from. They were often saved for posterity by coastal residents who reclaimed pieces from wrecks or sunken ships. In particular, objects donated to the museum in its early years lack a lot of information.

One example of these anonymous figureheads is a fine half-length portrait of a male figure. The only information registered is that it was donated to the museum by a Mrs Lund in Nyhamnsläge in 1949. Someone later added *Wynyard 1855* to the index card, but there is no indication where the information came from. One snow, *Wynyard*, is found in several editions of Lloyd's Register of Ships. Its home port was Seaham Harbour, in the coal-mining district of County Durham, England. The harbour was built by the local industry's biggest stakeholder, Charles Vane, to ship coal and as a complement to the larger harbour in Sunderland, five miles to the north. The snow was one of six vessels owned by Martin Moore in Sunderland, who was involved with coal shipping. In other words, he was a mid-sized shipowner and also had a seat in Sunderland's local government. How can we know now that the figurehead actually has something to do with the *Wynyard* found in Lloyd's register?

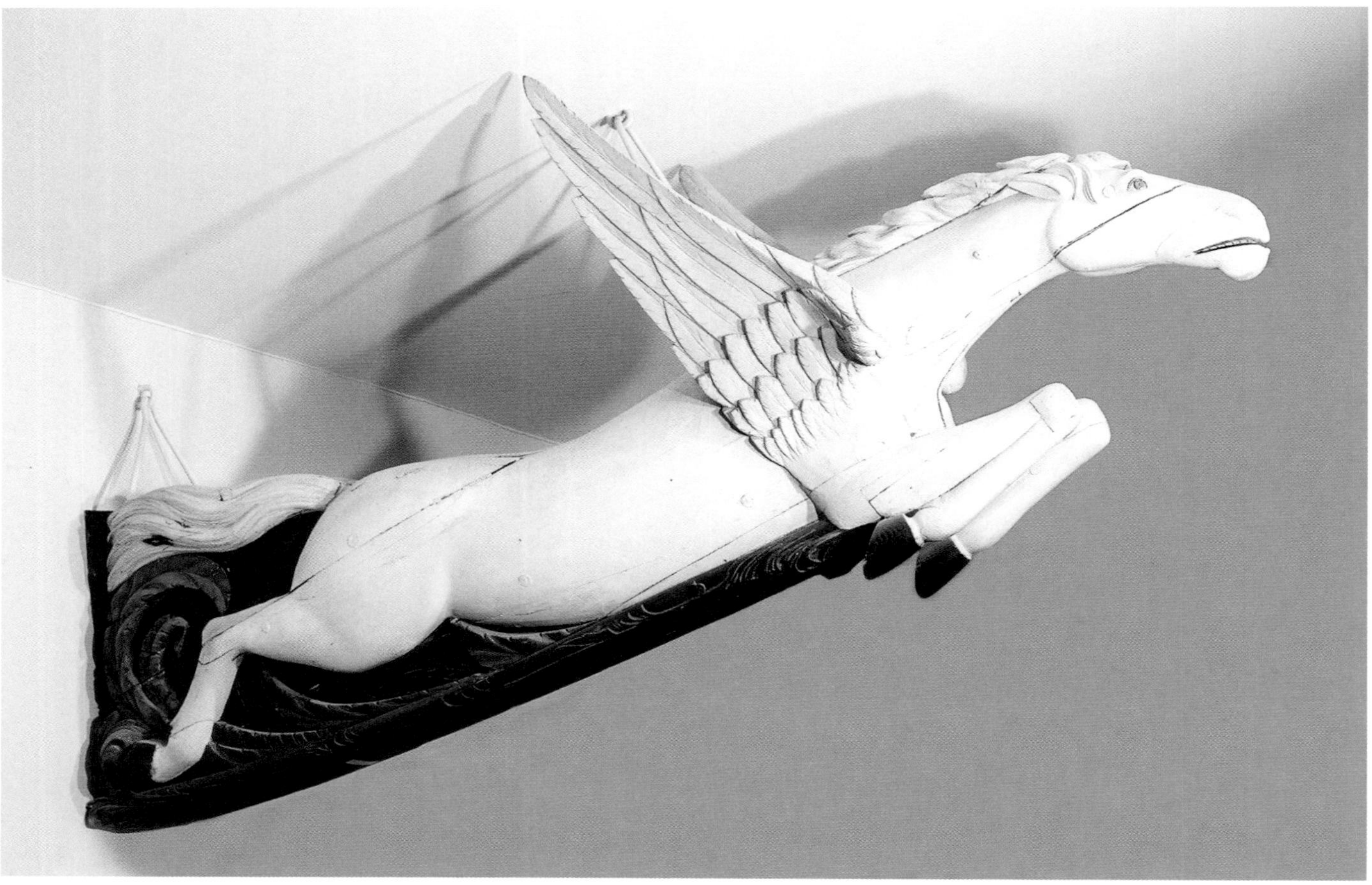

The figurehead is of some help here. Charles Vane's estate, located midway between Sunderland and Seaham Harbour, has the same name as the snow. When the snow was built, it had just been inherited by Vane's son, Frederick William Stewart, the young Marquess of Londonderry. Judging from a portrait of him, he is the same man depicted in the figurehead. Frederick William Stewart is described as being far more socially involved than his father, who was generally considered severe and as lacking an understanding of the common folk. It would be especially interesting in this case to take a closer look at the circumstances behind the choice of figurehead ornamentation.

In the 1850s, a new type of sailing vessel became increasingly popular. The clipper, built in the United States, took over much of the merchant ship market, which had been dominated by England. The early clippers were made of pitch pine, with a tonnage of around 1,000 tonnes. The museum has a figurehead from one such clipper, *Chrysolite*, built in New Brunswick, Canada in 1858 for Potter & Co in Glasgow.[12] It served emigrant traffic between London and New Zealand. The figurehead, a woman draped in a floor-length gown with her hair up and wearing a diadem, is traditionally posed – her gaze directed upward and forward, one hand to her side, the other across her chest. Below, towards the stem attachement is a coat of arms. We do not know who carved it. There is some information available about sculptors working in the New Brunswick area, but they were numerous. By this time, the town had developed a significant shipbuilding industry with related workshops for sculpting. It is believed that US and Canadian sculptors were initially commissioned but gradually began to carve on spec and thus use motifs that had wider appeal, such as bald eagles, Indians and Creole women and men.

The figurehead for the clipper *Pegasus* holds a special place at the Maritime Museum. It has featured in the museum's exhibitions ever since it opened in 1938. The figurehead originally sat on the four-masted iron-hulled barque *Pegasus*, built in Liverpool in 1884 for the textile manufacturer Charles Webster Corsar's shipping company Flying Horse Line, whose vessels all featured the winged horse as their figurehead. Only when the English developed the technology to build iron hulls could they build clippers that were as fast. These were also much larger and had far more capacity than the American models. For example, *Pegasus* had twice as much tonnage as *Chrysolite*. As a result, the English were able to reclaim their position in global trade.

With *Pegasus*, we see ornamentation that for the first time is apparently a trademark. Other vessels, such as *Wynyard*, linked economic affairs with the name of the ship. In the 19th-century's age of sail, goods were delivered with ever increasing speed, and people even competed to arrive first in the markets with their cargo of wool or tea. It is easy to imagine that it became important to showcase one's brand in port, and what more suitable location was there than the vessel stern that extended across the wharf?

There are still many unwritten, intriguing stories among the ornaments from vessels that took part in the expanding global trade of the 19th-century.

CONTRIBUTORS

ANNIKA BÜNZ holds a PhD in archaeology. In her dissertation *Upplevelser av förhistorier, analyser av svenska arkeologiska museiutställningar* (Experiences of prehistories, analyses of Swedish archaeological museum exhibitions) (2015) she made detailed and systematic analyses of the way people and their worlds are depicted in archaeological exhibition narratives. She is currently a researcher at Gothenburg University's Department of Cultural Sciences on a project examining how Nordic maritime museums depict navigation and seafarers.

LARS EINARSSON is Head of Department at Kalmar County Museum and Project Manager for the Kronan Project. He has devoted much of his professional life to investigating what life was like both on naval vessels and on land during the wars of the 17th century. Einarsson is also the author of the book *Regalskeppet Kronan: historia och archaeology ur djupet* (The Royal Ship Kronan: history and archaeology from the depths).

NIKLAS ERIKSSON is Associate Professor in archaeology and a researcher at Stockholm University's Department of Archaeology and Classical Studies. Eriksson specialises in maritime archaeology, and his research has to a large extent been based on shipwrecks. The work has led to the doctoral dissertation *Urbanism under Sail: An Archaeology of fluit ships in Early Modern Everyday Life* (2014). He is also the author of *Riksäpplet: arkeologiska perspektiv på ett bortglömt regalskepp* (Riksäpplet: archaeological perspectives of a forgotten royal ship), in which he writes about how the historical description of the fleet's defeat during the Scanian War affected maritime archaeology in connection with *Riksäpplet* and other wrecks.

FRED HOCKER holds a PhD in anthropology, and is currently Head of Research at the Vasa Museum and the Swedish National Maritime and Transport Museums. He has experience of many maritime archaeological excavations in the US, Northern Europe and the Mediterranean, and since 2003 he has led the research into the *Vasa*. Hocker is furthermore the author of *Vasa: A Swedish Warship.*

LEOS MÜLLER is Professor of History at Stockholm University's Department of History. Müller's special area is maritime history, and he is Director of the Centre for Maritime Studies (CEMAS), as well as being one of the editors and authors behind the book *Sjövägen till Sverige. Från 1500-talet till våra dagar* (The sea route to Sweden. From the 16th century to the present day). Müller's current research concerns Swedish navigation and trade during the 18th century – above all the relationship between foreign policy and economic development.

JOHAN RÖNNBY is Professor of Marine Archaeology at Södertörn University's School of History and Contemporary Studies, and is also Director of Södertörn University's Maritime Archaeological Research Institute (MARIS). His research has concerned various shipwrecks in the Baltic Sea, as well as Viking lake dwellings, ports and prehistoric landscapes under water, coastal landscapes and man's interaction with water.

INGRID ULFSTEDT is a Curator at the Swedish National Maritime and Transport Museums, with a special interest in the histories our objects represent. Ulfstedt's principal field of work is collections of objects.

NOTES AND REFERENCES

THE FIGUREHEAD'S ADVENTURES THROUGH TIME AND SPACE

NOTES

1. in Pinckney 1940: 39–40
2. Rittmeister 1958
3. Ernst Plate in Wietek 1961: 5
4. Marshall 2003: 1
5. Hamilton 1984
6. There are exceptions, where the shipowner has reintroduced the figurehead on modern vessels, above all the Norwegian shipping company Messrs Fred Olsen being known for this (Henningsen 1959–60: 341–343, Weightman 1974: 194, Norton & Olsen 2008).
7. Lisberg Jensen 1984: 4, 6
8. Hunter 2007: 21
9. Bennett 1995: 2
10. Pearce 1999: 18–19
11. Svanberg 2017: 156–157, see also Thor Tureby & Johansson 2017
12. Lidchi 1997: 160
13. Svanberg 2017: 158, Thor Tureby & Johansson 2017
14. 2006: 217, see also Basu & Macdonald 2007: 2
15. see, for example, Pinckney 1940, Poulsen 1974, Norton 1976, Costa: 1981, Stammers 1983: 5, Kristensen 2001: 10, Hunter 2007: 20, Gøthesen 2010: 9
16. Bjering 1963: 42, Costa 1981: 8
17. Cordingly 2001:175
18. Hallén & Rüster 1975: 7–8, see also Stammers 1983: 5
19. e.g. Stammers 1983: 5
20. Costa 1981: 8, see also Bjering 1963: 45
21. see, for example, Carr Laughton 1925, Pinckney 1940, Poulsen 1945, 1975, 1976, Rogers 1946, Christensen 1952, Henningsen 1959–60, Brewington 1962, Stackpole 1964, Bøggild 1979, Hansen 1979, Costa 1981, Stammers 1983, Hamilton 1984, Thomas 1995, Riise Kristensen 2001, Marshall 2003, Pulvertaft, David 2009, 2011
22. see, for example, Humbla 1933, Leander 1943, Cederlöf 1965, Halldin 1968, Reyde 1969, Färnström 1993, 1994
23. *Sculpture Journal* 24.2, Kelly 2015: 136
24. Cordingly 2001
25. Bjering 1963: 47
26. Hunter 2007: 21
27. Cordingly 2001: 175, see also Soop 1978
28. Bjering 1963: 53
29. Cordingly 2001: 172
30. Cordingly 2001: 174–175
31. Cordingly 2001: 177
32. Weightman 1974: 194
33. Norton 1976: 66
34. Text in the 'Models Room', Royal Danish Naval Museum in Copenhagen 11.08.2015, see also Bjering 1963: 47.
35. Hallén & Rüster 1975: 13
36. Bjering 1963: 99–100
37. Lisberg Jensen 1984: 12
38. Bjering 1963: 99–100
39. Wietek 1961: 27
40. Costa 1981: 10
41. Rogers 1946: 2
42. Costa 1981: 99, see also Pinkney 1940, Hallén & Rüster 1975: 77
43. Stammers: 1983: 5
44. Bjering 1963: 42
45. Marshall 2003: 4
46. Costa 1981: 8
47. Poulsen 1974: 75–76
48. Hamilton 1984
49. Stammers 2005: 7
50. Marshall 2003: 1
51. Marshall 2003: 1
52. Gøthesen 2010: 7–8
53. Bjering 1963: 50
54. Cordingly 2001: 177
55. Bjering 1963: 50
56. Lindqvist i Färnström 1994: 2
57. Cordingly 2001:178
58. Färnström 1994: 9
59. Bjering 1963: 62
60. see, for example, Humbla 1933, Leander 1943, Cederlöf 1965, Halldin 1968, Reyde 1969, Hallén & Rüster 1975: 35–57, Färnström 1993, 1994, cited literature in Bjering 1963
61. Bjering 1963: 62
62. Norton 1976: 10
63. Bjering 1963: 62
64. Bjering 1963: 58
65. Hallén & Rüster 1975: 80
66. Lisberg Jensen 1984: 10
67. Bjering 1963: 58
68. Norton 1976: 10
69. Hallén & Rüster 1975: 80
70. Lisberg Jensen 1984:10–11
71. Kelly 2015: 136
72. Peters 2013
73. from Nils Holgerssons underbara resa genom Sverige. Lagerlöf 2011 [1906–1907]
74. Wietek 1961: 21–22
75. Wietek 1961: 21–22
76. see, for example, Poulsen 1945: 12–14, Alexandersson 2018: 58, Palmsköld 2018: 91–93
77. Humbla 1933: 95–96
78. Svärdskog 2004: 6
79. Svärdskog 2001, 2004
80. *Svenska Dagbladet* 17.01.2008
81. The Maritime Museum, Stockholm, Identification Number: S 2734.
82. *Sex & the Sea* at M/S Maritime Museum, Helsingör 09.09.2016–03.09.2017
83. Wietek 1961: 23
84. Wietek 1961: 24
85. Wietek 1961: 27
86. Marshall 2003: 1
87. Costa 1981: 101
88. Shirazi 2014: 21–23
89. Lisberg Jensen 1984: 12

REFERENCES

Alexandersson, Henrik 2018. 'Från stävprydnad till minnesföremål' in ed. Annika Bünz *Galjonsfigurens många öden och äventyr. Studier av berättande-, minnes- och kunskapande praktiker.* Lund: Nordic Academic Press. pp 43–65.

Basu, Paul, Macdonald, Sharon 2007. 'Introduction: Experiments in Exhibition, Ethnography, Art, and Science' in ed. Sharon Macdonald & Paul Basu *Exhibition Experiments.* Malden, Osford, Victoria: Blackwell Publishing. pp 1–24.

Bennett, Tony 1995. *The Birth of the Museum. History, Theory, Politics.* London, New York: Routledge.

Bjering, K. B. 1963. 'Stävdekorationer genom tiderna. Galjonsbilder, ornament och namnbräden i Hallands museum.' in *Halland 1963.* Year 46. Årsbok för kulturhistoria och hembygdsvård i Hallands län.

Halmstad. pp 41–111.
Bøggild, Hansaage 1979. *Skibsudsmykninger.* Rønne: William Dams Boghandel A/S.
Brewington, M.V. 1962. *Shipcarvers of North America.* Barre: Barre Publishing.
Carr Laughton, L. G. 1925. *Old Ship Figure-Heads & Sterns.* London: Halton & Smith, New York: Minton, Balch & Company.
Cederlöf, Olle 1965. 'Johan Törnströms galjonsbilder' in Årsbok för Marinmusei Vänner. pp 525
Christensen, Erwin O. 1952. *Early American Wood Carving.* Cleveland & New York: The World Publishing Company.
Cordingly, David 2001. *Heroines & Harlots. Women at Sea in the Great Age of Sail.* London, Baskingstoke & Oxford: Macmillan.
Costa, Giancarlo 1981. *Figureheads. Carvings on Ships from Ancient Times to Twentieth Century.* Lymington: Nautical publishing company.
Färnström, Maud 1993. 'Två Karlskronaprofiler: Törnström och Chapman.' in Årsbok för Marinmusei Vänner.
Färnström, Maud 1994. *Johan Törnström 1744–1828.* Karlskrona: Naval Museum.
Galionsfiguren 1961. *Altona Museum, Hamburg.* Special exhibition JulySeptember 1961.
Gøthesen, Gøte 2010. *Gallionsfigurer på norske skip.* Oslo: Sem & Stenersen.
Halldin, Gustaf 1968. 'Amiralitetsbildhuggaren Johan Törnström' in Årsbok för Marinmusei Vänner. pp 17–51.
Hallén, Tore, Rüster, Reijo 1975. *Galjonsbilder.* Stockholm: Rabén & Sjögren.
Hamilton, Georgia W 1984. *Silent Pilots. Figureheads in Mystic Seaport Museum.* Mystic: The Museum.
Hansen, Hans Jürgen 1979. *Galionsfiguren. Illustrated General Catalogue.* Oldenburg, Munich, Hamburg: Stalling.
Henningsen, Henning 1959–60. 'Gallionsfigurer' in *Søens Verden* ed. Kaj Lund. Year 1. pp 325343.
Humbla, Philbert 1933. *Johan Törnströms galjonsbilder.* Off-print from *Konstrevy* No. 33.
Hunter, Richard 2007. 'Figureheads' in *The Oxford Encyclopedia of Maritime History,* Hattendorf, John B. editor-in-chief. Oxford & New York: Oxford University Press. pp 20–21.
Kelly, Julia 2015. 'Introduction: sculpture and the sea' in *Sculpture Journal* 24.2. pp 135–139.
Lagerlöf, Selma 2011 [1906–1907]. *Nils Holgerssons underbara resa genom Sverige.* eBook. Stockholm: Bonnier Carlsen.
Leander, Ulla 1943. 'Galjonshuggaren Johan Törnström. 1743–1828.' Offprint from *Blekingeboken. Årsbok för Blekinge hembygdsförbund och Blekinge museum.* Karlskrona: Blekinge Museum and Local Folklore Association.
Linnestad, Bjørn, Olsen, Olav 2008. *The Fred. Olsen & Co. Figureheads.* Messel Forlag.
Lisberg Jensen, Ole 1984. *Galjonsfigurer och fartygsornament in Sjöfartsmuseet, Göteborg.* Gothenburg: Maritime Museum in Gothenburg, publication series No. 4.
Marshall, Gordon de L. 2003. *Ships' Figure Heads in Australia.* Kalamunda: Tangee Pty Ltd.
Norton, Peter 1976. *Ships' Figureheads.* Newton Abbot, London, Vancouver: David & Charles.
Palmsköld, Anneli 2018. 'Museet och galjonsfiguren' in ed. Annika Bünz *Galjonsfigurens många öden och äventyr. Studier av berättande-, minnes- och kunskapande praktiker.* Lund: Nordic Academic Press. pp 67–96.
Pearce, M. Susan 1999. 'Presenting Archaeology' in *Making Early Histories in Museums.* Nick Merriman (ed.) London: Leicester University Press. pp 12–27.
Peters, Andrew 2013. *Ship decoration 1630–1780.* Barnsley S Yorkshire: Seaforth publishing.
Pinckney A. Pauline 1940. *American Figureheads and their Carvers.* New York. W.W. Norton & Company.
Poulsen, Hanne 1945. *Gallionsfigurer og skibsornamentik på Bangsbo.* Fredrikshavn: Bangsbo Museum.
Poulsen, Hanne 1974. 'Gallionsfigurer og anden dansk skibsornamentik indtil 1850' in *Handels- o Søfartsmuseet på Kronborg Årbog 1974.* pp 74118.
Poulsen, Hanne 1975. *Danske gallionsfigurer efter 1850. Gallionsbilledhuggerne H.J. og W.E. Møen.* Yearbook. Published by Selskabet Handels- og søfartsmuseets venner. Handels og Søfartsmuseet på Kronborg. pp 42–83.
Poulsen, Hanne 1976. *Gallionsfigurer og ornamenter på danske skibe og in danske samlinger.* Copenhagen: Rhodos.
Pulvertaft, David 2009. *The Warship Figureheads of Portsmouth.* Stroud, Glocestershire: The History Press.
Pulvertaft, David 2011. *Figureheads of the Royal Navy.* Barnsley S Yorkshire: Seaforth Publishing.
Riise Kristensen, Jens 2001. *Flådens ansigt. Galionsfigurer fra Den Danske Marine.* Stenstrup: Skib Forlag.
Reyde, Folke 1969. 'Johan Törnströms-biografien av 1856. Några synpunkter' in Årsbok för Marinmusei Vänner. pp 14–29.
Rogers 1946. 'Ships figureheads. Selections from the Collection of The Mariners' Museum.' A reprint from *The Shipyard Bulletin.* February Vol XI No. 4.
Sculpture Journal 2015. 24.2.
Soop, Hans 1978. *Skulpturer. Regalskeppet Wasa.* Stockholm: Liber.
Stackpole, Edouard A. 1964. *Figureheads & Ship Carvings at Mystic Seaport.* Mystic, Connecticut: The Marine Historical Association.
Stammers, Michael. K. 1983. *Ships' Figureheads.* Shire publications Ltd.
Svanberg, Fredrik 2017. 'From totality to infinity' in red Christina Johansson & Pieter Bevelander *Museums in a time of migration.* Lund: Nordic Academic Press. pp 155–168.
Svenska Dagbladet 17.01.2008 www.svd.se/fagelskramma-var-dyrgrip [09.12.2017].
Svärdskog, Karl-Eric 2001. *Jenny Lind and the Clipper Nightingale Figurehead.* Portsmouth: the Portsmouth Marine Society.
Svärdskog, Karl-Eric 2004. *Jenny Lind – från Näktergal till fågelskrämma.* Gothenburg: Swedish Nightingale förlag.
Thomas, P. N. 1995. *British Figurehead & Ship Carvers.* Mount Pleasant, Beamish Lane, Albrighton, Wolverhamton: Waine Research.
Thor Tureby, Malin, Johansson, Jesper 2017. 'The Making of Cultural Heritage and Ethnicity in the Archive' in ed. Christina Johansson & Pieter Bevelander *Museums in a time of migration.* Lund: Nordic Academic Press. pp 169–195.
Weightman, A.E. 1974. 'Figureheads' in *The Decorative Arts of the Mariner.* Red Gervis Frere-Cook. London: Cassell & Co pp 193–204.
Wietek, Gerd 1961. *Galionsfiguren. Ships' Figure-Heads.* Hamburg: printed by Th. Dingwort.

SWEDEN AND THE SEA – FROM THE 16TH CENTURY TO THE PRESENT DAY

REFERENCES

Ekström, Simon, Müller, Leos & Nilson, Tomas, *Sjövägen till Sverige: från 1500-talet till våra dagar*, Universus Academic Press, Malmö, 2016.

Fritz, Martin, 'Shipping in Sweden, 1850–1913', *Scandinavian Economic History Review*, 28: 1980/2, pp 147–160.

Glete, Jan, 'Bridge and bulwark: the Swedish navy and the Baltic, 1500–1809', in Göran Rystad, Klaus-R Böhme & Wilhelm M. Carlgren (eds.), *In quest of trade and security Vol. 1*. Probus, Stockholm, 1994, pp 9–59.

Müller, Leos, *Sveriges första globala århundrade: en 1700-talshistoria*, Dialogos, Stockholm, 2018.

Sjögren, Hans, *Spelet in Saléninvest: staten, bankerna, ägarna och ledningen 1973–97*, Ekerlid, Stockholm, 1999.

Tenold, Stig, Iversen, Martin Jes & Lange, Even (eds.), *Global Shipping in Small Nations: Nordic Experiences after 1960*, Palgrave Macmillan, Houndmills, Basingstoke, 2012.

Östlund, Joachim, *Saltets pris: svenska slavar in Nordafrika och handeln in Medelhavet 16501770*, Nordic Academic Press, Lund, 2014.

THE DANISH MONSTER

NOTES

1. Landström 1961, pp 12–13.
2. Adams & Rönnby 1996, 2013, 2019; Rönnby & Sjöblom 2015.
3. Einarsson, L. & Wallbom, B., 2001, 2002; Einarsson, L. & Gainsford, M. 2007; Einarsson 2012; Rönnby 2015
4. Sjöblom 2015.
5. Adams 2013.
6. Glete 2000, 2002.
7. Adams & Rönnby ibid.; Rönnby & Sjöblom ibid.
8. Sjöblom ibid.
9. Landström ibid.; Möller Jensen 2007
10. Eriksson 2016; Lavér 2017.
11. Mandeville 1484.
12. Möller Jensen 2007, p. 198.
13. Möller Jensen ibid.
14. Svanberg & Qvarnström 1998.

REFERENCES

Adams, Jonathan, *A Maritime Archaeology of Ships: Innovation and Social Change in Medieval and Early Modern Europe*, Oxbow Books, Oxford 2013.

Adams, Jonathan & Rönnby, Johan, *Furstens fartyg: marinarkeologiska undersökningar av en renässanskravell*, Maritime Museum, Stockholm, 1996.

Adams, Jonathan & Rönnby, Johan, 'One of His Majesty's 'BesteKraffwells': The Wreck of an Early Carvel-built Ship at Franska stenarna, Sweden', in *The International Journal of Nautical Archaeology*, vol. 42.1 (2013), pp 103–117.

Adams, Jonathan & Rönnby, Johan, 'The Consequences of New Warships: Medieval to Modern and Our Dialectical Relation with Things', in Johan Rönnby (ed.), *On War on Board: Archaeological and Historical Perspectives on Early Modern Maritime Violence and Warfare*, Södertörn Academic Studies, 2019 (in prep.).

Christiansen, A. E., 'Skipsristningene i Sæby kirke', *Handels- og Søfartsmuseet årbog*, Copenhagen, 1969.

Einarsson, Lars & Wallbom, Björn, *Marinarkeologisk besiktning och provtagning för datering av ett fartygsvrak beläget i farvattnen syd Saxemara, Ronneby kommun, Blekinge län*. Report. Swedish Archaeology Commission (UV)/Marine Archaeology *2001*, Kalmar County Museum, Kalmar, 2001.

Einarsson, Lars & Wallbom, Björn, *Fortsatta marinarkeologiska undersökningar av ett fartygsvrak beläget vid St. Ekö*. Report. Swedish Archaeology Commission (UV)/ Marine Archaeology *2002*, Kalmar County Museum, Kalmar, 2002.

Einarsson, Lars & Gainsford, Matthew, *Rapport om 2006 års marinarkeologiska undersökningar av Skeppsvraket vid St. Ekö. Saxemara, Ronneby kommun, Blekinge län*, Kalmar County Museum, Kalmar 2007.

Einarsson, Lars, *Rapport om hydroakustisk kartering och kompletterande dendrokronologisk provtagning och analys av vraket vid St. Ekö, Ronneby kommun, Blekinge län*, Kalmar County Museum, Kalmar 2012.

Eriksson, Bo, *Monster: en världshistoria om det skrämmande*, Natur & kultur, Stockholm, 2016.

Landström, Björn, *Skeppet: en översikt av skeppets historia från den primitiva flotten till den atomdrivna ubåten med rekonstruktioner i ord och bilder*, Forum, Stockholm, 1961.

Landström, Björn, *Vägen till Indien: upptäcktsresor till lands och sjöss från expeditionen till Punt 1493 f. Kr till upptäckten av Godahoppsudden 1488 e. Kr.*, Forum, Stockholm, 1964.

Lavér, R., 'I gapet på Gribshunden (1495). Skulpturer och deras bildspråk under medeltiden', Bachelor's thesis in archaeology, Stockholm University 2017.

Linderson, Hans, 'Dendrokronologisk analys av Ekö-vraket, Saxemara i Blekinge skärgård', i *Nationella Laboratoriet för Vedanatomi och Dendrokronologi Rapport 2012:1*, Lund 2012.

Mandeville, Sir John, *Johannes von Montuilla Ritter: Reysen unde wanderschaff. Translated by Otto van Diemeringen*, Johann Prüss, Strasbourg, 1484.

Møller Jensen, Janus, *Denmark and the Crusades, 1400–1650*, Brill, Leiden, 2009.

Jahrehorn, Max, *Föremål från ett medeltida skeppsvrak St. Ekö, Ronneby kommun. Konserveringsrapport*, Kalmar County Museum, Kalmar 2009.

Glete, Jan, *Conflicts and the Transformation of Europe*, Routledge, London, 2000.

Glete, Jan, *War and the State in Early Modern Europe: Spain, the Dutch Republic and Sweden as Fiscal-military States, 1500–1660*, Routledge, London, 2002.

Svanberg, Jan & Qwarnström, Anders, *Sankt Göran och draken*, Stockholm, Rabén Prisma, 1998.

Rönnby, Johan (ed.), *Gribshunden (1495): Skeppsvrak vid Stora Ekön, Ronneby, Blekinge. Marinarkologiska undersökningar 2013–2015. Blekinge museum rapport 21*, Blekinge Museum/Södertörn University 2015.

Rönnby, Johan & Sjöblom, Ingvar, 'Havsguden från Västervik: om skeppet Neptunus och vasakungarnas nya stora kravlar', in Veronica Palm (red,), Västerviks historia II, Västervik Museum 2015.

Wegener Sleeswyk, André, 'The Engraver Willem A Cruce and the Development of the Chainwale', in *The Mariner's Mirror*, Vol. 76:4 (1990), pp 345–361.

Sjöblom, Ingvar, 'Identifiering och historiska sammanhang', i Johan Rönnby (ed.), *Gribshunden (1495): Skeppsvrak vid Stora Ekön, Ronneby, Blekinge. Marinarkologiska undersökningar 2013–2015*. Blekinge Museum Report 21, Blekinge Museum/ Södertörn University, 2015.

THE ROYAL SHIP KRONAN - PALACE AND WAR MACHINE

NOTES

1. Einarsson, 1990; Einarsson, 2016.
2. Glete, 1999, p. 19.
3. Lundmark, 1931, p. 77.
4. Soop, 2007, p. 136.
5. RA, Am.koll. specialräkn. 1651, Vol. 1170–71.
6. See, for example, Franklin, 1989.
7. Soop, 2007, p. 129 ff.
8. Laughton, L.G. Carr, 1991[1925], p. 71.
9. Fox, 1980.
10. Fox, ibid. p. 133; Eriksson, 2017.
11. Einarsson, 1997, p. 209 ff.
12. Hocker, 2006, p. 461 ff.
13. Soop, 1992 [1986].
14. Tångeberg, 1993.
15. E.g. Fox, ibid.: Franklin, 1989; Lavery, 1984.
16. Danish National Archives, Copenhagen, Naval Department. General Commissariat's archive *Instruktionscopiebog 1685–1716.*

REFERENCES

Einarsson, Lars, 'Kronan: Underwater Archaeological Investigations of a 17th Century Manofwar', in *The International Journal of Nautical Archaeology and Underwater Exploration*, Vol. 19, No. 4, 1990.

Einarsson, Lars, 'Artefacts from the Kronan: Categories, Preservation and Social Structure' in Mark Redknap (ed.), *Artefacts from Wrecks: Dated Assemblages from the Late Middle Ages to the Industrial Revolution*, Oxbow Books, Oxford, 1997.

Einarsson, Lars, *Regalskeppet Kronan: historia och arkeologi ur djupet*, Historiska Media, Lund, 2016.

Eriksson, Niklas, *Riksäpplet. Arkeologiska perspektiv på ett bortglömt regalskepp*, Nordic Academic Press, Lund, 2017.

Fox, Frank, *Great Ships: The Battlefleet of King Charles II*, Conway, London, 1980.

Franklin, John, *Navy Board Ship Models 1650–1750*, Naval Institute Press, London, 1989.

Glete, Jan, 'Hur stor var Kronan?: något om stora örlogsskepp i Europa under 1600-talets senare hälft', in *Forum navale*, 1999 (No. 55), pp 17-25.

Hocker, Fred (ed.), *Vasa I. The Archaeology of a Swedish Warship of 1628*, Medströms Bokförlag, Stockholm, 2006.

Laughton, L.G. Carr, *Old Ship Figure-heads & Sterns*, Dover Publications, London, 1991 [1925].

Lavery, Brian, *The Ship of the Line. Vol. 2, Design, construction and fittings*, Conway Maritime, London, 1984.

Lundmark, Efraim, 'Jost Schutze: En bild- och vapensnidare i Stockholm på 1600-talet', in *Personhistorisk Tidskrift*, Year XXXII, Booklets 1–2, Stockholm. 1931.

Soop, Hans, *The Power and the Glory: the Sculptures of the Warship Wasa*, 2. [rev.] pr., Royal Academy of Letters, Stockholm, 1992 [1986].

Soop, Hans, *Flytande palats: utsmyckning av äldre svenska örlogsfartyg*, Atlantis, Stockholm, 2007.

Tångeberg. Peter, *Rapport över färgundersökningarna på regalskeppet Vasa 1990–1993*, stencil, Lästringe, 1993.

UNPRINTED SOURCES

Swedish National Archives (RA), Stockholm: Admiralty Board, specialräkn. 1651, Vol. 1170-1171.

Danish National Archives, Copenhagen: Naval Department. General Commissariat's archive. *Instruktionscopiebog 1685–1716.*

BETWEEN LION FIGUREHEADS AND LIONS OF FASHION

NOTES

1. Noldus 2005.
2. See overview in Eriksson 2014a.
3. Weber, 1978; Schama, 1987; De Fries & Van Der Woude, 1997; Israel, 1998.
4. Se Schama, 1987, p. 311; Eriksson, 2014a, p. 173
5. Zettersten 1903, p. 569; Eriksson 2014a, 156.
6. Eriksson, 2014, Chapter 7.
7. Magnusson & Nordin, 2015, pp 66–67.
8. Eriksson 2017a, p. 16.
9. Soop 2007, pp 81–83.
10. Lundmark 1931, also von Corswant-Naumburg, 1999.
11. Eriksson, 2017b, Chapters 5 and 7.

REFERENCES

Corswant-Naumburg, Inga von, *Huvudbaner och anvapen under stormaktstiden*, diss. Stockholm University, Visby, 1999.

De Vries, Jan & Woude, Ad van der, *The First Modern Economy: Success, Failure, and Perseverance of the Dutch Economy, 1500–1815*, Cambridge University Press, Cambridge, 1997.

Eriksson, Niklas, *Urbanism under Sail: An Archaeology of Fluit Ships in Early Modern Everyday Life*, Södertörn Doctoral Dissertations 95, 2014a, 2014.

Eriksson, Niklas, 'Seglande halsgavelhus: om skulpturer på flöjtskepp i Sverige och Nederländerna under tidigmodern tid', in *Historisk tidskrift*, 2014b, 134.3, pp 385–408.

Eriksson, Niklas, 'Svärdet (1663–1676): ett regalskepp byggt på holländskt manér, in *META historiskarkeologisk tidskrift*, 2017a, pp 7–26.

Eriksson, Niklas, *Riksäpplet: arkeologiska perspektiv på ett bortglömt regalskepp*, Nordic Academic Press, Lund, 2017.

Israel, Jonathan Irvine, *The Dutch Republic: its Rise, Greatness, and Fall : 1477–1806*, Oxford University Press, Oxford, 1998.

Lundmark, Efraim, 'Jost Schutze: en bild- och vapensnidare i Stockholm på 1600-talet', in *Personhistorisk tidskrift*, 1931, Nos 1–2, pp 69–76.

Magnusson, Börje & Nordin, Jonas, *Drömmen om stormakten: Erik Dahlberghs Sverige*, Medström, Stockholm, 2015.

Noldus, Badeloch, *Palats och herrgårdar: nederländsk arkitektur i Sverige = Stadspaleizen en Buitenplaatsen: Nederlandse bouwkunst in Zweden*, Ambassade van het Koninkrijk der Nederlanden (Netherlands Embassy), Stockholm, 2002.

Soop, Hans, *Flytande palats: utsmyckning av äldre svenska örlogsfartyg*, Signum, Stockholm, 2007.

Schama, Simon, *The Embarrassment of Riches: an Interpretation of Dutch Culture in the Golden Age*, Collins, London, 1987.

Weber, Max, *Den protestantiska etiken och kapitalismens anda*, Argos, Lund, 1978.

Zettersten, Axel, *Svenska flottans historia: åren 1635–1680*, Norrtelje tidnings boktr., Norrtälje, 1903.

SHIPS' ORNAMENTS IN THE MARITIME MUSEUM'S COLLECTIONS

NOTES

1. Förteckning över överförda föremål från Karlskrona örlogsvarv till Sjöhistoriska museet 1912. Maritime Museum's drawings archive.
2. Panofsky, Erwin: 'Iconography and Iconology: An Introduction to the Study of Renaissance Art', in *Meaning in the Visual Arts*, London, 1955, pp 51–81.
3. Galjonsfigurens många öden och äventyr. Studier av berättande-, minnes- och kunskapspraktiker, ed. Annika Bünz, Lund, 2018.
4. Ritningar till galjonsbild i Sjöhistoriska museets arkiv, *Galjons Bild för 70 Canons Skeppet Konung Gustaf den III.* Chapman Collection, Maritime Museum's drawings archive. SB 1384b and ÖR 5936.
5. Drawing for stern ornament. In the literature the drawing has been attributed to Jean Bérain and has been deemed to be the model for decoration of the stern *of Konung Karl.* Drawing in National Museum's archive, NMH THC 3851.
6. Chapman Collection, Maritime Museum's drawings archive.
7. *Ritning till ett Linie Skepp om 62 Canoner.* Dated July 1768 and signed Fred. H. Chapman. Maritime Museum's archive. ÖR 1173. See also Daniel G. Harris: *Fredrik Henrik af Chapman: den förste skeppsbyggnadsarkitekten och hans verk.* Translation Roderick Klintebo, Karlskrona, 1998. Harris states that the drawing he includes on p. 53 must have been done by Jean Eric Rehn. The drawing is to be found in the Chapman Collection in the Maritime Museum's drawings archive, but is unsigned. ÖR 1033:17
8. *Cestiuspyramiden i Rom 1756.* Drawing by Jean Eric Rehn in the National Museum's archive. NMH 111/1995.
9. Drawing *Linie skepp nr 1 Kronprins* Gustaf *Adolph,* Johan Tobias Sergel, odaterad. National Museum. NMH 837/1875. Drawing *Gallionsbild föreställande Moder Svea,* Johan Tobias Sergel, undated. National Museum. NMH 836/1875.
10. Harris, ibid., p. 59
11. Maritime Museum, Object Nos O 00144, O 00145 and SM 30166
12. Maritime Museum. Object No. S 5435

IMAGE RIGHTS

Cover: Ola Torkelsson

pp. 1–171: Ola Torkelsson
p. 172: ©The Maritime Museum, Sweden. Photo: Anneli Karlsson
p. 179: Annika Bünz
p. 182: The Naval Museum, Sweden. Photographer unknown
p. 186: Ola Torkelsson
pp. 187–189: The Maritime Museum, Denmark
p. 190: The Vasa Museum, Sweden. Photo: Karolina Kristensson
pp. 193–194: The Maritime Museum, Sweden. Photo: Anneli Karlsson
p. 197: The Maritime Museum, Sweden. Photo: Cecilia Nordstrand
p. 198: Johan Rönnby
p. 199: Drawing by Arthur Nordén
pp. 200–201: Ingemar Lundgren
p. 202: From *The Travels of Sir John Mandeville* (1484)
p. 203 (u.): Wikimedia Commons
p. 203 (b.): Marie-Lan Nguyen. Wikimedia Commons
p. 204 (l.): Woodcuts by William A. Cruce. Ashmolean Museum Oxford
p. 204 (r.): Engraving by Israel Meckenem. Wikimedia Commons
p. 205 (l.): From the *Aeneid*, translated by Octavien de Saint-Gelais
p. 205 (r.): From Conrad Grünenberg's manuscript describing his pilgrimage to the Holy Land (1486). Wikimedia Commons
pp. 206–207: Patric Leo
p. 208 (u.): Saint-Bavo's Cathedral, www.lukasweb.be – Art in Flanders. Photo: Hugo Maertens
p. 208 (b.): Ship in Säby church, Denmark. Source unknown
p. 209: Artist unknown. Hillerød Museum, Denmark. Wikimedia Commons
p. 210: ©Kalmar County Museum, Sweden. Photo: Rolf Lind
p. 212: ©Kalmar County Museum, Sweden. Photo: Pierre Rosberg
p. 213: ©Kalmar County Museum, Sweden. Photo: Lars Einarsson
p. 214 (u.): ©Kalmar County Museum, Sweden. Photo: Pierre Rosberg
p. 214 (b.): The Maritime Museum, Sweden / The Swedish National Maritime and Transport Museums. Photo: Anneli Karlsson
pp. 215–217: ©Kalmar County Museum, Sweden. Photo: Rolf Lind
p. 218: ©National Maritime Museum, Greenwich
p. 221: The Naval Museum / The Swedish National Maritime and Transport Museums. Photo: Mattias Billemyr
p. 222: Niklas Eriksson
pp. 224–225: Engraving by Wolfgang Hartmann 1650. Stockholm City Museum
pp. 226–230: Niklas Eriksson
p. 232: The Swedish National Maritime and Transport Museums. Photo: Karolina Kristensson
pp. 235–239: The Maritime Museum, Sweden. Photo: Anneli Karlsson
p. 241: Private
p. 242: The Maritime Museum, Sweden. Photo: Cecilia Nordstrand
p. 244: The Maritime Museum, Sweden. Photo: Johan Jonson
p. 245: The Maritime Museum, Sweden. Photo: Cecilia Nordstrand
p. 245: The Maritime Museum, Sweden. Photo: Cecilia Nordstrand
p. 247 (l.): The Maritime Museum, Sweden. Photo: Cecilia Nordstrand
p. 247 (r.): The Maritime Museum, Sweden. Photo: Oliver Lindkvist
pp. 248–249: The Maritime Museum, Sweden. Photo: Cecilia Nordstrand
p. 251: The Maritime Museum, Sweden. Photo: Oliver Lindkvist
p. 252: The Swedish National Maritime and Transport Museums. Photo: Karolina Kristensson

INDEX

Vessels in italics

Abraham Rydberg 250
Abrahams offer 223
Af Chapman 243, 245, 246
af Chapman, Fredrik Henrik 180, 244, 245, 246, 250
Andersen, Ludvig Emil 188
Äpplet 211, 214
Augustus, Roman Emperor 236
Ax:son Johnson, Axel 7, 188
Ax:son Johnson, Margaret 7
Beecq, Jan Karel 219
Bérain the Elder, Jean 244, 245
Bjering, K. B. 177, 180
Blixen, Karen 15, 188
Brandskogsskeppet 199
Britannia 214
Broström, Dan 196
Bull 246, 252
Carl XIV Johan 250
Carolus XI 245
Charles II, King of England and Scotland 218
Charles IX, King of Sweden 234, 236, 240
Charles X Gustav, King of Sweden 214, 220
China of Bath 40
Christian I, King of Denmark 204, 205, 208
Christian II, King of Denmark 208
Christian V, King of Denmark 219, 220
Christianus Quartus 220
Christianus Qvintus 220
Chrysolite 253
C.L. Bahr 162
Columbus, Christopher 205, 208
Cordingly, David 175, 176, 180
Costa, Giancarlo 175, 177, 183
da Vinci, Leonardo 199
Dahlbergh, Erik 228
David and Goliath 226
de Saint-Gelais, Octavien 205
De Voss, Jacob 228
Delfinen 250
Didius Julianus, Roman Emperor 236
Dorothea, Queen of Denmark 208
Draken 250
Ehrensvärd, Carl August 244, 246
Elina 86
Engelska fregatten, see also *Postiljon* 220
English Frigate, see also *Postiljon* 220
Esplendian 246
Estonia 191
Euridice 246, 250
Fäderneslandet 246
Felix 196
Fenix 211, 213
Forssberg, Anna Maria 241
Friedrich Wilhelm IV 90
Galten 250
Georg 150
George II, King of Great Britain 178
Glete, Jan 191
Gnarly 199
Gosselman, Christ. 244, 245
Göthe, Gustaf 250
Götheborg 181, 191
Gotlandia 227
Gribshunden 16, 203, 208
Griffen 199, 203, 204, 205, 208
Grünenberg, Conrad 205
Gustav Adolf, King of Sweden 234, 236, 240, 241
Gustav II Adolf, King of Sweden 234
Gustav III, King of Sweden 244, 246
Haakon Haakonson 32
Hallén & Rüster 175, 180
Halvmånen 226
Hartmann, Wolfgang 223
Heidegger, Martin 183
Henry the Navigator 205
Janszoon Liorne, Pieter 223
Järnvågen 219, 220
John III, King of Sweden 234
John, King of Denmark 203, 204, 205, 208
Johnson, Axel 196
Jonah and the Whale 226
Jutholm wreck 227
Karl, Duke of Södermanland 234
Katarzyna Jagiellonka, Queen of Sweden 234
Kelly, Julia 180, 181
Konung Gustaf III 246
Konung Karl 245
Kristian IV 220
Kristian V 220
Kronan 5, 16, 191, 205, 211–220, 228, 231
Kronprins Gustaf Adolph 246
Kung Adolph Fredrik 245
La Santa Maria 205
Lady Stanley 44, 82
Lammet 227
Lejonvraket 223
Liljevalch, Fredrik 252
Lind, Jenny 182
Lion wreck 223
Lisberg Jensen, Ole 177, 180
Longfellow, Henry Wadsworth 174, 178
Louis XIV, King of France 192
Lundell, Ulf 191
Machiavelli, Niccolò 199
Mars 191, 205
Marshall, Gordon 178, 183
Mary Ann 252
Mary Rose 205
Masreliez, Louis 246
Møinichen, Claus 218, 219, 220
Moore, Martin 252
Moser, Stephanie 175
Naseby 214
Neptunus 211, 219, 220
Nerpin, Henrik 245, 250
Nightingale 182
Noah's Ark 226
Nordin, Lars 245, 250
Norton, Peter 180
Notke, Bernt 205, 208
Odysseus 203
Onkel 48
Örnskjöld 40
Oseberg Ship 175
Østergaard, Ingeborg 188
Oxenstierna, Axel 192
Panama of Blyth 134
Pauline et Marie 18
Pegasus 252, 253
Peters, Andrew 181
Pilot of Montrose 154
Postiljon, see also *English Frigate* and *Engelska fregatten* 220
Poulsen, Hanne 178
Prince Royal 214
Rehn, Jean Eric 244, 245, 246
Riksäpplet 228, 230, 231
Riksnyckeln 228
Rosen 226, 227
Royal Catherine 214
Royal Charles (1673) 214
Royal Charles (1680) 214
Royal George 178
Royal James 214
Royter, O.M. 196
Rydberg, Abraham 250
Saga 182, 245, 250
Schütz, Henrik 245
Schütz, Jost 213, 227, 228
Septimius Severus, Roman Emperor 236
Sergel, Johan Tobias 244, 246
Sheldon, Francis the Elder 211, 214, 220, 228, 230, 231
Solen 226
Soop, Hans 228, 241
Spökskeppet 223, 226
St. Andrew 214
Sten Sture the Elder 204, 205, 208
Stewart, Frederick William 250, 253
Stockholm 245
Sundvall, Carl August 245, 250

Svanen 199, 226
Svärdet 211, 219, 220, 227, 228, 230, 231
Tessin the Elder, Nicodemus 245
Tessin the Younger, Nicodemus 244, 245
The Prince 214, 218
The Rose 226, 227
Thor 245, 250
Tiberius, Roman Emperor 236
Törnström, Emanuel 243, 245, 250
Törnström, Johan 176, 180, 245, 246
van de Velde the Elder, Willem 219
van de Velde the Younger, Willem 218, 219
Vanadis 245, 250
Vane, Charles 252, 253
Vasa 16, 176, 191, 205, 211, 218, 219, 228, 233–241
Vasa, Gustav 191
Vintunnan 226
Wasen 234, 236
Webster Corsar, Charles 253
Werngren, Nils 252
Wietek, Gerd 177, 181, 183
Willerup, F.C. 180
Winblad, Ulla 183
Wrangel, Carl Gustav 228
Wynyard 250, 252, 253
Zygmunt I, King of Poland 234
Zygmunt III Waza, King of Poland, Grand Duke of Lithuania 234, 236, 240

IMAGERY OF THE SEA
Figureheads and Symbols

The essays are based on the seminar
Imagery of the Sea, held at Engelsberg Ironworks
in Västmanland, Sweden, Spring 2019.

Edited by:
Kurt Almqvist, President, Axel and Margaret Ax:son Johnson Foundation for Public Benefit
Svante Helmbaek Tirén, Curator

Photographer: Ola Torkelsson
Translation: William Sleath and Susan Long
Design: Patric Leo
Layout: Patric Leo and Petra Ahston Inkapööl
Pre-press and coordinator: Italgraf Media AB, Sweden
Print: Narayana Press, Denmark
ISBN 978-91-985237-9-9

BOKFÖRLAGET STOLPE

AXEL AND MARGARET AX:SON JOHNSON FOUNDATION
FOR PUBLIC BENEFIT